Travels into Print

Travels into Print

Exploration, Writing, and Publishing with John Murray, 1773–1859

INNES M. KEIGHREN,
CHARLES W. J. WITHERS,
AND BILL BELL

The University of Chicago Press

CHICAGO AND LONDON

INNES M. KEIGHREN is a senior lecturer in human geography at Royal Holloway, University of London. CHARLES W. J. WITHERS is the Ogilvie Professor of Geography at the University of Edinburgh. BILL BELL is professor of bibliography at Cardiff University.

The University of Chicago Press, Chicago 60637
The University of Chicago Press, Ltd., London
© 2015 by The University of Chicago
All rights reserved. Published 2015.
Printed in the United States of America

24 23 22 21 20 19 18 17 16 15   1 2 3 4 5

ISBN-13: 978-0-226-42953-3 (cloth)
ISBN-13: 978-0-226-23357-4 (e-book)
DOI: 10.7208/chicago/9780226233574.001.0001

Library of Congress Cataloging-in-Publication Data

Keighren, Innes M., author.
Travels into print : exploration, writing, and publishing with John Murray, 1773–1859/Innes M. Keighren, Charles W. J. Withers, and Bill Bell.
pages ; cm
Includes bibliographical references and index.
ISBN 978-0-226-42953-3 (cloth : alk. paper) — ISBN 978-0-226-23357-4 (e-book)
1. Travelers' writings—Publishing—England—History—18th century.
2. Travelers' writings—Publishing—England—History—19th century.   3. John Murray (Firm)   4. Travel—History—18th century.   5. Travel—History—19th century.   6. Travel writing—History—18th century.   7. Travel writing—History—19th century.   I. Withers, Charles W. J., author.   II. Bell, Bill, 1961– author.   III. Title.
Z286.T8K45 2015
070.50942'09033—dc23
2014029845

♾ This paper meets the requirements of ANSI/NISO Z39.48–1992 (Permanence of Paper).

It is far easier to travel than to write about it.
—David Livingstone (1813–73), African explorer

Contents

Preface and Acknowledgments

This is a book about books—books of travel and of exploration that sought to describe, examine, and explain different parts of the world, between the late eighteenth century and the mid-nineteenth century. Our focus is on the works of non-European exploration and travel published by the house of Murray, Britain's leading publisher of travel accounts and exploration narratives in this period, between their first venture in this respect, the 1773 publication of Sydney Parkinson's *A Journal of a Voyage to the South Seas, in His Majesty's Ship, the* Endeavour, and Leopold McClintock's *The Voyage of the 'Fox' in the Arctic Seas* (1859), and with the activities of John Murray I (1737–93), John Murray II (1778–1843), and John Murray III (1808–92) in turning authors' words into print. This book is also about the world of bookmaking. Publishers such as Murray helped create interest in the world's exploration and in travel writing by offering authors a route to social standing and scientific status—even, to a degree, literary celebrity. What is also true is that the several John Murrays and their editors, in working with their authors' often hard-won words, commonly modified the original accounts of explorers and travelers, partly for style, partly for content, partly to guard the reputation of author and of the publishing house, and always with an eye to the market. In a period in which European travelers and explorers turned their attention to the world beyond Europe and wrote works of lasting significance about their endeavors, what was printed and published was, often, an altered and mediated version of the events of travel and exploration themselves. With particular reference to questions of authorship and the authority of what was being claimed in print, *Travels into Print* is a study of the relationships between the facts of travel and of geographical exploration and how the published versions of those travels came to appear in print.

Geographical exploration, travel writing, and book history are each topics of considerable importance, perhaps especially so from the later Enlightenment to the middle years of the nineteenth century in which period European scientists, individual travelers, and public audiences turned their attention to the nature of the world beyond themselves. Study of these topics has generated widespread interest in and across several fields. Geographers and historians of science have stressed the importance of the published accounts of voyages of exploration and travel to the emergence of modern science and to modern ideas about the dimensions and the content of the world. Historians of cartography and of the visual arts before photography have scrutinized the images produced by these explorers and travelers, seeing in their sketches and maps not only attempts at convincing depiction but also expressions of anxiety about the problems that new and diverse geographies and peoples posed for notions of authenticity and for the credibility of author and artist. Literary scholars and historians of the book have turned to the different forms of travel writing, to the exploration narrative as a genre, and to the production and edition history of travel texts. In one way or another, then, books printed and illustrated are no longer seen as simple bearers of geographical or historical truths but have themselves become the objects of scholarly enquiry. *Travels into Print* is intended as a contribution to these fields and as a demonstration of the fruitful links that can come from examining books as objects of knowledge from these different perspectives.

Our work was greatly aided by the acquisition, in 2006, of the John Murray Archive (JMA) by the National Library of Scotland, and the move north from the Murray offices at 50 Albemarle Street, London, to Edinburgh, of this extensive and unique collection of material. The JMA is one of the largest and most important publisher's archives in Britain. Our book is based on a detailed analysis of the JMA—principally of its rich manuscript materials, the correspondence files of letters into and out from the several John Murrays, and of the production and financial records and ledger volumes, as well as of the printed books themselves. Where relevant, we have made use of other publishers' archives, and manuscript and other sources, in order to illumine the story of exploration's authorship and authentication. The research on which this book is based was initially supported, from 2008 to 2010, by a research grant from the UK Arts and Humanities Research Council for a two-year project titled "Correspondence: Exploration

and Travel from Manuscript to Print, 1768–1846" (AH/F009364/1). This
research project ended with a conference held at the Institute of Geography, University of Edinburgh, and in the National Library of Scotland in
April 2010, and we are grateful for the comments and suggestions made by
delegates concerning our ideas and for the stimulus of the papers delivered
at this meeting. Further work from 2010 has been supported by the University of Edinburgh, Royal Holloway, University of London, and the British
Academy.

The idea of correspondence that lies at the heart of *Travels into Print*
embraces three related themes and sets of ideas, each of which provides a
major thread running through and across the chapters of this book. The
first is epistolarity: the cultures and practices of letter writing as evident in
works of travel. More generally, how did explorer-authors write? For whom
did they write? The second is epistemology. Explorers and travelers have to
convince their readers—and publishers, as well as their publisher's editors
and literary advisers—of the truth claims of what they were writing about.
Seeing things for oneself is a different route to truth than being told by others. One's preparedness as a traveler in a strange land to believe things told
by others depends strongly on trust in the teller, not just in the knowledge
being imparted. How did explorer-authors justify the claims they made in
and of their works and, even, of themselves? Simply, by what means did
the written book claim to correspond with the real world it purportedly
depicted? The third is editing. This term embraces processes of authorial
mediation by the publisher. At different times, in different ways, and for
different reasons, the Murrays and their editorial employees amended authors' words. We show here how common this later redaction by publishers
of explorer's "in-the-field" writing was and what the resultant effects on the
final printed work were. Editing also encompasses books' edition history
and knowing why certain books were reissued or revised, or published at a
different price or in a different format. Our initial period of interest, covering the non-European travel and exploration material in the JMA between
1768 and 1846, reflected the foundation of the house of Murray in 1768
and James Cook's pioneering voyage that year into the Pacific, accompanied by Sydney Parkinson, and the synthesis of British polar exploration
by John Barrow of the Admiralty in his *Voyages of Discovery and Research
within the Arctic Regions* (1846). As the research began to focus in forming
this book, our period altered to reflect the first Murray publication in 1773
arising from the Cook-Parkinson *Endeavour* expedition, and moved into
the 1850s to encompass further examples of explorers' book writing and

publisher's bookmaking by looking at the works and words of David Livingstone, Charles Darwin, and Joseph Hooker among others. Our terminus in 1859 is not simply arbitrary—it reflects a particular moment in British exploratory culture as an era of large, often Admiralty-sponsored voyages of science and territorial investigation gave way to increasingly individual and touristic travel.

We have incurred many debts in undertaking this work and it is a pleasure to acknowledge them. By far our greatest debt is to David McClay, curator of the John Murray Archive at the National Library of Scotland (NLS). Without his support, courteous guidance through the material, and gracious responses to a barrage of questions and requests during the years spent consulting JMA materials, this book would be very much poorer. We also owe thanks to other NLS staff, notably to Rachel Beattie, Kenneth Dunn, Chris Fleet, and George Stanley, and we are grateful to the NLS for its reproduction of images from the Murray manuscripts and printed works and for the permission of the trustees to include them. We also acknowledge the support of Virginia Murray and John R. Murray of the house of Murray for their interest in the project and, particularly, for supplying from private family collections the illustrations of John Murray I, John Murray II, John Murray III, John Barrow, Alexander Burnes, and the drawing room scene from 50 Albemarle Street, London, which appear in the color plates. Other illustrative material was supplied by the Royal Geographical Society (with the Institute of British Geographers) and by the National Portrait Gallery in London.

David McClay, Virginia Murray, William St. Clair, and Bill Zachs, author of the definitive work on Murray I and his foundation of the publishing dynasty, kindly read the typescript in its near-finished form and made suggestions that greatly improved the final version. The cumulative bibliographical work relating to the non-European books of travel published by Murray in our period of concern has benefited from the assistance of Helen Beaney, Department of Early Printed Books and Special Collections, the Old Library, Trinity College Dublin; Tricia Boyd, Centre for Research Collections, University of Edinburgh Library; Sophie Connor, Rare Books, Cambridge University Library; Timothy Cutts, Rare Books Librarian at the National Library of Wales; Mastan Ebtehaj, Middle East Centre Library, St. Anthony's College, Oxford; Samantha Gibson, the London Library; Paul Hambelton, National Library of Scotland; and Jeremy Hinchliff, Balliol Library, Balliol College, Oxford. We thank the librarians and archivists in the Bodleian Library, Oxford, the British Library, the National Maritime Museum, the Centre for the History of the Book at the University of Edinburgh, and the University of Edinburgh Centre for Research Collections for their assistance. Conference audiences at meetings in Cape Verde, Edinburgh, Las Vegas, London, Manchester, Munich, Seattle, Toronto, and Valencia helped us refine and defend our interpretations. For additional insight, we thank Benjamin Colbert, Felix Driver, Mike Heffernan, Nigel Leask, David Livingstone, Fraser MacDonald, Sarah Millar, Miles Ogborn, Jonathan Wild, and Karina Williamson. We have benefited enormously from the reports of the readers appointed by the University of Chicago Press. In Abby Collier, Christie Henry, Ryan Logan Smith, and Yvonne Zipter of the Press, we could not have wished for better editors and publishers: encouraging, forgiving, patient. If they and others have amended our words, it has always been for the better.

As we were finishing this book, news came through of the untimely death of Susan Manning, Grierson Professor of English at the University of Edinburgh and director of the Institute for Advanced Studies in the Humanities there. Susan Manning discussed the ideas contained in this book with each of us at one time or another. This book is dedicated to her memory.

CHAPTER ONE

Exploration and Narrative: Travel, Writing, Publishing, and the House of Murray

Between the publication in 1773 of Sydney Parkinson's account of the first Pacific voyage of James Cook and the high latitude sensationalism of the 1850s surrounding John Franklin's death in the Arctic, geographical exploration and travel tales captivated public audiences and journalistic commentators alike. Everywhere, it seemed, was coming under the explorer's gaze. Everywhere, too, explorers and travelers were turning their observations on the far away and the unknown into print. Such geographical enquiry and the narratives written about it were to have lasting effects. Reflecting on them in 1858, social theorist and commentator Harriet Martineau noted how "one of the discontents of our saucy modern days is at the smallness of the globe we live on: Between the recent discoveries in astronomy, on the one hand, and the prodigious achievements in geographical exploration on the other, together with the saving of time from steam-travelling, we seem to have obtained a command over the spaces of the globe which considerably diminishes the popular reverence for the mysteries of our planet."[1]

A century and a half after Martineau's reflections, and nearly 250 years after Cook, Vancouver, and other European explorers were at work in the Pacific, the issues of geographical exploration and published narratives that then shrank the world and expanded the minds of contemporaries are now the subject of rich interdisciplinary enquiry. At heart, these enquiries are about the relationships between global exploration, literary production, and the literate public during what one scholar has termed the "Age of Wonder."[2] The focus of these interests in the geographical "opening-up" of the world through exploration and travel rests in the events themselves and in their consequences. Since few could share the individual explorer's experiences, knowledge about the events crucially lies in investigating the accounts printed of them. For most people, exploration was typically experienced at a remove, on the page, not at first hand. And that meant that the facts of exploration had to travel into print. Exploration and travel was hazardous enough: knowledge secured through hardship and privation seemed, somehow, doubly important if hard won from nature's grasp. On return, then, the facts of the explorer's voyaging or land-based travels had to undertake a further and no-less-awkward journey—the "voyage into narration"—if they were to be known about and become the basis for new knowledge.[3]

The move into print was far from a straightforward recounting of events or the cataloging of simple travel facts. Numerous events could and did get in the way of bringing words to book. Authors could amend their field notes, and so could construct new meanings and chronology. Editors and publishers might alter their authors' words in order to meet perceived audience demands. Later editions could supplement, abridge, or recast initial accounts. Explorers' actions and intentions might be given different interpretations under review. Although the explorer-author may have sought exactness in what he or she recounted, authors' written claims to what some called "plain and unvarnished truth" were often founded on others' verbal testimony in the field or on but fleeting observation of the phenomena and locale in question. Either way, what was written might reflect only limited first-hand experiences. If, on return, explorers' words were for one reason or another modified as their notebook jottings were transformed into print, or as rough sketches became seemingly authoritative maps and illustrations, so their books might be only partly the work of single authors. As Ian MacLaren has it, "The first-person authority of the sojourner was thus compromised as it was augmented."[4] Truth telling in exploration writing was far from plain and seldom unvarnished.[5]

Travel and its social and political consequences were central to the making of "globalization" and "imperialism" and to our understanding of periods and categories such as Enlightenment and the romantic era.[6] The terms involved—"explorer" and "exploration," "travel" and "travel writing," even "geography" and "geographer," "print," "authorship," and "book"—were complex, their interrogation always warranting further study. As a result, travel and travel writing have been examined as part of the processes of (unequal) global exchange. What has been seen by some as the interests of the colonizing state in securing useful knowledge about the colonized periphery and its political status has by others been interpreted as unequal "global pillage."[7] But the too-simple assertion that colonization and its antecedent exploration were singular, one-way phenomena, and that exploration narratives straightforwardly reveal the process of discovery or culture contact, does not stand up to the evidence. Disputes between government officials, missionary organizations, and travelers in the field would suggest that the phenomena referred to as exploration and colonialism were commonly ambivalent, anything but simple and cooperative. Accounts of knowledge produced "from the margins" have helped recast and decenter understandings of scientific exploration as more than simply contact with a view to conquest: empirical exploration as the calling card to imperial exploitation.[8] Literary students of travel writing have spilled much ink examining the genre's different narrative styles and interpretative approaches.[9] Historians have turned their attention to explorers' notebooks and in-the-field writing practices, and historians of the book have revealed the many "instabilities of print"—in form, meaning, and interpretation—involved in understanding the book as an object of, and means to, communicative action in science, literature, and exploration.[10] This list might be extended *ad libitum* since, as we show, there is also a comparable critical literature—notably within the history of the book, attaching to notions of authorship, to editing, reading, and reviewing, and within geographical and historical studies—to understanding the actions and works of individual explorers.

In this book, we address and explore the complex relationships between travel and exploration and the resultant narratives in print. We examine published narratives of travel and exploration and the manuscripts and other in-the-field writings on which they were based to see how they were undertaken. We disclose and examine the claims made by authors and others with regard to the truth and authority of their accounts. We scrutinize the various processes involved—writing and rewriting, editing and so on—that brought such books into being. We look at the ways authors and others shaped the material form and the epistemological content of such books, and at the ways readers helped shape reactions to texts, truth claims, and authors alike. These are the issues making up what we term travel into print, and we expand on them here and in what follows with reference to the publishing house of John Murray.

This agenda is not to overlook the comparable movement of print into travel. While the writing of explorers and travelers can be seen as post factum accounts of their experiences and observations, there is a sense in which one person's travel writing created an appetite for others' journeys and critical accounts, especially where geographical and scientific questions were dealt with inconclusively or simply left unanswered. With reference to British narratives of Arctic exploration, for example, Janice Cavell has shown how the search for the Northwest Passage became almost a mania in the nineteenth-century British press, as one work after another was circulated, commented on, and challenged.[11] Such texts were not simply reflections on past encounters or unanswered questions but often acted as encouragements to further investigation. The a priori discourse of travel writing could help to frame the shape and texture of travelers' experiences in the field. Certain methods of observation—often, although not exclusively, learned from already existing printed works—were replicated, mimicked, adapted and modified, or simply rejected. Even texts from the later eighteenth century onward bear the traces of previous accounts on which they comment, sometimes to confirm, sometimes to contradict: travel narrative as a genre was, in one sense, highly accretional in texture, "composed" in numerous ways and part of multiple conversations between exploration and authorship as complex processes of attribution and authority. By the mid-nineteenth century, the genre of travel writing had come to adopt and adapt stock methods of narrating and illustrating the experience of travel as a narrative form. Publishers would consolidate what they saw as their literary property by incorporating lists of other publications from that firm, and the advertisement of other works through the formula "by the author of" was another way in which the commercial, the textual, and the status of certain authors can be seen to intersect. Explorers may have sought status by "filling in" the world's "blank spaces," but travel writing was hardly ever a blank canvas.[12]

By the beginning of the nineteenth century, there was a vast repertoire of travel writing (as well as a growing number of scientific authorities) on which contemporary authors could draw as literary models. Contemporaries and modern scholars have recorded a remarkable rise in the production of travel narratives from about 1800. In 1815, the editor and publisher John Scott observed that "our book-shelves groan with the travels of persons who have suddenly arisen from almost every class and profession of life, to go their ways into almost every other country, as well as into every parish of their own."[13] For the geographer Julian Jackson writing in 1835, the commonplace nature of books of travel had by then been attended by a decline in their style: "Never perhaps were books of travel so much read as now. This has induced every tourist to give to the world the account of his rambles under all kinds of titles. The corruption of that kind of writing has followed as a natural consequence."[14] Among modern scholars, Benjamin Colbert has provided statistical evidence of an increasing interest in travel writing in Europe in the early nineteenth century, and in his analysis of trends in nineteenth-century British publishing, Simon Eliot shows a marked general increase in the production of works of travel throughout the early nineteenth century.[15] While religious tracts and commentaries dominated the early nineteenth century, and adult and juvenile fiction prevailed from midcentury, texts of travel and geography remained, almost consistently, the second-largest genre. Notwithstanding the inconsistent categorization of travel texts in contemporary catalogs and indices, Eliot's statistical survey has shown that, between 1814 and 1846 at least, the portmanteau category of geography, travel, biography and history accounted for 17.3 percent of British book production (of which approximately three-quarters were texts of travel), narrowly trailing religion at 20.3 percent.[16]

Given the number and variety of travel accounts published in the period between Cook's endeavors and Franklin's death and given their importance and the range of modern scholarship paid to them, these concerns with the authorship and publishing of exploration's print culture across the best part of a century might seem to be not a matter of "focus" at all, being too large, too ambitious, and too inchoate to be achievable. Our endeavors are given particular direction because they are focused on the authors, books, and activities of one publishing firm in particular, the London-based company of John Murray.

The house of John Murray was begun in 1768—the year in which Cook encountered the Pacific—by the Edinburgh-born John Murray (1737-93; plate 1). Under the guidance of his son, John Murray II (1778-1843; plate 2), and the latter's son, John Murray III (1808-92; plate 3), the firm became a major publisher, bringing into print the works of Jane Austen, Lord Byron, George Crabbe, Humphry Davy, Charles Darwin, Washington Irving, Charles Lyell, and Walter Scott, among other significant literary and scientific figures. Official accounts of British publishing have tended to emphasize the "literary" over other generic forms in their attempts to glamorize house history, and this has in the past been as true in the case of Murray as those other firms whose lists were equally as diverse.[17] Travel publishing was to become a major part of the Murray firm's raison d'être.

Under the guidance of John Murray II, in particular, the firm's involvement in the publication of narratives of travel and exploration between the late eighteenth century and the mid-nineteenth century (and later, too) was so significant that one modern commentator reckons it, and not literary publication, to be the firm's "greatest contribution to the advancement of knowledge and of human understanding of the world."[18] We examine how the Murray firm worked to solicit and to publish books of non-European exploration between its first venture in this respect, the posthumous work of 1773 by the Edinburgh-born Sydney Parkinson who sailed with Cook on the *Endeavour*, and the account of 1859 by Leopold McClintock who, with others, undertook to search for John Franklin's lost polar expedition. The importance of the firm's involvement in publishing English-language works of travel and exploration in this period was strengthened by the fact that it was, from 1813, official publisher to the Admiralty (the commanding authority of the British Royal Navy), and to the Board of Longitude, and so undertook the publication of most British narratives of Arctic and African exploration, in addition to other works, in the first half of the nineteenth century. Many of these expeditions were sanctioned by John Barrow, second secretary to the Admiralty (plate 4), an important mediator between explorers and the recorded words on their travels, and between those words and the printed versions of them as published by Murray.[19] From 1831, the house of Murray published the Royal Geographical Society's *Journal*, and this and other outlets for texts of travel and exploration, notably the Murray-published *Quarterly Review* from 1809, allowed Murray books to be reviewed and debated in the public world of literary and scientific periodicals.

Before turning to examine the activities of this publishing firm and of its many authors as they first explored and later wrote, it is helpful to review those questions of travel, travel writing, truth, and authority in geographical exploration that are central to our aims and that were of concern to the workings of the house of Murray.

TRAVEL, TRAVEL WRITING, AND EXPLORATION
NARRATIVE IN INTERDISCIPLINARY PERSPECTIVE

Not the least of the many issues to do with travel and travel writing is knowing what is the object in question: "From the amount of critical attention and the number of labels applied to travel writing in recent years, one may well wonder whether critics are discussing the same object."[20] For Jan Borm, the wide range of terms in use, such as "travel book," "travel journal," even "traveler's tale," and categories such as travel writing and travel literature, may be reduced not to matters of absolute definition but to differences between the travel book as a predominantly nonfiction genre and travel writing as a description, "an overall heading for texts whose main theme is travel."[21]

Narratives of exploration may be bracketed with travel books given their avowedly factual emphasis, but such narratives were often the consequence of formally sanctioned institutional imperatives, frequently the results of an organized expedition, and not simply the result of individual travelers. In contrast to travel, exploration commonly had a predetermined and significant end in mind and, often, a significantly more consequential beginning in official enquiry. Exploration usually also had a lasting public "afterlife" as the results were debated in scientific institutions as well as in the periodical and newspaper press.[22] The term "narrative of exploration" is too often generic. In what follows, we address the form and purpose of narrative as a literary style and its use by Murray in titles, alongside other terms such as "journal," "letters from," and other constructions used to frame particular notions of travel and exploratory encounter. Exploration, moreover, was usually aimed at uncovering and documenting the hitherto unknown (to the practitioner-authors and their audiences at least). In this sense, it differed from those instances of travel whose purpose lay more in recounting the traveler's own experiences than in the scientific importance of the facts documented. This is to suggest a difference within the predominantly factual category books of travel between travel accounts, where the author's circumstances are to the fore, and exploration narratives wherein the author's experiences and actions, including the fact of being the author at all, were sometimes reduced in scope, occluded by the new facts or interpretations being brought to the reader's attention. To see travel writing as simply a description of texts whose main theme is travel is inadequate. Travel writing is an analytic and interpretive category whose study involves the textual and stylistic analysis of works of travel and of exploration and, particularly, of authorship, the style of writing, its underlying purpose, and the power of such writing to delimit, explain, or misrepresent the objects of its attention.

These distinctions have been the subject of enquiry from several scholarly quarters. Historians of science stress the importance of voyages of exploration and land-based factual travel from the late eighteenth century and of their published accounts to the emergence of modern science, both in opening up the world and in developing systematic and instrument-based methodical approaches to its examination, classification, and representation. What Harry Liebersohn has seen between the late eighteenth and the mid-nineteenth centuries as a "distinctive era" for scientific ethnography and travel for reasons of that period's related technological and political changes was also a profoundly textual era, evident in the production of new geographical accounts of the world so revealed and in the description and circulation of natural and human specimens whose characteristics provided raw material for Europe's scientists.[23]

Historians of art have explored the practices of representation associated in this period with the visual depiction of new knowledge, peoples, and cultural productions and of natural phenomena and places. Narratives of exploration were, commonly, illustrated accounts. Illustration in different senses reinforced the printed word—in the specific exemplification of a plant or mineral specimen, the visual depiction of topography, portrayal of the peoples encountered, or scenes showing the explorer or his or her ship at work in foreign surroundings. Scientific illustrators strove for objectivity in several ways. Pictures of the author-explorer included as frontispiece or elsewhere and showing him or her to be distinguished in bearing, or posed in military uniform, could help reinforce the text and the author's standing as credible or intrepid. For pictures as for printed words, however, travel did not always make for truth: there was no simple "voyage into substance" for the illustrated travel narrative.[24] Quite apart from the epistemological problems of securing a deeper understanding of nature's relationships or culture's workings when on the move and with limited time, imaging the world in the years before the camera presented similar problems to writing the world.[25] Field sketches might be amended as they became engravings; botany developed particular practices of type-specimen depiction; geology adopted graphic styles to show what could only be inferred; ethnographic scenes stylized moments of encounter instead of the longer-run processes of negotiation and exchange that followed first contact and from which more certain knowledge might follow.[26]

Geographers and cartographic historians have shown mapping to be a powerful means to scientific enquiry and of empirical and imperial authority from the later eighteenth century. Maps were a form of symbolic territorial inscription with the capacity to "write in" or, of indigenous inhabitants, "write over" or even altogether "write out" native knowledge. Maps might reflect what was encountered, but they also constituted particular visions of the world: they are never mirrors of it.[27] The place of maps as integral to explorers' narratives and particularly maps' relationship with the printed text—whether the map may itself be seen as a form of writing as it was also an aid to travel and how its material in-the-field production and later inclusion was part of an individual book's history—have not been the subject of detailed investigation. Maps were an important motive for, and consequence of, exploration and travel for several of the Murray firm's authors, but they were seldom a reliable guide. Geographers have also scrutinized travel texts for what they reveal about different places, about traveling as a process, and about the discursive qualities of travel writing, not least for Africa and in relation to gender.[28] In such work, human geographers share the concerns of cultural theorists and anthropologists with culture and literary criticism in comparative perspective and with travel writing as a form of ethnographic (mis)representation.[29] Others have considered the epistemic practices of geographical writing, often with a historical focus, or have addressed the traditions of textual practice in books of geography and the complexities of geographical authorship and editing.[30]

Historians of the book have revealed the centrality of the history of print to the history of culture, and the importance of understanding the practices of printing, reading, and reviewing in determining the meanings and the reception of ideas. For Elizabeth Eisenstein, for example, printing brought both fixity and standardization: "The fact that identical images, maps and diagrams could be viewed simultaneously by scattered readers constituted a kind of communications revolution in itself."[31] Similar claims for the fixity of print have been made by Walter Ong in his attention to reason and discourse in early modern method.[32] Robert Darnton's development of the "communications circuit" in 1982 was one way to model the cultural presence of print, from author to publisher, from publisher to printer, shipper and bookseller and, from there, to readers and so back to the author, at a time, even then, when Darnton himself feared that book history might become "interdisciplinarity run riot."[33] Later work has challenged and developed these claims, although interdisciplinarity has remained a constant. Adrian Johns's *The Nature of the Book* (1998) took issue with Eisenstein's emphases, for instance, seeing instead multiple local differences among printers, publishers, and authors over their respective credibility, stylistic variations in how printed works looked, and, particularly, connections among the place of production, its constituent social relationships, and the making of printed knowledge. Darnton's work has been extended by others who saw in it too little attention to bibliography, including edition history.[34] Reading and the ideas of audience, of interpretive communities, and the meanings of reception have been given renewed attention by scholars such as Roger Chartier, Robert Darnton, Wolfgang Iser, and others who have shown books and print to be notoriously leaky vessels for the circulation of ideas. As Leslie Howsam shows in her "field guide" to studies in the history and culture of the book, book history is not the exclusive preserve of

historians, literary scholars, or bibliographers (although they have dominated to date) but is interdisciplinary territory open to different fields and
to the relationships between them.[35] Books *in* history, as "containers" of
history—of science, of empire, of exploration—have themselves become
the objects *of* historical enquiry, neither objects of "fixity" in a technical or
an interpretive sense nor simply "representative" of such things as exploration, travel, and science but vital means by which our knowledge about
them exists at all.

Quite where such interdisciplinary perspectives and what others have
termed "cross-border approaches to travel writing" may be leading and with
what result is hard to know, just as is tracing the origins of such work—to
the crisis of representation in the humanities, to the powerful colonial
critiques and postcolonial counterarguments prompted by Edward Said's
Orientalism (1978), to Said's work on "traveling theory," to the humanities'
engagement with "the spatial turn," or to the more evidently materialist
hermeneutics of some book historians and historians of science.[36] The study
of books and of books of travel, including narratives of exploration and of
travel writing, is now so strongly interdisciplinary that it has become almost
transdisciplinary, a body of scholarly enquiry in its own terms. From one
point of view, this is to be welcomed: it will be clear how much our focus
on John Murray's authors and books rests on perspectives gleaned from the
scholarly fields summarized above, and it is certainly not our place to pro-
scribe intellectual cohabitation. But, from another, there is a danger to the
understanding of books as material objects, written artifacts, and cultural
transactions if such transdisciplinary concerns neglect the insight of specialists: of literary scholars about genre, style, and textual criticism; of book
historians on edition history, the economics of the book trade, materiality,
and audience; of geographers on place and space; and of historians of science on the nature and mobility of textual knowledge as a form of communicative action between places and over time. What is also apparent is how
each of the central terms—"exploration," "knowledge," "book," "map,"
and so on—is a contingent, not an essential, category. Even the description
"explorer," for example, had numerous meanings, most persons who merit
the term in later analysis not having been so described in their lifetimes:
"discoverer," "navigator" or, for the French, *voyageur naturaliste*, were the
more commonly used contemporary epithets.[37] Understanding the nature
of travel and exploration and the move of travel into print in historical context requires attention to the contemporary meaning of these contingencies
rather more than adherence to modern explanatory and theoretical cate-

gories. This collective engagement within and above different disciplines throws into relief questions about authoring and authorship, trust, credibility, and truth telling and about the strategies and procedures by which authors undertook to write their books of travel, and through which, as we shall see, John Murray turned the words of others into print.

EXPLORATION, TRAVEL, AND TRUTH

One critical matter arising from the study of travel writing and exploration is the relationship between travel and truth. There is an established body of work concerning the proven travel liar, persons whose claimed travel never actually happened or for whom exaggerations in fact or style were part of a satirical or other literary strategy.[38] But notions of truth and trust permeate travel accounts and exploration narratives in more complex ways than outright falsehood. Because printed narratives about travel did not always correspond to the events themselves and because authors, editors, and publishers could and did amend text and image before publication, it is vital to interrogate the actual practices of exploration writing—how and where such writing took place, what the authors' purpose was, and so on—and to know the reasons behind later emendations of it. In several senses, this is a matter of correspondence.

Correspondence is at one level a matter of written culture and communication. In this period, the world of learned literate enquiry was one of correspondence and epistolarity—people wrote to one another constantly. Writing, especially letter writing, was a form of literate sociability as well as of polite enquiry and information dissemination. It was a means by which in letters of introduction, for example, one's criterion as a man or woman of status was established, among distant contacts, or one's standing as a "Man of Letters" was made and reinforced. As Martin Lyons and others have shown, correspondence was a vital via media by which knowledge was exchanged and the credibility of the author or bearer of the letter was established.[39] Correspondence in our period of concern and for several Murray authors was also more than a private culture of epistolary exchange in that some authors chose to title their work of travel "Letters from. . . ." Whether this was because the work in its initial formulation was based on letters written from the place of travel, and later presented unexpurgated in that form, because use of the term "letters" gave the printed work an air of immediacy and insight into the private world of the correspondents, or because to be lettered was to be highly literate, or for other reasons, is often difficult to know. The use of the epistolary form was by the mid-eighteenth century a firmly established genre, one that served to personalize and authenticate the author as author. At the same time, it had an obvious appeal to the reader, who was invited either to occupy vicariously the position of direct addressee or, at least, to take the perspective of a privileged insider. The fact of letter writing as a form of in-the-field travel writing also suggests a certain authorial strategy and hints, too, at a certain social and physical space that afforded authors an opportunity to write—to retrieve narrative from memory—in a particular form and a particular place.[40]

Correspondence is importantly also an epistemological matter: simply, what is written about should correspond in appropriate ways to the object or phenomenon being described. Correspondence in this sense is a matter of credibility of and for the author and a matter of the chosen writing strategy: how can you assure the reader (who, it may be presumed, has not seen or, at least, is not familiar with the objects or places in question) that you, the author, are telling the truth, that such an event really did occur, that the account as written corresponds to the actual experiences of travel and exploration? Correspondence in this more material sense presumes, of course, that a real world of visible phenomena and material conditions exists to be written about and the author has a mimetic or metonymic power to be able to do this. In the exegesis of printed travel narratives, however, this is centrally a matter of trust and epistemology: how and what did authors write in order to convince audiences of the truth of the tale told?[41] As we shall show, the strategies that Murray's authors employed to demonstrate their veracity varied in relation to genre, social status, and the geographical region being written about; the means by which credibility was secured and trust earned were multiple and often contradictory.

As Frederic Regard notes in discussing examples of British explorers' "first contact" accounts, "all reports—log entries, journals, retrospective narratives, fictional reelaborations—were *narratives*. Exploration accounts, as well as ethnographic descriptions or anthropological studies, were, and still are, *literary artefacts*."[42] In his work on the ways in which historiographical writing mimics literary and, in particular, fictional form, Hayden White has shown how seemingly innocent accounts of the past have resorted to what he calls "the tropics of discourse." That it is possible to describe a wholly imaginary world in the style of an actual travel account—and that historians interrogate the construction of travel narrative in different ways—serves to indicate how the stylized artificiality of the genre has become at once apparent and a source for discursive critique.[43] It is precisely because of their status as, first, forms of discursive practice and, second, as material artifacts that the composition and making of exploration and travel accounts are important objects of research: how is truth, or what passes for truth, registered in the author's words; what are the evidence bases ("I saw this," "I heard it from a reliable witness," and so on); and who, if not the author alone, is involved—and how—in the making of these written objects. As Adriana Craciun observes, these issues are of importance precisely because "the social and collaborative dimensions of authorship, knowledge-making, and material textuality predominate across many interdisciplinary fields."[44]

Historians of science have shown for authoritative scientific writing how the truthfulness and trustworthiness of narratives of enquiry and of their authors was established in several ways.[45] One way was an insistence on the value of immediate encounter, for the author to give assurances about having seen the phenomena or place in question at first hand. In such a strategy, accompanying evidence in the form of a map or sketch "drawn at first hand" could be marshaled as textual corroboration. Instrumental measurement might be invoked both as a sensory extension of the author's personal encounter (in-the-field "readings" as a form of in-the-field writing) and as a form of warrant about standardization and exactness that would itself travel. Temperature, for instance, might be given as numerical figures derived from using and reading a calibrated thermometer rather than via the less precise textual descriptor "very hot today," or the recording of geographical bearings would adopt standard forms of distance and direction as opposed to phrases such as "about ten miles from the coast." So, too, the truth claims of the narrative or trust in the author might be established by the explorer-author assuring the reader that he or she had been told something by a credible and reliable source. Here, first-hand experience is replaced or supplemented by reliance on the social status of the informant. Credibility rooted in the testimony of significant others is potentially a matter of epistemology *and* of morality.

This is, emphatically, not to argue that travelers and explorers and, later, Murray authors in the later eighteenth and nineteenth centuries explicitly identified and were governed by those self-same maxims for the evaluation of secure knowledge that Steven Shapin has identified in discussing truth making, testimony, and "epistemological decorum" in seventeenth-century natural philosophy.[46] Equally emphatically, it is to insist that analysis of books of travel and of travel writing needs to pay closer attention to the criteria of credibility by which truthfulness and trust, in the account and in the author, was secured. As one leading historian of the book has it, "The making of credibility should be our subject in analysing the importance and development of the book."[47] Print itself could act as a guarantor of authority, of course, not least so when it issued from the house of Murray, which over time, through its close association with Admiralty officialdom as well as its stable of celebrated explorers and literary authors, acquired the aura of respectability and of celebrity by association. Whatever the perceived reputation of Murray authors and of the imprint, several factors might militate against the perceived reliability of reportage from the field. Questions of credibility could be compromised by the need for physical security. Global travel and exploration is intrinsically dangerous. Long-distance oceanic navigation has its attendant risk of shipwreck. Unaccompanied travel in African deserts or in parts of the Near East and Central Asia was no less dangerous: there, Christian white men who were often regarded with suspicion, especially if dressed in military uniform, risked robbery or worse. Travel always prompts questions of safety (fig. 1).

Such perils were often answered, by many British and other land-based travelers in non-Christian lands, at least, by their adoption of disguise. In

FIGURE 1 The dangers of travel exemplified. This illustration—*Sangah in the Vale of Dogalee*—highlights the dangers of travel for beasts of burden and humans alike. Source: Godfrey Mundy, *Pen and Pencil Sketches* (London: John Murray, 1832). By permission of the Trustees of the National Library of Scotland.

its more practical senses—adopting native dress, for example, or learning a language and traveling with locals—disguise was, self-evidently, a matter of security and of sound common sense.

Disguise is likewise a matter of correspondence and, thus, of authorial credibility. In disguising oneself, however well-intentioned one's motives, the author runs the risk of being taken for what he or she is not and of soliciting information by underhand means. The demands of practicality need to be balanced by those of propriety: while travelers regularly had to write ahead to secure safe passage, and so corresponded in an epistolary sense, many did not correspond in their physical appearance or in their stated handwritten objectives with who they said they were and why they were traveling. In mapping Tibet and the Himalaya in the later nineteenth century, for example, British-trained pundits had no need to disguise their skin color or to adopt native costume but every need to mask their instruments from prying eyes and to record their movement by measuring secretly the routes and distances traveled. In debates in the early 1830s that followed accounts by British and French explorers over reaching the fabled city of Timbuktu, John Barrow chastised the French (whose representative, in disguise, had got there ahead of the British officer who traveled and died in full uniform) over the French explorer's lax ethical standards (and, by implication, the moral code of the whole French nation). How, as Barrow argued in fulminating against the French, could one trust the truth claims in what was so obviously achieved by deceit?[48] Disguise, a form of travel lying, could take many forms but it always had moral, intellectual, and practical consequences, for the authority of the traveler-explorer, for those he or she traveled among, and for later readers.

Assessment of travel accounts can also reveal the facts of authors' dependence on indigenous inhabitants in several ways. This is likewise a matter of pragmatic and epistemological significance. In many instances, travel was only possible because of local knowledge, native assistance with food, water, and lodgings, help with foreign customs, and sometimes arcane linguistic translation or in restoring sick travelers to full health. Such engagement with local knowledge systems and with the bearers of that knowledge could present problems for authors who were later keen to be seen to document things from first-hand experience or whose own standing depended, for their testimony to be believed, on their informants being of an equivalent or better social position. This returns us to arguments about the connections linking truth, credibility, and gentlemanly social status. It is also to recognize what we may think of as a "double movement" in many explorers' narratives: between a declared dependence on indigenous agents for practical matters and an epistemological dependence on indigenous knowledge that was, commonly, undeclared because such knowledge came from and was rooted in a culture deemed inferior to that of the European traveler-author. Either way, explorers and scientists in the field certainly depended on such "go-betweens," local authorities or figures who made themselves into vital intermediaries for the accumulation of knowledge in ways that were indispensable to the explorers' own authorial and authoritative claims, even if it was not acknowledged as such. Too great a reliance on such brokers could reduce one's claims to be an authoritative figure; too little, and one might not come home to turn the encounters into print.[49] The delivery of information was often part of a fiscal exchange, and in this sense, native witnesses might themselves be thought of as authors, explorers merely occupying the position of in-the-field readers or, more often than not, listeners. Native authorities also often engaged in cartographic acts, providing for their interlocutors rudimentary diagrams and directional aids.[50] Such matters of contact and codependence must be recognized, in the field and elsewhere. In scrutinizing the authoring of Murray's books of travel and exploration, such issues mean we must consider the extent to which explorers' texts were collaborative, even intercultural works, and, even, acts of cultural appropriation: John Barrow, we shall suggest, was no less a go-between, albeit of a very different sort, than indigenous intermediaries.

If dependence on human guides in the field might prove a matter of life or death or, more mundanely, result only in correct directions and temporary assistance, travelers could always seek guidance in print. From the later eighteenth century, a growing number of how-to travel guides in print sought to regulate travel and the traveler. Such guides have a long history, but from the later seventeenth century they were associated with developments in natural philosophy regarding that combination of plainness in speech and print, adoption of mathematics, and regulation in method and experimental procedure, now conventionally labeled the Scientific Revolution.[51] Empirical information on what, how, and why to observe became an important evidence base for what explorers and travelers could state as truthful and trustworthy facts. Training the eye meant training the hand: exploratory observation and the nascent travel fact required deliberative inscription to become a fact as writing and the notebook replaced memory as a guide to secure knowledge.[52] Eighteenth-century natural philosophers in the laboratory and field regularly cautioned about the need for accurate observation and on-the-spot recording: "The rationale for travel note-taking derived from the twin dangers of an unruly observation in the field and an unreliable memory."[53]

Printed guides for travelers were of different sorts albeit they had a shared end in view: that travel should be easier and safer and that travelers should become effective instruments—authors to be relied on as persons against whom others could assay truth claims. We may, loosely, distinguish between the moral, practical, and epistemological emphases of such travel guides: advice about comportment and manners being part of the first; matters of clothing, preparation, and in-the-field safety in the second; and issues of observation, writing, and reliance on and tolerance of other human instruments and precision devices making up much of the third. Although such guidebooks did not necessarily travel into the field, others did: hand guides to identifying, collecting, and pressing plant specimens, for example, had to. Books on the move directed travel just as specimen guides directed later exploration as to its nature and purpose.[54] Such texts were in effect guides to the traveler as a regulated moral and, even, a political-cum-scientific instrument. Questions of epistemology and truth telling in print were ineluctably linked to the status of one's informant, the social standing of the author, or the warrant by association that came with being officially sanctioned to have undertaken the travel or the exploration by a government or a scientific body (fig. 2). Such differences were a feature of Murray

FIGURE 2 The lone traveler—as well as the larger and scientific expedition—had to be properly equipped with the right instruments to survive, to explore, and to write. Source: Francis Galton, *The Narrative of an Explorer in Tropical South Africa* (London: John Murray, 1853). By permission of the Trustees of the National Library of Scotland.

publications: John Bettesworth's 1783 *The Seaman's Sure Guide*, for example, had much in common, despite its rhetorical style and more precise intended audience, with Sir John Herschel's edited *A Manual of Scientific Enquiry; Prepared for the Use of Her Majesty's Navy: and Adapted for Travellers in General* of 1849. So, too, the Murray *Handbook* series, begun in 1836 and which aimed at the discerning traveler but not at men of science, shared much with Sir Richard Colt Hoare's *Hints to Travellers in Italy* of 1815. Even so, regulating the traveler in these ways did not necessarily mean that their systematized observations and inscribed words about travel facts made it safely into print or, as the examples of Cook, Laing, Lander, Moorcroft, and Franklin attest, that the explorer made it home at all.

TRAVELS INTO PRINT, MAKING PRINT TRAVEL

The matters arising from Marie-Noëlle Bourguet's term "voyage into narration" center on that sequence of stages or process through which, as Felix Driver puts it, "the explorer in the field was translated into the published author."[55] As William Sherman notes, in order for exploration and travel narratives to take their final form, they must undertake two journeys, the one geographical, and the other authorial.[56] For Ian MacLaren, the process of making the one into the other is more complex than a simple binary model. Several stages are involved in an individual's exploration writing. "The first of those stages is the field note or log book entry, which is written *en route*. It marks the first effort by the traveller (who may or may not be travelling in order to write) to mediate experience in words." The second stage is the journal, "the writing-up of the travels either at their conclusion or following a stage of them." There is, MacLaren notes, often a rhetorical or lexical shift between field notebook and journal as the latter reviews, edits, and puts into narrative form, often with a degree of assured reflection, the uncertainties and perhaps abbreviated language of the first. This is to be explained by an awareness of audience: "Writers' awareness of readers vitally conditions the narrative, in terms of the way events are structured, plotted and phrased." The third stage is that of draft manuscript, the fourth and last that of publication, the difference between them, in MacLaren's terms, being again one of audience: "If the world ought to hear about one's travels, is the traveller the one to tell the world? Many travelled and explored brilliantly but did not write in a fashion that either they or a publisher considered sufficiently literary to lure the interests and purses of a readership."[57] In a later refinement, MacLaren advanced a six-part model for the evolution from traveler to author in print: from "field notes, diary, log book, containing entries made regularly if not daily while *en route*"; to a "retrospective journal, report, or letter, completed by explorer/traveller upon return home or at the end of a stage of a journey/voyage"; to the draft manuscript, "with or without interlineations and other indications of revisions"; to the printed book, in various formats and editions; to a scholarly edition; and, finally, to a further scholarly or trade edition.[58]

MacLaren's conceptual models, and Craciun's attention to "the manifold social agents and contingencies involved in bringing exploration writings into public circulation," help clarify the key issues we address here with respect to the house of Murray: of "author" as a category, but of "authorship" as an accomplishment and of "authoritativeness" as the criteria by which the explorer-author might be believed and as a matter of the rhetorical style he or she brought to the narrative.[59] They also highlight the importance of the material form of explorers' writing (if less readily the related issue of where such different stages of writing were done and if and how location matters to content and style). The difference in textual form (certainly in physical form) from field notebook to published book is too often taken to be suggestive of fixity of purpose and of meaning: authorial intent, we are being invited to consider, is at its clearest in the final published book stage. But as MacLaren, Craciun, and others show, printed publication did not equate to a single shared meaning. As MacLaren remarks, "That travel and exploration writing seldom receives discussion about the status of the text itself is worrying given the weight of ideological consideration to which books in this genre have been subjected, perhaps especially by students of imperialism and colonialism."[60]

In the field and elsewhere, writing and authoring could take different forms. Editing was not something done unto authors but part of "authorship," what one scholar has called "author-mongering."[61] All this being so, who, indeed, was the "author," whose was the "authoritative" voice? Edition history is no less a matter of correspondence: how much did later editions correspond in form and content to the first published edition? Publishers— always mindful of audience—make the final decision on content, print run, volume size, and costs, not authors. How was the book put together? Ought we even to think of a "final" text? For Jerome McGann, describing what he has called "the socialization of texts," there could, in the nineteenth century, never be any fixed version of a text, especially when multiple versions were in circulation. Against those textual critics who look for the authoritative version, on the one hand, and those who would argue that print endows the word with extreme stability on the other, McGann presents the printed word as more unstable than is generally thought.[62]

Several brief illustrations will serve to illuminate these points and lead us toward their fuller assessment with respect to the authors and books associated with the house of Murray. The words and images of George Back's account of his expedition to Hudson Bay on HMS *Terror* in 1836–37, for example—"an eleven months' harrowing voyage of nearly no geographical significance"—were, given Back's temporary and debilitating illness, amended on his return by a friend, by John Barrow, and by a Murray editor in order to make the facts of the travel seem more arduous, the author's actions more adventurous and perilous.[63] Analysis of the diary entries, the journal account, and the published version of John Franklin's first polar narrative, published by Murray in 1823 and documenting Franklin's expedition between 1819 and 1822, has revealed that, more mindful of the truth, Franklin "made a conscious effort to produce a more accurate and favorable account for publication than the one provided by his diary entries." Specifically, Franklin was more complimentary to a native tribal chief on whom he and others had relied: a fact overlooked by later commentators keen to see Franklin as an explorer in imperial mode (even though his first expedition was, as was Back's, a failure overall).[64]

Such Arctic travels into print had African counterparts. In 1820s West Africa, the explorer Hugh Clapperton undertook what has been termed his "in the raw" (that is, his in the field) writing in two formats, a log and a journal, altering the second more or less daily from the records contained in the first. The result was a fair copy text that was, in turn, later amended, partly by Clapperton, partly by Barrow.[65] For the celebrated African explorer John Hanning Speke, whose exploits with Richard Burton in the late 1850s investigating the source of the Nile led both to great public acclaim and bitter personal rivalry, the shift from explorer to author took place in the offices of the Edinburgh-based publisher William Blackwood & Sons. Speke's reputation was forged in Africa, but it was later confirmed and made public only through textual emendation in Scotland. The problem lay mainly with Speke's dreadful field notes and writing—"He writes in such an abominable, childish, unintelligible way" noted his publisher, John Blackwood—as well as with his innate sympathy for many of the African peoples encountered. The printed result, *Journal of the Discovery of the Source of the Nile* (1863), was neither wholly Speke's in style or substance nor at all his in imperialist sentiment. Speke the explorer was translated into Speke the author, but not in a manner of his choosing: "By emphasizing Speke's unique role as a successful explorer and Imperial adventurer . . . Speke the inarticulate traveler was transformed into an articulate, saleable commodity. . . . by reiterating standard views of Africa which fit generalized, European notions of the area."[66] The involvement of editors and publishers in effecting the evolution or translation from explorer to author could be complicated by the translation of an explorer's words from one language to another, with consequent changes in content and meaning. New English translations in the 1850s of Alexander von Humboldt's *Personal Narrative*, which Murray published in several volumes between 1814 and 1825, were undertaken with a view to "modernising" both the author and his text for new audiences: Humboldt's narrative became a "text in flux."[67]

Certainly illuminating, these illustrations are also instructive. MacLaren's cautionary remarks—that "the findings in the course of one book or of one explorer are not necessarily pertinent to any other case," and "nor should the availability of publishers' correspondence with authors necessarily serve to undermine the status of the published text itself"—are well made.[68] Different editions present a dilemma in terms of interrogating the "original" version.[69] Can we retrieve "authorial intent" when different people were at work in writing and making a book, when different editions may have altered both content and format and, vitally for our understanding of explorers' words, when we, as researchers, face "the question of basing arguments on printed versions when in so many cases the manuscript diary or journal on which the published account was presumably based is extant, waiting impatiently to reward its study"?[70] Authors may have had the truth in view and editor-publishers had audiences in mind. But differences in the evidence and interpretation of the reception of their printed works—from in-text marginal notes, written accounts regarding private reading, and printed public reviews—need also to be borne in mind, for it was in such ways and spaces that the reading public made sense of, and believed (or not), these accounts of travel and other achievements in geographical exploration.[71] With these issues in mind, let us return to the house of Murray and to its books of non-European travel and exploration.

THE HOUSE OF MURRAY AND BOOKS
OF TRAVEL AND EXPLORATION

The history of the house of Murray, particularly of Murray I, has been documented extensively elsewhere and so is not rehearsed here.[72] Rather, we outline the firm's engagement with works of travel, document the number of non-European works of travel the house published between 1773 and 1859, and present the thematic elements of our study. This is not to lose sight of the fact that the firm's establishment and development coincided with the emergence of identifiably modern practices in bookmaking and bookselling. The period was characterized by a spirit of commercial ingenuity that saw books treated increasingly as commodities directed to specific audiences.[73] Publishers sought to satisfy a reading public whose number and wealth were growing as a function of Britain's economic development, although it is impossible to know with certainty the size and composition of the audience for books, either at the end of the eighteenth century, or in the first half of the nineteenth century, when technological advances brought down the cost of book production to the extent that new titles became more widely available to a mass reading public and subscription, circulating, and public libraries brought reading matter into the hands of less privileged readers.[74]

Between 1773 and 1859, the house of John Murray published around four hundred books of travel, topography, and exploration. For several reasons, this figure is necessarily approximate. As William Zachs has shown, many of the firm's books in its early years were copublished.[75] This total does not include works of formal geographical instruction, such as geographical gazetteers, but it does include works of topographical description, such as John Wood's 1769 *A Description of Bath*. The total includes works of practical guides to scientific travel, including John Bettesworth's *The Seaman's Sure Guide, or, Navigator's Pocket Remembrancer* (1783) and John Herschel's edited *Manual of Scientific Enquiry* (1849), many of whose contributors were Murray authors in their own right. It does not include that particular genre of guide to travel, the *Handbook* series, which was begun by Murray II in 1836 (based on his own experiences as a traveler in Holland in 1829) and which eventually ran into several hundred volumes.[76] Nor does the total include works such as Lord Byron's *Childe Harold's Pilgrimage* (1812–18), since, although it is a kind of late Grand Tour travelogue, we have here excluded narrative poems.

The house of Murray published 239 works of non-European travel and exploration in the period between Sydney Parkinson's account in 1773 and Leopold McClintock's Arctic work of 1859. These texts are listed in the appendix (see pp. 227–85). We use this listing as a database in what follows, referring for ease of use to the short title and to the identifying number of the text from this list. Again, some notes of caution are justified in interpreting this total. The appendix includes works whose subject matter incorporates some travel beyond Europe even although their main focus lay with travel in Europe. Some titles, notably within the era of Murray I, were copublished, or published by others before the house of Murray became involved in a later edition or a change of format. Some were published several years after the events they described. Some texts went into at least a second edition. The total is arrived at from assessment of the works published by John Murray I, examination of the annual publication lists produced by John Murray II, analysis of the John Murray Archive, and examination of union catalogs and contemporary advertisements. As Leslie Howsam notes in her experiences determining the Kegan Paul imprint, it is possible that a meticulous bibliographer would find more for Murray's non-European travel list.[77] The appendix should, therefore, be considered a work in progress. From this total of 239 works, a number of features emerge with respect to the house of Murray's publishing of non-European travel in the eighty-six years from 1773.

The first is chronology. Between Parkinson's narrative of 1773 and John Taylor's *Letters on India* in 1800 (app.: 1–17), Murray had a share in only a dozen or so non-European travel books. From the second decade of the nineteenth century, however, the firm's publication of non-European travel and exploration narratives gathered pace: from three or four such works annually in the early 1810s, for example, to a "peak" of eleven such works in 1819, ranging from Thomas Bowdich's African narrative to John Ross's Arctic account, and from Edward Sabine's *North Georgia* chronicle to Abraham Salamé's Algerian enquiries (app.: 48–58). In the 1820s, Murray published fifty-eight such works, but only about half that number in the 1830s (app.: cf. 59–116 [1820s] to 117–45 [1830s]). In the 1840s, the firm returned to earlier levels in this respect, with fifty-five non-European works in that decade (app.: 146–201).

In part, this phasing is easily explained: after 1815, as war with France ended, restrictions on European travel were eased, and, for global discovery especially, notably in Africa, East Asia, and the Polar regions, British naval officers on half pay became available for voyages and land travels on behalf of the Admiralty. This trajectory, as William St. Clair argues, reflected the "rapid expansion of new writing on foreign travel, voyage, [and] exploration" that followed the conclusion of the Napoleonic Wars: "Military men on reconnaissance, naval men on patrol, scientific explorers, industrial exporters, travelling gentlemen, and Christian missionaries . . . [all produced] diaries which told of their adventures and their discoveries."[78] The increase in the 1820s was despite the 1825 banking crisis that bankrupted the Edinburgh publisher Archibald Constable and forced the firm of Murray to "retrench" in the period 1826–27.[79] The downward trend of the 1830s reflects the aftereffects of a series of failed speculative enterprises in South America (which were largely responsible for the 1825 panic), and a tailing-off in the initially enthusiastic assaults on the Northwest Passage, with the result that public interest in, and Murray's output of, travel texts dipped. The 1830s presented a series of challenges for the book trade in Britain as a whole, mostly related to Britain's political and economic instabilities. George Paston estimates that "at this time not more than one in fifty [books published] paid its expenses": one reflection of this shift in fortunes was that Murray increasingly required its authors to assume the "sole cost and risque [*sic*]" of publication.[80] In fact, the financial arrangements the Murrays struck with their authors for books of travel and exploration followed a fairly standard model, with the publisher taking all of the risk and the author either half share or two-thirds of any profits. The Murrays would occasionally vary these arrangements, sometimes by giving advances on any share of profits, for first and subsequent editions, as was the case with David Livingstone's *Missionary Travels* (1857; app.: 235), or, exceptionally, for individual authors, tried-and-tested friendship might weaken long-established business models: "My business is to pay authors and not be paid by them," Murray III wrote to Woodbine Parish in 1855, "for this reason my rule is not to publish any work except at my own expense and risque and when I make an exception it is only as a special favour to oblige a friend, such an exception is the reprint of your Buenos Ayres" (app.: 137).[81]

A second topic is the format of the books. Table 1 shows trends in Murray's books of non-European travel and exploration by format between 1773 and 1859. As will be clear, the growth in numbers of non-European works in the 1820s was evident also in the rise of the quarto volume (a relatively large, expensive format), a fact discussed by Janice Cavell for Murray and others' narratives of polar exploration but which was not alone associated with accounts from that region, as works on Egypt, Australia, Southern Africa, and China among others reveal (app.: 59, 62, 64, and 67).[82] Format reflected publishers' sense of their market. To judge from one commentator, the shift away from the quarto format in the 1830s in Murray's non-European travels, here for an octodecimo volume of the Landers' Niger travels in the firm's Family Library series, was welcome:

> Of all books, voyages and travels are, generally speaking, among the most read and the most relished. To this the varied nature of their contents chiefly contributes; for the matter-of-fact incidents which are narrated, give them the stamp

TABLE 1. Books of Non-European Travel and Exploration Published by John Murray between 1773 and 1859, by Decade and Format

Decade	2°	4°	8°	12°	18°	Total
1770s	0	2	0	0	0	2
1780s	0	2	2	3	0	7
1790s	0	2	5	0	0	7
1800s	0	2	3	0	0	5
1810s	0	20	17	0	0	37
1820s	0	22	33	1	2	58
1830s	0	0	25	2	2	29
1840s	1	1	46	7	1	56
1850s	0	0	35	3	0	38
Total	1	51	166	16	5	239

Note: Data derived from the appendix. Book sizes are abbreviated thus: 2° = folio; 4° = quarto; 8° = octavo; 12° = duodecimo; and 18° = octodecimo.

of history, while their novelty of scene and subject gives them the interest of romance. The contents of these little volumes, which a few years since would have been ushered into the world in a goodly and ponderous quarto, will amply repay readers of all classes for the time spent in their perusal, and we trust that they will find numerous purchasers to recompense Mr. Murray for his enterprise in securing them for the Family Library—the first, if not the best, of those numerous libraries which have recently sprung up so prolifically for the cheap dissemination of amusement and instruction among all ranks of society.[83]

The vast majority of Murray's authors were men, and British men of a certain social status and position, often within the British Admiralty, or associated with the diplomatic service or Britain's overseas trade, or, simply, possessed of independent means and with the desire to travel. Of our sample, only thirteen Murray authors between 1773 and 1859 were women, Maria Graham standing out as the exception in this small group as the author or editor of three works (app.: 6, 75, 78, 79, 91, 113, 130, 148, 161, 163, 170, 172, 199, 210, 225, and 230). Finally, the geographical coverage of Murray-published non-European books of travel and exploration shows that all regions of the world were the subject of attention, although closer scrutiny of the appendix reveals trends within the chronology and geography of publishing: no polar work before Barrow's 1818 *Chronological History* (app.: 43), then a concentration of such work from the 1820s; much interest in Africa, especially northern and western Africa, in all periods; the Near East and Egypt likewise a source of continued interest; South American works concentrated in the 1820s.

In what follows, we are principally interested in what lies behind these numbers, the periodicity of publishing, and the varied formats used—interested, that is, in what the appendix does not show, namely, how the firm worked to solicit its books and, particularly, how it negotiated with its authors over their production. Our focus is on the acts of exploring and the writing of works of travel and exploration, what may be thought of as the connective tissues between authoring, authorizing, and authoritative content. To do this has involved detailed study of the surviving manuscript evidence—notably, but not only, the correspondence between author and publisher over the house of Murray's narratives of travel and exploration—together with study of the printed works themselves. From analysis of this material on explorer-author matters and from assessment of such financial records from the ledger volumes that bear on the authorial process and the production of individual books and so illuminate the author-publisher relationship as a whole, it is possible to discern how Murray, the firm's readers, and its many authors worked to produce printed words from field notes, even, for some, when such notes had been lost and where memory had to stand in for inscribed record, or where Murray altered authors' words before they made it into print. Where possible, and with respect to some individual works, we have placed the Murrays' endeavors in context with other publishers' work at this time: our detailed study of questions of correspondence and authorship precludes lengthy comparative analysis. In his history of Longman, for example, Asa Briggs claims that "travel was always as important an element in the Longman lists . . . as it was in those of Murray," but offers no detailed assessment in support of his claim.[84] Travel and geographical research was, at 6 percent, a small part of Kegan Paul's output. It was more significant from the 1870s, and, in the 1880s, it was the travels of Sir Richard and Lady Burton that dominated their list.[85]

Our principal focus on processes rather than numbers and formats, on authorship rather than on cost or aspects of the technology of publishing, is mindful nevertheless of the firm's development as a publishing house. Medical texts and poetry were the dominant categories of Murray I's publications, these subjects representing approximately 23 percent and 10 percent, respectively, of the total number of books published in the period 1768-93.[86] Travel and topography together were only Murray I's tenth most common genre (equal with drama) and equated to just a little over 3 percent of books published in this period (a figure that is even smaller when topographical texts and geographical grammars are excluded).

In 1783, Murray I established the monthly *English Review*, which built on his experience in producing two earlier periodicals, the *Repository: or Treasury of Politics and Literature* (founded in 1770) and the *London Medical Journal*, established in 1781. In 1791, the firm published Isaac D'Israeli's celebrated miscellany, *Curiosities of Literature*, inaugurating a friendship that was continued by Murray II.[87] D'Israeli was one of the first of Murray's authors who, along with John Barrow, John Wilson Croker, William Gifford, Henry Milton, John Gibson Lockhart, and others, would become the firm's trusted "readers," evaluating the literary merits of proposed new works as they arrived at the firm. The house of Murray's social and literary circle included literary authors who helped secure the firm's reputation: Lord George Byron, the first two cantos of whose *Childe Harold's Pilgrimage* were issued by Murray in 1812; and, although not famous in her lifetime, Jane Austen, whose *Emma* was published by Murray in 1815; and many others.[88] In the firm's drawing room at 50 Albemarle Street, Murray entertained "persons of the very highest rank for Literature & talent," a group that Sir Walter Scott subsequently styled the "four o'clock friends" (plate 6).[89] Although it is the literary and poetical members of these soirées who have tended to attract the attention of the firm's chroniclers, Murray II's undertakings in travel literature began to take off after this period and to become the firm's financial backbone.

John Murray II, in particular, managed the publication of books on parts of the world in relation to market demand: authorship reflected matters to do with audience as well as with the facts of writing in the field, truth telling, and those other elements experienced by explorers and travelers who became "authorized." John Murray III's interests in the earth sciences helped him bolster the standing of the firm, which issued several leading works of science: Charles Lyell's three-volume *Principles of Geology* (1830–33) and Charles Darwin's *Journal of Researches* (1845) and *On the Origin of Species* (1859). The Murray *Handbook* series, not considered here, followed Murray III's own European continental travels during the early 1830s.[90] The resultant publication in 1836 of *A Hand-Book for Travellers on the Continent* (issued under the imprint of John Murray and Son) marked the firm's entry into the production of books specifically aimed at the then-emerging tourist market.[91] Murray's handbooks came to establish a reputation for comprehensiveness and accuracy: as one contemporary had it, "he [the reader] trusts to his Murray as he would trust his razor, because it is thoroughly English and reliable."[92] Murray recruited and carefully selected authoritative travel writers for his Handbooks and non-European texts, a retinue that included John Wilkinson for Egypt and Richard Ford for Spain. Like Murray II, Wilkinson and Ford were fellows of the Royal Geographical Society (RGS; Murray was a founding member), and Wilkinson and Ford authored other works under Murray's aegis.[93]

The house of Murray's place in the publishing landscape was aided further by its establishment in 1809 of the *Quarterly Review:* "a Tory counterblast" to Archibald Constable's *Edinburgh Review*.[94] The *Quarterly* helped secure income and influence as well as expanding Murray's circle of friends and authors.[95] Where the *English Review* was commercially significant for the house of Murray, the *Quarterly Review* with monthly sales of over twelve thousand afforded both an important critical space and an alternative to rival outlets: John Barrow, writing to Murray in 1816 over several literary matters, reported "I have a letter from Edinburgh in which I am told that the Quarterly is working its way in that auld toun in spite of its rival [*Edinburgh Review*], and what is still better, in spite of metaphysics and metaphysical doctors."[96] Barrow and John Croker, respectively, second secretary and first secretary to the Admiralty, were regulars at Murray's literary salons (see plate 5). Barrow and Croker shared interests concerning Britain's maritime policy, its exploratory and imperial agenda, and its international scientific standing. Both men served on the council of the Royal Society, both facilitated vital links between Britain's scientific, maritime, and publishing establishments, and both were recruited by Gifford to contribute to the *Quarterly*, something they did prodigiously: Barrow composed around two hundred articles; Croker was involved, as author or coauthor, in 270 items.[97]

The *Quarterly* was perhaps the principal popular medium for the communication of Barrow's exploratory and imperial vision, ambitions given additional impetus following the conclusion of the Napoleonic Wars in 1815.[98] The necessity to find a meaningful role for the approximately twenty thousand naval officers laid up on peacetime half pay was pressing. For Barrow, the solution was obvious. In his editorial introduction to the Murray-published *Narrative of an Expedition* (1818; app.: 47, and see chap. 2, pp. 41–45), undertaken by James Kingston Tuckey (who was in no position to adjudicate on written accounts of his own words, having died on the expedition), Barrow sketched his overall vision with characteristic patriotic rhetorical flourish: "To what purpose indeed could a portion of our naval force be, at any time, but more especially in a time of profound peace, more honourably or usefully employed, than in completing those *minutiæ* and details of geographical and hydrological science, of which the grand outlines have been boldly and broadly sketched by Cook, Vancouver,

Flinders, and other of our own countrymen."[99] Where Barrow read drafts and reviewed published versions, Byron lauded his publisher. Writing to Murray from Venice in 1818, Byron wryly congratulated Murray—whom he now nicknamed the Admiral—on his relationship with Barrow:

> Along thy sprucest bookshelves shine
> The works thou deemest most divine—
> The "Art of Cookery," and mine,
> My Murray.
> Tours, Travels, Essays, too, I wist,
> And Sermons to thy mill bring grist;
> And then thou hast the "Navy List,"
> My Murray.
> And Heaven forbid I should conclude
> Without "the Board of Longitude,"
> Although this narrow paper would,
> My Murray.[100]

Barrow's influence on Murray's publishing of works of exploration and travel appears in several guises throughout our book, precisely because it points to the complex explorer-author-publisher relationships we examine here. As others have disclosed, Barrow's influence was especially evident in relation to polar exploration. In 1817, Barrow's article in the *Quarterly* was, notionally, a review of Edward Chappell's *Narrative of a Voyage to Hudson's Bay* (1817), a text he dismissed as "slight" and "extremely ill advised."[101] Barrow devoted most of the remaining twenty-four pages to the possibilities for Arctic exploration that recent intelligence concerning the "disappearance of an immense quantity of arctic ice" suggested.[102] The scheme, setting out the scientific and commercial benefits of high-latitude exploration, and particularly the desirability of settling the longstanding question of the existence or otherwise of the Northwest Passage, owed a considerable (and largely unacknowledged) debt to the whaler-navigator William Scoresby, who had proposed such an undertaking first to Joseph Banks and then to Barrow. As Fergus Fleming notes, Barrow simply published "Scoresby's plans as his (anonymous) own," an act his position made possible, if not morally defensible.[103] As Kim Wheatley shows, Barrow used the *Quarterly* more generally to propose geographical theories, advance his views on Arctic and British exploration, and, commonly, to offer lengthy commentaries on the personnel and issues he took to be leading figures or failures in advancing Britain's interests in these respects.[104] Barrow's 1817 article, and others that followed on the same subject, were important prompts to British Arctic exploration from 1818 and served to spark popular interest "on a mass scale."[105]

The house of Murray was at the heart of these polar exploratory enterprises and the narratives that emerged from them because John Murray II was at the heart of the institutional networks which sustained them. But Barrow's role as go-between was far from unique, and the firm's publishing of overseas travel for other parts of the world and at different times was shaped by other circumstances. As a further and final illustration in introducing what was involved in the making of exploration authorship, let us consider the fractious genesis of the first such book the house produced, Sydney Parkinson's posthumous *A Journal of a Voyage to the South Seas* (1773; app.: 1).[106] Parkinson, a talented watercolor artist specializing in botanical subjects, was initially commissioned by Joseph Banks to depict material collected during Banks's 1766 expedition to Newfoundland and Labrador. Banks then recruited him to serve as his botanical artist on an expedition to the South Seas—what became James Cook's *Endeavour* expedition. From 1768 to 1771, Parkinson completed more than a thousand drawings of landscape and natural history subjects and additionally compiled vocabularies of Pacific languages. Parkinson died on January 26, 1771, before reaching home, having contracted malaria and dysentery at Batavia.

One of the stipulations of the expedition's sponsorship was that "all journals and diaries would be surrendered at the end of the voyage" and submitted, under the direction of the Admiralty, to a "professional author," in this instance John Hawkesworth, who would compose the official expedition narrative.[107] Given that Parkinson was not officially a member of the *Endeavour*'s crew, but part of Banks's personal retinue, a dispute ensued concerning the copyright of Parkinson's personal journal of the expedition, particularly because it contained much colorful detail not otherwise recorded and retained by Cook and Banks. Here, in its first overseas travel text, the house of Murray was, as part of a larger publishing syndicate (see app.: 1), exposed to questions of authority over others' in-the-field work of exploration. Banks, as Parkinson's employer, felt that the right to publish the papers rested with him. Parkinson's brother, Stanfield, felt otherwise and arranged independently for the book's publication. To prevent this, Hawkesworth took out an injunction. This was only lifted in June 1773 on the publication of Hawkesworth's three-volume account, a narrative that lacked almost totally the thrill and novelty that its source material contained.[108] The Hawkesworth narrative has been described as "prolix, abstract, and much given to philosophical digression," not at all what the reading public expected from an exotic excursion in the South Seas.[109]

Parkinson's volume appeared later that year, notionally edited by Stanfield and sold by Murray and five other London booksellers. The volume also contained a lengthy and scurrilous preface (signed by Stanfield but authored largely by William Kenrick, a writer and satirist whom Stanfield had employed) that detailed the long debate over the rights to Parkinson's papers and possessions.[110] The preface depicted Banks and John Fothergill, a Parkinson family friend who had mediated in the dispute, in an unfavorable light. Fothergill was scandalized by Stanfield's abuse of his good offices and purchased some four hundred remaining copies of the book. Into these copies, Fothergill inserted a twenty-two-page pamphlet offering his explanation of the dispute.[111] Following Stanfield's death, Fothergill secured the rights to Parkinson's journal. This was, at Fothergill's request, republished in 1784, after Fothergill's own death, complete with his riposte to the earlier preface.

At a time when "Polynesian affairs were . . . the rage," Murray's decision to take a share in Parkinson's volume was based on a desire to satisfy "a reading public avid for news of far-off lands."[112] As we have discussed, travel writing was then being given considerable momentum by the combination of scientific discovery, public interest, and political competition: there was, by century's end, scarcely a location on the globe that seemed not to have been "sighted, explored, traded with, taken possession of, catalogued or written about by British travellers."[113] As one contemporary had it, "The press almost daily teems with some new travels, journies [sic], or voyages."[114] The public interest (and potential profitability) associated with travel texts is evidenced by the "notoriously extravagant" £6,000 fee paid to Hawkesworth for his production of the *Endeavour* narrative (a volume that also drew together a number of earlier, and related, expeditionary accounts).[115] Importantly, too, Hawkesworth's text "flouted many of the conventions of eighteenth-century travel writing" in that it was the work not of an eyewitness but of a literary compiler and editor working at a geographical and temporal remove from the events described. That the book failed, in this respect, to "satisfy a freight of generic and moral expectations" assumed of travel literature dented its credibility in the view of some, but it seems not to have affected its popularity to many of the reading public.[116] For contemporaries, Parkinson's eyewitness account offered a valuable corrective to Hawkesworth's interpretation (misrepresentation, in the views of some) of the *Endeavour* expedition. For modern researchers, its story foreshadows many of the issues that the Murray firm faced in later years as it brought exploration narratives into print.

What follows examines the nature of exploration, the processes involved in books' making and authors' shaping, and the reception given to Murray's travel and exploration narratives in relation to the issues identified above and for 239 non-European books of travel and exploration. The structure of this book broadly follows the stages of books' travel into print as explorers, authors, and publisher would have encountered them: the undertaking of the work in the field, the move of manuscript and in-process works of exploration from travel to print, from published work and the processes involved in preparing the book for production, to audience reaction. In chapter 2, we turn away from the publishing firm to consider the motives behind exploration. Since many works of exploration were prompted by formal government enquiry and by the scientific questions of the day, it is pertinent to ask if notions of authorship and exploration were shaped even before the events themselves took place. Chapter 3 addresses questions of authorship and authorization in looking at the ways in which explorers became authors. Chapter 4 discusses the routes to truth in travel and exploration writing. In chapter 5, we consider the nature of the books themselves and how Murray sought to manage and to present the information the firm's explorers and travelers unearthed. Chapter 6 looks at Murray's books in the marketplace and at their reception and review. Chapter 7 draws together our evidence and examines its wider implications. The terms used by MacLaren, Craciun, and others, such as "explorer-author," are subject to interrogation as we examine how these works of exploration and travel were put together and became printed books. Our attention to the issues of exploration and authorship and to bookmaking as processes involving different actors means that several themes, and key illustrations of them, appear in more than one chapter. We have been cautious not to extrapolate too greatly from the particular examples we offer about the nature of non-European exploration and travel writing in the period under review. It will be clear, nevertheless, that the house of John Murray was at the center of the production of travel narratives. Our interrogation of the John Murray Archive discloses how exploration and travel moved from the field to the page, how books made authors of explorers, and how truth was made and credibility attained in and through print culture in late eighteenth- and early nineteenth-century Britain. Other scholars working at the intersection of book history, authorship, and geographical and historical change have sought to elucidate the "operations that made it possible to set the world of the written word in order."[117] *Travels into Print* is just such a study—in showing how writing the world involved putting others' words to order.

CHAPTER TWO

Undertaking Travel and Exploration: Motives and Practicalities

Writing to John Murray II in 1826 about the reception of his just-published *Journal of a Third Voyage for the Discovery of a North-West Passage* (app.: 96), William Parry alluded to the several demands of being a naval officer, geographical explorer, and polar author: "It is very gratifying to my feelings to be assured that my Book is well spoken of by those whose judgment is the most to be valued on such subjects." But, he continued, "Too much is, in these book-making days, expected of Naval Officers in this respect; for they are accustomed to *act* more than to *write*—but both are expected from us now."[1]

Parry was not alone in his feelings about the demands of seamanship and authorship. For him and other naval officers engaged in scientific work, exploration depended on writing and mapping. Writing was required as a record of exploratory undertaking, as testimony to a voyage's significance, and as a record against which that and other information could be tested. A ship's logbook was a record of authority and observational accuracy, in intent at least: something to be given up on return as an indication of command, a reflection of success, and a record of one's ability to write and record accurately. Mapping, too, was a form of inscription that sought accuracy and yet was often hindered by the very geography it sought to represent. Like his fellow polar explorer-authors, and others with military backgrounds—future Murray authors recounting tales of sub-Saharan Africa or of mapping and diplomacy in Central Asia—Parry's exploration and writing reflected official instructions from the Admiralty or from respective colonial secretaries of state. For those authors and their superiors, travel and exploration had clear motives. Because it did so, it is legitimate to ask whether certain practical consequences followed— observation and recording at regular intervals, writing not just in the field but at certain times of day and in certain ways, in order to ensure accuracy of account and to emphasise the scientific content and the longer-term benefits rather than recount the immediate personal circumstances of traveling in unpredictable and dangerous environments. Formal instructions, we must suppose, presumed formal purpose and with particular ends in view for the exploration. But that is quite different from supposing either that such directives necessarily determined a formal approach to the nature and style of in-the-field writing about exploration or that we should presume there to have been what we might think of as an authorial regimen, of moments set aside for observing, recording, writing things down.

For other Murray authors, the motives for being a traveler and a writer stemmed not at all from others' institutional instructions-cum-official imperatives for the advancement of geographical and scientific knowledge or in extending Britain's commercial reach. Other, more personal, reasons were advanced: curiosity about a place or people; interest in addressing the "novelty" of a region; a desire to see things for one's self, to test others' words at first hand, to make a name for one's self, or, simply, to heed the call to adventure. Motives could be, and often were, multiple in nature. William Henry Edwards's interest in South America had been piqued by "the graphic illustrations in school Geographies, where men riding rebellious alligators form a foreground to tigers bounding over tall canes, and huge snakes embrace whole boats' crews in their ample fold": his Amazonian travels aimed at being an "unpretending volume" in which he felt able to admit to "the absence of some of the monsters which did *not* meet his curious eye."[2]

Consider, too, George Featherstonhaugh, author in 1844 of *Excursion through the Slave States* (app.: 166) and one of several travelers to North America who became a Murray author. Featherstonhaugh's travels throughout the southern United States in the mid 1830s were motivated both by a personal commitment to reveal the condition of the slave-owning states in the face of a perceived lack of such information within Europe and from a professional interest in America's geology and natural history following his appointment in 1834 as the first US government geologist. Gentlemanly travel by public stagecoach with the first aim in view conflicted with his ability to collect specimens and inspect the country rock at first hand: "Travelling in a public vehicle would seem to present singular impediments to a correct investigation of the geology and natural history of a country, as no doubt it does; and if I had not been already familiar with the structure of the Alleghany ridges immediately west of the *Blue ridge*, I should have regretted the very limited opportunity now afforded me of forming accurate opinions." As he moved from place to place, obtaining "rational and obliging answers" from "very civil people," so Featherstonhaugh was known by different sobriquets: "Mr Wright [his host] seeing me curious about rocks and shells, always called me *Doctor;* most of the people . . . with whom I had formed an acquaintance, called me *Colonel;* and some of the blackeys that waited upon me, called me *Judge*."[3] Only from 1844 could he term himself "author": the British-born Featherstonhaugh's strongly held abolitionist views hindered publication until, back home, he felt able to "express with perfect freedom any opinion that were [*sic*] on the side of humanity, of rational liberty, and the moral government of mankind."[4]

The different motives and experiences illustrated in the examples of Parry and Featherstonhaugh speak to the concern of this chapter with the motives and practicalities of travel, specifically to the related questions of "Why travel at all?" and "Did travel influence writing in the field?" The fact that Parry's work was published and that he became a distinguished polar author (and, in writing, revealed himself to be a distinguished polar explorer) stemmed, as for all Murray narratives of travel and exploration, from the publisher's judgment as to the merits of the manuscript, guided by his perceptions of the potential and the opinions of audience and his in-house readers. But the processes of becoming an author, with knowing how and why Murray and others judged the merits of texts, how putative authors were authorized, and how their narrative was regarded, and in what sense, as authoritative—questions we address in the chapters following—are altogether different from knowing why people traveled and with knowing how such travel acted on their writing in the field.

In beginning his *Life in Abyssinia* (app.: 218), Mansfield Parkyns observed that "a book of travels should be either a scientific work or an entertaining one."[5] For the modern researcher, it is appropriate to ask how far, if at all, prior motive was a determinant on travelers' writing in ways that made the work the one or the other. Did the different facts of travel in different places impede or facilitate writing? For Richard Wilbraham, writing was initially a private matter, latterly a solace to the loneliness of travel in the Caucasus: "My journal was, at first, kept for the amusement of my own family: latterly, travelling as I was without the society of any European, it became quite a companion to me."[6] For others, traveling under orders,

writing in particular forms almost became an obligation, to be done despite
the rigors and the loneliness. For Francis Bond Head, who covered over
three thousand miles on horseback across the pampas of South America
in a matter of weeks in undertaking a survey of gold and silver mining
prospects, "rough riding" meant "rough writing"—and reviewers loved his
Rough Notes (1826; app.: 95) for that reason: "The reader is fairly instigated
to stop and take breath," noted one, "as if he had been *bodily* accompanying
the author."[7] Head, author of other books for Murray and one of the firm's
readers, was ever after known as Galloping Head. Head's narrative reflected
his mode of travel—fast moving, interrupted only when necessary, a breath-
less encounter for author and reader alike. His case, and those of Parry and
Featherstonhaugh, highlight the need to think about writing in the field
and to consider the relationships between the motive and the practicalities
of writing before the "final" work was brought home, there to be reflected
on, amended, and authorized.

Placing travel writing and travelers in a typology is, as a practice, almost
as long established, and as difficult, as travel itself. To Lawrence Sterne, author of the popular *Sentimental Journey* (1768), "the whole circle of travelers
may be reduced to the following Heads: Idle Travellers; Inquisitive Travel-
lers; Lying Travellers; Proud Travellers; Vain Travellers; Splenetic Travel-
lers," and to what he called "the Travellers of necessity, The delinquent and
felonious Traveller, The unfortunate and innocent Traveller, The simple
Traveller, and last of all (if you please) The sentimental Traveller," in which
category Sterne placed himself.[8] Such an essentially moral categorization
is not especially useful with respect to questions of motive for the many
Murray authors of non-European exploration and travel under review here,
quite apart from the fact that Sterne's taxonomy is highly satirical. Neither
is that simpler and more recent distinction between "the adventurous" and
"the less venturesome" traveler since, although the first included missionaries, circumnavigators, and scientists, many in the latter category traveled, as
did Featherstonhaugh, with mixed motives in mind.[9] For Nigel Leask, the
emphasis of some contemporary land-based travelers on "abiding curiosity," their motives rooted in antiquarian interests and in the aesthetics of
antique lands and peoples, certainly produced narratives self-consciously
subjective in tone. But many such romantic-period land-based travelers,
whether to the Near East or to South America, were also sustained by scientific and rational motives akin to their ocean-going counterparts, albeit
that the language of the text and the attention paid to the travel experience
sometimes occluded the factual content.[10]

Not all of Murray's explorer-authors disclosed their motives, nor did they make clear how they had written in the field. And given what we shall see of authors' writing on return and of editorial redaction, we must be wary over such statements offering either a true or a full explanation. Our attention is to different sets of examples: exploration narratives prompted by formal instruction over scientific and imperial ends; travel accounts motivated by novelty and utility; and the works of travelers whose travel and writing aimed at the correction of others' accounts. We look also at the house of Murray's involvement in producing how-to guides to travel and at what these works suggest about guidance on the nature of writing as part of the travel experience. Our intention is neither to propose a typology nor to compare examples against the emergence of normative standards of practice in travel and travel writing but, simply, to explore the different ways explorers and others thought about being in, and writing in, the field.

"THE PRACTICAL ILLUSTRATION OF GEOGRAPHICAL SCIENCE": EXPLORATION AND WRITING UNDER INSTRUCTION

Of the 239 non-European travel accounts and narratives of exploration under review, thirty-four works either declared themselves as or can be shown to be the result of some degree of formal instruction issued before the travel itself.[11] Such directives were issued by the British Admiralty, alone or in association with the Colonial Office of the British government or, occasionally, with the Royal Society.[12] Each of these thirty-four works was a narrative of scientific exploration and, even, of British geographical and imperial aggrandizement. The relative attention given to different regions of the world in the years between 1813, the date of the publication of John Macdonald Kinneir's *Geographical Memoir of the Persian Empire*, and James Clark Ross's *Voyage of Discovery and Research in the Southern and Antarctic Regions* in 1847 (app.: 26 and 187, respectively), is suggestive of a "core" geography to Britain's imperial and scientific interests in this period: principally the polar regions, western and northern Africa, and Central Asia but not yet central and southern Africa or the Near East. The polar regions account for fourteen of the thirty-four texts (all but one of the fourteen being concerned with Arctic exploration), Africa's exploration eight, Central Asia six, South America and Australia (together with Oceania) two each, with Arabia (including Yemen and Oman) and China each being represented by one work.

As with the overall totals and for any typology, a note of caution is necessary. The accounts in 1824 and 1826 by Maria Graham (later, Lady Maria Calcott) were facilitated by the official naval duties of her first husband, not by any brief given directly to her. John Macdonald Kinneir's *Geographical Memoir of the Persian Empire* (1813), Mountstuart Elphinstone's *An Account of the Kingdom of Caubul* (1815), and John Malcolm's *Sketches of Persia* (1827) were each the consequence of Britain's diplomatic and political interests in Central Asia following Malcolm's appointment as ambassador to Persia in 1799 (app.: 26, 31, and 100, respectively). Perhaps understandably, given diplomatic sensitivites, the several authors are silent about the formal terms of their appointment. Similarly, the works of Alexander Burnes in 1834 and in 1842 (app.: 126, 156) were the result of extensive travel in Afghanistan, Central Asia, and British India's northwest frontier between 1824 and 1842 and from a heady mix of cartography, military surveillance, and imperial espionage that, while hardly hinted at in his printed narratives, is all too evident from surviving manuscript evidence pertaining to "secret committees."[13] Other works followed from association with their superiors' formal explorations: this is the case for Lyon's *Private Journal* (1824; app.: 83) and Wood's *Personal Narrative* (1841; app.: 155) and their connections with William Parry's polar work and Burnes's Afghanistan and Central Asian travels, respectively. Lyon was published because of his journal's ethnographic focus on "Esquimaux" (i.e., Inuit) cultures rather than Parry's official and scientific account in his *Journal of a Second Voyage*. Wood's motive lay in extending geographical information about the Oxus River and from what he saw as necessary correctives to Burnes's inflated claims about the navigability of the Indus and the possibilities for Britain's commercial and political advance into northwest India.[14]

The association between formal instruction and exploration is apparent in several texts. In Phillip King's *Narrative of a Survey of the Intertropical and Western Coasts of Australia* (1827; app.: 99), we are told, as was he, the purpose of his travels: "The principal object of your mission is to examine the hitherto unexplored Coasts of New South Wales, from Arnhem Bay, near the western entrance of the Gulf of Carpentaria, westward and southward as far as the North West Cape; . . . as the chief motive for your survey is to discover whether there be any river on that part of the coast likely to lead to an interior navigation into this great continent"[15] (fig. 3).

Guidance was given on "the most important subjects, on which it will be immediately your province, assisted by your officers, to endeavour to obtain information on any occasion which may offer"; the categories included climate, topography, animal, vegetable, and mineral resources, the peoples

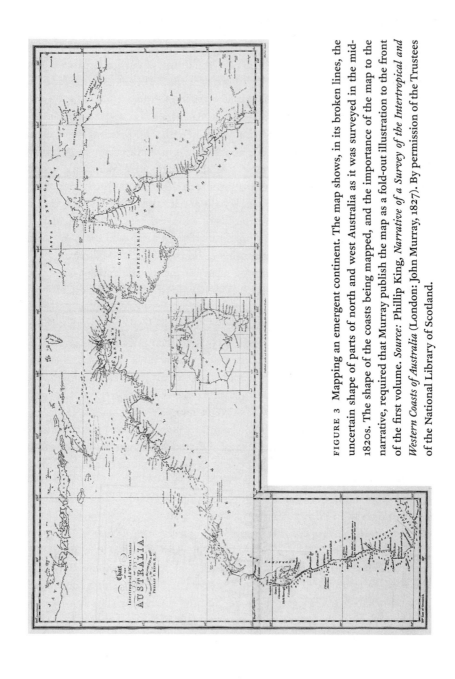

FIGURE 3 Mapping an emergent continent. The map shows, in its broken lines, the uncertain shape of parts of north and west Australia as it was surveyed in the mid-1820s. The shape of the coasts being mapped, and the importance of the map to the narrative, required that Murray publish the map as a fold-out illustration to the front of the first volume. *Source:* Phillip King, *Narrative of a Survey of the Intertropical and Western Coasts of Australia* (London: John Murray, 1827). By permission of the Trustees of the National Library of Scotland.

and culture of the region, and the native languages spoken. Colonial Secretary Bathurst's memorandum of instruction closed with guidance over writing and record keeping: "You will not fail regularly to keep a journal of your proceedings, and to note down your observations, as they from time to time occur, transmitting home by every opportunity intelligence of the progress which you have made, and of the leading events which may have befallen you."[16] But in the field and especially on his return, King's authorship of his finished narrative involved the editing of his observations and giving priority to the charts over the textual narrative: "Since the return of the Expedition, my time has been occupied in arranging the narrative and divesting it of such parts as were neither calculated to amuse the general reader, nor to give information to the navigator; but this has been so much impeded by the more important employment of constructing the Charts of the Survey, as to defer until the present season the publication of the events of a voyage that was completed nearly three years ago."[17] Such evidence provides a clear indication of the role of memory in authors' writing and, in the case of King at least, clear statements about the work's content and the delay between the travel events and the travel writing. The *why* of his travel is clear; its *how*, specifically how and when he wrote in the field, much less so. This pattern is apparent in the case of several others to whom formal instruction was given.

James Kingston Tuckey's *Narrative of an Expedition to Explore the River Zaire* (1818) illustrates the intent and precision of contemporary instructions and the difficulties faced by modern researchers in knowing exactly what such instructions actually meant for writing and recording on a day-to-day basis (app.: 47). The work, motivated by Britain's concern for "the advancement of physical and geographical knowledge" (which had been "retarded in its progress" by wars with the French), incorporated a lengthy "Memorandum of an Instruction" from John Barrow, which was included in the book itself. Barrow emphasized the importance of journal writing and the desired completeness of journal entry to Tuckey and to his scientific companions (the Norwegian botanist and geologist Professor Christen Smith, Mr. Tudor the anatomist, and John Cranch, natural historian):

> It is almost unnecessary to observe to you, how important it will be to keep a journal of your proceedings. In this journal all your observations and occurrences of every kind, with all their circumstances, however minute, and however familiar they may have been rendered by custom, should be carefully noted down; and although the gentlemen employed in the several departments of science, will each be instructed to keep their respective journal, it will not,

on that account, be the less desirable that you should be as circumstantial as possible in describing, in your own, the general appearance of the country, its surface, soil, animals, vegetables, and minerals; every thing [*sic*] that relates to the population; the peculiar manners, customs, language, government, and domestic economy of the various tribes of people through which you will probably have to pass.[18]

Barrow further instructed Tuckey to "employ one or more of the respectable inhabitants to proceed with you as interpreters" and "collect from them every information they may be able to give of those parts of the continent through which the river descends." Particular attention was to be paid to noting the river's features as well as to recording the "latitude and longitude of every spot remarkable." Detailed instrumental observation was crucial, with "variation of the compass . . . [to be] taken and stated down" and recordings made of the "dip of the magnetic needle." For that purpose, the Admiralty had "directed a very excellent dipping needle, by [Thomas] Blunt, to be supplied."[19] As expedition leader, Tuckey controlled the working practices of his scientific companions: "The proper times of his [Cranch's] going on shore for the purpose of collecting, will of course be regulated by your orders, and be such as not to interfere with the general convenience of the expedition."[20] Finally, Barrow cautioned, the notebooks had both to be in summary form and in multiple copies as a safeguard:

And as all of them have been given to understand that their journals are, in the first place, to be transmitted to the Admiralty, you are to call upon them, whenever an opportunity may occur, to send, along with your own, a copy or an abstract of these journals, according as you may deem the occasion that offers for a conveyance to be a safe or a doubtful one: and to prevent as far as possible, your and their labours from being lost to the world, it is strongly recommended, that triplicates at least be kept of all the journals, and that each person carry about with him a brief abstract of his observations, in order that, in the event of any accident befalling his journals, he may still preserve the abstract to refresh his memory.[21]

This insistence on detail, guidance over corroboration, usage of instruments, triplication of written record, and production of an abstract from memory (and as an aid to it should accident befall either the narrative or its author) leaves no doubt about the importance and nature of writing in exploration. But if carried out to the letter, such instructions might also have constrained expeditionary conduct: they seemingly left little time to do anything other than measuring, observing, and noting down, rather than writing at length. In such cases, especially where exploration required collection, measurement, and comparative inquiry on the move, as it did for Tuckey and others, the requirement to write down the facts of a bewilderingly diverse and unknown natural world may have hindered the efficiencies of travel and, in turn, affected the written record. This is why field writing so often took note form. Note taking in the field and its later recall, editing, and expansion such as that undertaken by King, Tuckey, and others was doubly necessary: the need to cover a reasonable amount of territory in pursuit of one's identified objectives (failure to do so might substantially limit the worth of the final written account) meant that, in the field, writing had to be in note form for fear that recording might be diminished by the facts of travel and because memory had to be relied on as a key resource for later reflection.[22] Neither should the practical constraints imposed by access to paper, pencils, ink, and quills be overlooked in this respect: borne of experience in his own African travels, Francis Galton offered a section titled "Writing Materials" in his *Art of Travel* (1855), including guidance on the use of "sympathetic," or invisible, ink.[23]

As it turned out, James Tuckey and his companions had no opportunity for post-travel reflection: nearly all of those on the expedition died of fever within weeks of starting out. Tuckey's hard-won words outlived him. Barrow used his authority within the Admiralty and his connections with Murray II to edit the work, making clear to Murray his intentions in that respect in a letter of May 8, 1817.[24] In this two-fold role as editor and, perhaps, to place his initial instructions (and himself) in a good light, Barrow altered the composition and arrangement of the materials at his disposal before passing the work to Murray. One subsequent reviewer tendered "thanks to the editor" for thus assembling the expedition's materials and putting them forward for public consumption in this way.[25] Another noted that the volume appeared "greatly amplified by the insertion of his [Barrow's] own reflections and [moderated by] deductions from the account given by the travellers," before concluding that the expedition members' accounts had as a result been "properly preserved in a distinct form."[26]

Tuckey's posthumous text almost outran Barrow and Murray both. Formal instructions insisted that Admiralty-backed explorers' journals became the Admiralty's property on an expedition's return: this requirement of journal writing extended not simply to its being undertaken but to its safe keeping as well and from a concern over what its content might reveal about how the exploration had been conducted and by whom. Barrow's actions were driven not only by a concern to order Tuckey's words regarding his explorations when the author could not, but also stemmed from concerns over a pirate edition of the expedition's travel: "It is too true that the rascals have some how or other got access to Tuckey's journal," Barrow wrote in a rage to Murray II, "through what channel I fear it will be utterly impossible to discover, or if we could discover, be able to prevent their going on; the extraordinary fact is that none but Common Artificers & Seamen returned, all the officers except the Master, the Surgeon and a Mate having died, and the Master, as he thought having secured all the Journals."[27] Despite despairing of finding "the vagabond who has stolen it," Barrow soon did so. "I have seen Mrs Tuckey"—Barrow does not tell us how or when—"and she tells me that Mrs Eyre the 1st purser's widow informs her that among her husband's things she found a copy of a Journal which she supposed to be her husbands." Barrow could do nothing formally since, in not being an officer, the purser was not subject to Admiralty injunctions over the keeping and transmission of journals: "As we did not uniform her late husband I fear we have no hold of her." But, Barrow informed Murray, if the purser's wife were to come before Murray, "I will endeavour to frighten her" over any attempt to use that unofficial manuscript journal as the basis to publication.[28] It may well be that Barrow was protecting his own financial interests, too: he was paid £350 for his editorial duties over Tuckey's *Narrative* and, by the end of 1818, had drawn a further £50 share of the profits.[29]

The importance of this example rests not in the fact that a senior Admiralty official bullied a sailor's widow into compliance over an unofficial exploration narrative (perhaps to protect his own interests in the official account) or that official instructions related to the control over the final field journals or, even, that authorizing explorers' texts generally required such personal intervention after the event. It rests in the fact that, however clear instructions might be over procedure about getting to the field and what to do once there, travel writing that resulted from being in the field could as easily be obscured by later reflection and redaction as by the business of travel itself. Later amendment was necessary precisely because of the often fragmentary form of in-the-field writing. On their return the facts of travel could have an authority beyond their content value, beyond what it took to produce them in the first instance, not by virtue of being the result of formal directive but because they had to be "worked up" beyond what was undertaken as writing on the move. Precise guidance as to purpose was no guarantee of success in the field, nor was it—at least as far as these examples show—a means to regulate the how and when of explorers' writing there. Nor, as Tuckey and his companions found, could it ensure that authors came home with their words at all.

The significance of official instructions regarding why to travel and what to investigate is most apparent for the polar narratives published by Murray between John Ross's *Voyage of Discovery* (1819; app.: 55) and Leopold McClintock's *The Voyage of the 'Fox' in the Arctic Seas* of 1859 (app.: 238).[30] In effect, the volumes between John Ross's 1819 account and that of George Back in 1838 (app.: 55, 136) represent for Britain the results of a range of combined imperial ambitions that centered on the Arctic: geographically, to record, map, and speculate on the shape and content of the Arctic and to learn more about the region's peoples; scientifically, to determine the Magnetic North Pole; and, commercially, to find the Northwest Passage.[31] James Clark Ross's *Voyage of Discovery and Research* (1847), with its emphasis on terrestrial magnetism and the southern polar continent, is the single Antarctic exception to this sequence of northern narratives (app.: 187). For almost twenty years, the Arctic—as an imaginative construct, the focus of a literary genre, and the locus of scientific enquiry—depended heavily on its presentation and negotiation in the studies of John Barrow, the Admiralty, and the house of Murray.

Janice Cavell has traced this "connected narrative" of Arctic exploration in British print culture with particular reference to Ross, Franklin, Parry, and McClintock. For her, the work of these men, produced as it was in different editions to ensure affordability and the widest popular interest, came together with the coordinating and reviewing activities of John Barrow to create a distinct literary culture for the British public; an Arctic metanarrative that treated the region as the epic locus of danger and mystery and cast British explorer-authors as patriotic heroes in the face of adversity.[32] If we look at the other end of the narrative, so to say, at the production and even their preprint production and not at their edition history and public reception, it is possible to think of the doing of the exploration that became Murray's Arctic publications between 1819 and 1838 as a similarly connected sequence precisely because the region was the object of sustained official scrutiny (fig. 4). In part, this is to combine Cavell's insights with those of Adriana Craciun and Kim Wheatley over the "polar nexus" of literary endeavor, Admiralty directives, Murray's publishing network, and intense public interest in the Arctic in the early nineteenth century.[33] In part also, it is to suggest that the emphasis on instruction as a consequence of this intense interest in the region afforded Arctic narratives a sort of accumulative

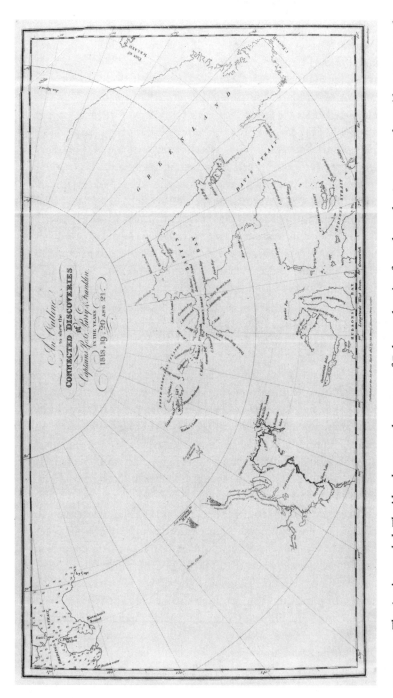

FIGURE 4 The Arctic revealed. The idea that textual accounts of Polar exploration from the early 1820s onward provide a "connected narrative" is endorsed in this map, "Connected Discoveries" of John Ross, William Edward Parry, and John Franklin between 1818 and 1821. Source: John Franklin, *Narrative of a Journey to the Shores of the Polar Sea, in the Years 1819, 20, 21, and 1822* (London: John Murray, 1823). By permission of the Trustees of the National Library of Scotland.

credit with regard to exploratory conduct. Because not all voyages in the region were successful, Arctic explorer-authors built in part on the instructive experience of their predecessors.

Individual explorers, of course, undertook their expeditions and wrote during and about their explorations in individual ways: ours is not an argument for a single and simple "Arctic rhetoric," although there are shared tropes in these authors' descriptions to do with Arctic aesthetics, the sublime, camaraderie, dependence on native interlocutors, and the rigors of daily life, including undertaking scientific measurements, in harsh environments (plate 7).[34] It is to suggest that in their work, notably in the work of Ross (1819), in the 1823, 1828, and 1829 volumes of John Franklin, and in the several voyages and volumes of William Parry, we have a body of Arctic exploration and of Arctic writing that builds cumulatively on individual experiences and others' endeavors (app.: 55; 74, 103, and 110; 66, 84, 90, 96, 107, 108, and 123, respectively). As Franklin remarked of Arctic discovery in closing his *Narrative of a Second Expedition* (1828; app.: 103), "Each succeeding attempt has added a step towards the completion of northern geography; and the contributions to natural history and science have excited a general interest throughout the civilised world."[35]

Over time such works, with others, helped provide for Murray a "model" of what best met the public's interest. Maria Graham read Parry's 1821 *Journal of a Voyage* en route to South America on board HMS *Doris*, describing his account to Murray II as "really good" (even though "I have not yet read it thoroughly because I have too little time at present every moment being filled up") and intimating that the ship's officers thought much the same.[36] Precisely because there was such interest, delays occasioned by authors were a source of anxiety to Murray II if he felt polar works might miss their moment in the public eye.[37]

The scientific and geographical motives behind Ross's 1818 voyage were clear enough: "Although the first, and most important, object of this voyage, is the discovery of a passage from Davis' Strait, along the northern coast of America, and through Behring's Strait, into the Pacific; it is hoped, at the same time, that it may likewise be the means of improving the geography and hydrography of the Arctic regions, of which so little is hitherto known, and contribute to the advancement of science and natural knowledge."[38] With this came a series of statements as to procedures: in its official instructions to Ross of March 31, 1818, the Admiralty directed "a great variety of valuable instruments to be put on board the ships under your orders," with instructions to Edward Sabine "to assist you in making such observations as may tend to the improvement of geography and navigation, and the advancement of science in general," with emphasis placed on magnetic observations, study of the tides, and depth sounding. Ross was further instructed "to make use of every means in your power" to collect and preserve natural specimens or to draw the larger ones and, finally, on return to England, to "demand from the officers and petty officers, the logs and journals they may have kept; and also from Captain Sabine, such journals or memoranda as he may have kept."[39] In their content and in their tenor, these instructions echo those given by Barrow to Tuckey.

The official instructions issued in May 1819 to Parry (who had earlier been in command of HMS *Alexander*, one of the four ships under Ross's charge) have word-for-word similarity in numerous respects, as well they might given the shared motive: attention to navigational practices on encountering continuous sea ice; concern over the safety of the crew, over the route home should a passage to Bering Strait be achieved, over the "great variety of valuable instruments to be put on board the ships," and over Sabine's assistance; and diligence with regard to the surrendering of all the written accounts at journey's end.[40] Likewise for his first Arctic land expedition, to the Coppermine River from 1819 to 1822, where "one of the principal objects of the Expedition is to amend the very defective geography of the Northern part of North America," the Admiralty's instructions to John Franklin insisted on instrumental measurement and a journal of his route and proceedings "describing therein every remarkable object that may occur." They further directed that the two midshipmen be employed "in assisting in all the observations" as well as in producing their own "accurate journals of all proceedings and occurrences" and, because Franklin and Parry were then both likely to be active in the same area at the same time and to much the same purpose, clear directives were issued that demanded they leave markers one for another.[41]

We are, then, consistently given insight into why Arctic travel was important, into what accounts of it should contain, and even over who, if not the author him- or herself, might assist in undertaking in-the-field writing. But we are seldom afforded in such instructions insight into *how* such writing was to be done. Despite the evidence of instructive rhetoric in common and shared motives, standard shipboard and expeditionary procedures, and control of the ships' journals at voyage's end—all suggestive of an accumulative culture of official instruction—we cannot readily glean the exact nature of on-board writing practices. Only occasionally do we get insight into what Parry observed to Murray II in 1826 over the requirements on naval men to both act and write. Noting how "my nautical education has taught me to act and not to question," John Ross was insistent that his 1819 text be read as the work of a seafarer; should his work prove disputatious (as, indeed, it would), he planned to remove himself from involvement in its implications: "My habits in literary composition are such, that I could not hope to put all these circumstances in a clearer point of view; and, as far as they partake of a controversial nature, it is not my business to enter into the discussion."[42] While this stands as an example of that authorial self-effacement and modesty that marked many of Murray's authors and others, it is also, for Ross at least, to discern connections between the nature of the exploration and the nature of the resultant text. In further introductory remarks, Ross hinted that the very nature of seaborne Arctic exploration was a root cause of his narrative style: the sameness of Arctic geography and voyaging, as well as his nautical training, determined a certain direct style and plainness of content:

> I have here attempted nothing beyond the journal of a seaman. If I had done more, I might have done worse; as I could not have hoped to add much elegance to the composition, nor much entertainment to the matter of a narrative, which was not productive of much adventure. From the nature of the service, we were almost always at sea, and were thus cut off from the sources of variety that are only to be found by frequent communication with unknown or interesting shores. If I have thus missed to give entertainment, I, however, trust, that I have diminished nothing from the utility of the statements to seamen, nor their authority to geographers.[43]

Ross's uneventful geography, factual narrative, and seaman's prose stands in contrast to Franklin's Coppermine River exploration. As mishap after mishap befell that expedition, so Franklin's narrative could not help but recount adventure in adversity: having lost some of his journal in a river crossing, Franklin's manuscripts of the affair, now stained and crumpled, survived only by being kept secure in his pocket.[44] Doing Arctic geography directly altered his writing (fig. 5).

For George Francis Lyon, Arctic authorship was a consequence of having maintained a private journal whose contents differed from official accounts.[45] Lyon was in command of HMS *Hecla* as part of Parry's second voyage in search of the Northwest Passage, between 1821 and 1823. Earlier, under Barrow's instructions, Lyon had undertaken exploration in the Sahara, the results of which were published by Murray in 1821.[46] Lyon's 1824 *Private Journal* following the Parry voyage is of interest for being just

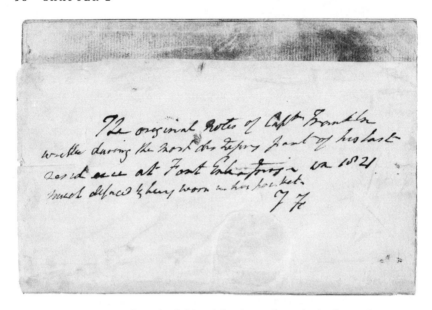

FIGURE 5 Manuscripts from the field and the rigors of travel. The figure shows some
of the stained and crumpled notes from Franklin's 1819-22 Coppermine Expedition,
which survived because they were pressed close to Franklin's body. The text reads:
"The original notes of Captn Franklin / written during the most distressing part of his
last / residence at Fort Enterprise in 1821 / much defaced by being worn in his pocket. /
J F." The account of this expedition was included in Franklin's *Narrative of a Journal to
the Shores of the Polar Sea, in the Years 1819, 20, 21, and 22* (London: John Murray, 1823).
Source: National Library of Scotland, MS 42237. By permission of the Trustees of the
National Library of Scotland.

that—private—and for its relationship to Parry's Arctic narrative of 1824.
In his preface, Lyon—in a modest self-positioning explored more fully in
chapter 4—makes clear that his journal was never intended to be published:
"It was written solely for the amusement of my own fire-side, and without
the most distant idea that it would ever see the light in any other shape
than that of its original manuscript."[47] Lyon had surrendered his manuscript journal, as required, at the voyage's end. Later close scrutiny by Barrow revealed, however, that it was both sufficiently different and interesting
enough to merit publication on its own terms, provided the overlap with
Parry's official account was excised:

> Being sent with the other journals to the Admiralty, in obedience to Captain
> Parry's instructions, my friend Mr. Barrow . . . advised me strongly to publish

it, on account of the number of little anecdotes it contained relative to the habits and disposition of a people entirely separated from the rest of the world,
and with whom we had for so great a length of time kept up an intimate and
constant intercourse.

He observed also, as an additional inducement, that Captain Parry, in his
authentic and official account of the expedition, had not deemed it fit or necessary to enter into many of those minute and peculiar traits which are requisite
for displaying the character of a strange people. Captain Parry's opinion on this
subject agreeing with Mr. Barrow's, I could no longer hesitate; and therefore,
after a few abbreviations, and the omission of some details of natural history,
and of scientific observations, I sent the original manuscript to the printer.[48]

Parry's 1824 work became the authentic and official narrative by virtue of its attention to science and his in-the-field authority, but the books
and the authors were closely linked in other ways. At sea, and in keeping
with official instructions, Parry had directed Lyon about what to sketch:
"I cannot deny myself the pleasure of bearing testimony to the obliging
readiness with which Captain Lyon has always attended to my suggestions
on this subject, as well as to the eagerness and assiduity with which he
seized on every opportunity of exercising his pencil, which so monotonous
and unpicturesque a voyage presented."[49] If there are in Parry's comments
echoes of Ross's remarks over the sometimes uneventful nature of Arctic
exploration, there is no doubt that on board and on shore, following orders influenced who did what in the production of the written and pictured
"raw materials" of exploration (fig. 6). Just as the travel itself reflected formal aims, the ends of journal writing in the field were in this sense clearly
shaped under the influence of official instructions. We know that Murray
made money from the result. From sales of the quarto format single volume
of Parry's 1821 *Journal of a Voyage*, which sold for £3 13s. and 6d. (app.: 66),
Murray had cleared a profit in excess of £2,300 by January 1823.[50] And we
know that the cheaper multivolume octodecimo publication (£1 for five
volumes) that appeared in 1828 (app.: 107) was welcomed as an inexpensive
edition: as one reviewer noted, "Perhaps none of our voyagers' narratives
possess such powerful interest as those which are here presented to the
public, in a popular and convenient style; and we consider Mr. Murray to be
doing an essential service to the literature of his country, by thus spiritedly
following up his plan of giving such works in this manner."[51] But the daily
practicalities of writing and recording that produced these books in other
places, of following these instructions to the letter, remain largely hidden
from view.

FIGURE 6 The art of polar exploration. George Francis Lyon's depiction of HMS *Hecla* and HMS *Fury* was undertaken on the orders of William Edward Parry, the expedition's leader. Notwithstanding that there may have been some embellishments made later by the engraver, Lyon was clearly an accomplished artist (see also plates 11 and 12). Source: William E. Parry, *Journal of a Second Voyage for the Discovery of a North-West Passage from the Atlantic to the Pacific* (London: John Murray, 1824). By permission of the Trustees of the National Library of Scotland.

NOVELTY, UTILITY, CURIOSITY, SECRECY: EXPLORATION, GEOGRAPHY, AND IN-THE-FIELD WRITING

The great majority of those travelers who became Murray authors were not guided by such formal instructions. Some, like Mansfield Parkyns, even lacked a firm sense of where they wished to go and for what purpose. For those who traveled with scientific aims in view, such as Joseph Jukes and Charles Lyell with North American geology as their focus (app.: 158, 175) or Joseph Hooker seeking to document Himalayan botany (app.: 220), such singular motivating purposes framed their approach to authorship and conduct in the field. Others wrote only on their return and then not from predetermined motives but from personal necessity. Hugh Tremenheere's account of the United States and Canada (app.: 212), for example, was put together entirely after his travels, partly because the work was based, as he was at pains to point out, on information received from "the proper authorities" and other "trustworthy sources of information."[52] For the impoverished American James Riley, writing to Murray II over "an account of my own shipwreck & suffering on the Western coast of Africa" (app.: 42), his motives for becoming an author rested in his unusual experiences and the novelty of his tale. Riley, whose case we describe more fully in chapter 3, had turned to authorship to survive: "My various sufferings & losses of property has put me under the necessity of resorting to other means than my former occupations by which to support my large & encreasing family."[53]

The great majority of Murray's travel authors were professional and literate men and women. Authorship reflected intellectual interests or professional connections with particular parts of the world rather than economic necessity. It is thus difficult to generalize on questions of motive although we can with assurance claim that, unlike explorer-authors traveling under official instructions or those having to write to make ends meet, travel writing for most was a choice, not a necessity. As John Ross found over the content and tone of some polar narratives following their authors' being "cut off from the sources of variety," as he put it, different geographies did influence explorer-cum-traveler authors in what and how they wrote.

This is not to argue for a deterministic relationship between the narrative and its locale. Rather, it is clear that for many traveler-authors, perhaps especially in the Near East, Central Asia, China, and South America, the very fact of encountering people and of writing about them as well as being dependent on them—virtually the antithesis of polar sea voyaging—presented travel writers with problems. These were not just problems of native credibility and authorial reliance on other's testimony. They revolved, rather, around the facts and implications of human history. Land travel, perhaps particularly in populated "antique lands," presented difficulties over what to record, what to omit, where to travel, and what to see given limited time and resources. Traveling in lands where Christians were treated with suspicion presented further difficulties in terms of the traveler's performance. And if one's motive required any form of covert surveillance, as was the case in Central Asia for John Macdonald Kinneir, John Malcolm, Mountstuart Elphinstone, Alexander Burnes, and William Moorcroft, to be seen writing openly was open to misinterpretation: there, to disguise oneself as something other than stranger, traveler, and writer was necessary. Alexander Burnes, who regularly changed his disguise in order to blend in with the local clothing styles and customs of the peoples he encountered, and who was repeatedly advised that his safety would "depend upon laying aside the name of European, at all events of Englishman," almost exposed himself as a foreigner and spy when he forgot to dress appropriately: "A Hindoo belonging to the minister went inside to announce our arrival, and, in the meantime, I rehearsed my tale, and drew on a pair of boots as well as for the uniformity as to hide my provokingly white ankles. My face had long been burned into an Asiatic hue and from it I feared no detection."[54]

What was written and why, perhaps even how and when, could thus be influenced by the constraints placed on travelers and explorers in different parts of the world. This was neither simply a matter of encountering human difference nor of having to protect one's identity in order to survive to write. Declaration of motive was not a sure guide, either, to the eventual conduct of the travel and the nature of the writing en route if the circumstances of travel meant that one's initial aims had to be amended in the field. In consideration, then, of how travel writings became printed words, it is to know, rather, how travel writing was undertaken when there were no formal directives as to motive and to know how the act of travel influenced the art and performance of writing before publishers and authors effected any post-travel amendments.

Murray's published works on China are particularly instructive in this regard. China was virtually unknown to western audiences: anything written from first-hand encounter was both novel and informative.[55] John Murray II and the author Sir Henry Ellis were at pains to stress this in advertising the latter's *Journal of the Proceedings of the Late Embassy to China* (1817; app.: 38; plate 8). Ellis was part of the British diplomatic mission to China under Lord Amherst in 1816: an earlier such mission, under George Macartney in 1793, had been a formative experience for the young John Barrow. "An Embassy to China is so rare an event in the history of Europe," noted Ellis, "that a correct narrative of the occurrences attending it possesses a degree of interest, almost independent of the mode in which the narrative itself may be executed."[56] This declaration is all the more important given the failure of this official mission and failure by the British to observe diplomatic conventions, specifically to kowtow with the forehead touching the floor as a gesture of respect. Asked to do so nine times, repeated British refusals were countered by the Chinese involving the restriction of their travel movements. Like Macartney's, the later mission nevertheless failed for reasons of mutual misunderstanding over trading policies rather than from these diplomatic slights alone.[57]

But because the British visitors were constrained by where they could travel, what they could see, and even to whom they might talk, writing about China was likewise limited: "Those who have talents for observation, or powers of description, may possibly find wherewithal to occupy the mind and the pen. Millet fields, willow-groves, junks, half-clothed inhabitants with little eyes and long tails, women with prettily-dressed hair but ugly faces, these are the daily and unchanging objects, and from these I cannot eke out any thing like interesting description."[58] His writing was tempered by his state of mind, and his state of mind by China's apparent monotony:

> I have now exhausted my recollections respecting China and its inhabitants; and have only to ask myself, whether, omitting considerations of official employment, my anticipations have been borne out by what I have experienced? The question is readily answered in the affirmative: curiosity was soon satiated and destroyed by the moral, political, and even local uniformity; for whether plains or mountains, the scene in China retains the same aspect for such an extent, that the eye is perhaps as much wearied with the continuance of sublimity as of levelness. . . . I have neither experienced the refinement and comforts of civilized life, nor the wild interest of most semi-barbarous countries, but have found my own mind and spirit influenced by the surrounding atmosphere of dulness [sic] and constraint.[59]

For John Davis, later a fellow diplomat and Murray author on China, prolonged residence there made no difference: "I have little pleasure in this country, except in the prospect of leaving it, and of again seeing my English friends."[60]

The house of Murray managed its authors and works on China in similar ways as it had its Arctic narratives. Informing Murray II that "I have lately heard from our friend Sir George [Macartney] who tells me he is going to print again," John Davis's purpose in writing in October 1822 was to propose a book on China given his own extended residence there. For reasons that are not clear, Davis's book was first published in 1836 by Charles Knight, Murray III producing an enlarged edition in 1857 (app.: 233).[61] Staunton, who had written on China in 1822 (app.: 72), called on the Murray firm years later, proposing that items be published on the country, given "the revival of the China question, on the occasion of the opening of the Trade."[62]

In contrast to the experience of traveler-diplomats in China, travelers to the Near East and Asia Minor from the later eighteenth century availed themselves of numerous guides both printed and human. For these regions, exploration and authorship was a matter of in-the-field testing and of an emergent critical methodology in addition to being the locus of romantic sensibilities. Travelers could test others' earlier travel accounts for accuracy and so aim to bring knowledge up to date, in which context canonical texts were subject to particular scrutiny and classical antiquities found in the field. Mapping, an important desideratum, required the use of mathematical instruments and a culture of regulated observation and measurement. Individual travelers could themselves be tested by the environment and by the circumstances of travel.

William Martin Leake, the Arabian traveler, antiquarian, and comparative philologist, was motivated by all this and more: "To the traveller who delights in tracing vestiges of Grecian art and civilization amidst modern barbarism and desolation, and who may thus at once illustrate history and collect material for the geographer and the artist—there is no country that now affords so fertile a field of discovery as Asia Minor." "Unfortunately," he further observed, "there is no province of the Ottoman empire more difficult to explore in detail." "Asia Minor," he noted, "is still in that state in which a disguised dress, an assumption of the medical character, great patience and perseverance, the sacrifice of all European comforts, and the concealment of pecuniary means are necessary to enable the traveller thoroughly to investigate the country, when otherwise qualified for the task by literary and scientific attainments, and by an intimate knowledge of the languages and manners of the people."[63] Before journeying, Leake traveled through the narratives of others in order to check locations, especially of sites recorded in antiquity, including geographer James Rennell's critical commentaries on classical Asia Minor.[64] Leake turned to Richard Pococke's *Description of the East* (1743), "the only traveller of the last century who has published his route with sufficient precision to be of any use to the geographer," but he also dismissed him: "He has been extremely negligent in noting bearings and distances."[65] Leake consulted the maps of the leading French mapmaker Jean-Baptiste Bourguignon d'Anville and more recent ones undertaken by the British hydrographer Francis Beaufort.[66] Once traveling (in a party of over thirty, "well armed, and dressed as Tatár Couriers"), Leake had a clear sense of what had been written by others, where to go and why, and he corrected others' textual descriptions and latitudinal positioning as he journeyed.[67] In all, and as the subtitle to his 1824 *Journal* indicates (app.: 82), his was a work of method in ancient and modern geography involving first-hand testing of classical accounts and adding new knowledge based on empirical measurement: "It is hoped that by these several means the future traveller will be furnished with an approximation that may assist him in ascertaining the real sites."[68]

Not everyone was as driven in their motive and method. Leake would go on to be a founder member of the Royal Geographical Society in 1830. Several Murray authors—Alexander Burnes in 1834, James Wellsted in 1837, and William Hamilton in 1844 (app.: 126, 139, 157, respectively)—were likewise closely associated as medalists or administrators with the RGS. Murray's eye for the market meant that the firm could, and did, use the status of such men to promote interest in their work and, also, to reorder their motives and words, even to the point of obscuring them altogether, in order to bring foreground novelty. James Wellsted's *Travels in Arabia* (1838) was the result of Wellsted's periodic land-based journeys undertaken during hydrographic surveys in the Red Sea and around Oman and Yemen from 1829. What Wellsted stressed in regard to the work's novelty—"many of the facts herein stated have never previously been made known to a European public"—was the science and accuracy of the coastal mapping and his descriptions of the Arabian interior and its peoples.[69] Murray II and III demanded a different emphasis because they knew that the public would be more interested in the latter, and so they published the travel narratives in the first volume, the science in the second. Seemingly, Wellsted acceded to this restructuring: "With respect to the form in which it is published I must confess that I would rather submit it to the judgment of others than my own—no one knows the public taste better than Mr Murray and there is no one whose opinion would be of more value."[70] One reviewer knew what had been done, and with what consequences:

> In Arabia, the place of honour is always given to age—not so in Albemarle Street or the Row. Among us, the great Sheikhs of publication, who recline voluptuously beneath their shady groves, while their literary herds browse in the desert, invariably give the preference to what is new; and, regardless of the sense, turn topsy-turvey whatever MSS. are placed in their hands, solely for the purpose of placing in the front whatever strikes the eye most with the glistening of novelty. . . . This disregard of chronological order releases us from the obligation of following very scrupulously in our author's track.[71]

For Edward Robinson, the American clergyman and scriptural geographer who traveled through the Holy Land in 1838 and in 1852, the sustaining motive for his travels lay in testing the Word of God. Murray published three books by Robinson: *Biblical Researches in Palestine, Mount Sinai and Arabia Petræea* (1841; which secured for Robinson the gold medal of the Royal Geographical Society in 1842); *Later Biblical Researches in Palestine, and the Adjacent Regions* (1852; a later edition of his 1841 work: app.: 153), and *Physical Geography of the Holy Land* (1865). Contemporaries regarded Robinson highly. For Josias Leslie Porter, fellow churchman, scriptural geographer, and Murray author, there was no one better: as Porter confessed in writing to Murray III in 1856 over preparations for the Murray handbook on the Near East, "I wish I had Robinson's new work, for he is the *only man* in whose accuracy I can depend; and his historical notes are invaluable."[72] Modern scholars have likewise praised Robinson's work.[73]

Here, we may note the emphasis placed by Robinson and his traveling companion, Eli Smith, on the means employed to meet their end in view.

> We early adopted two general principles, by which to govern ourselves in our examination of the Holy Land. The *first* was to avoid as far as possible all contact with the convents and the authority of the monks; to examine everywhere for ourselves with the Scriptures in our hands; and to apply for information solely to the native Arab population. The *second* was to leave as much as possible the beaten track, and direct our journies [*sic*] and researches to those portions of the country which had been least visited. By acting upon these two principles, we were able to arrive at many results that to us were new and unexpected; and it is these results alone which give a value (if any it have) to the present work.[74]

Robinson's avoidance of religious houses and monastic authority was in part a matter of Presbyterian polemics. As for Leake, it was also a methodological choice but in a different way: see things for yourself and rely on native authorities (Eli Smith was fluent in Arabic, both men in Hebrew). So, too, was the decision to leave established routes and the aim "to examine everywhere with the Scriptures in our hands." The two men also carried numerous books with them, in order to check in the field against others' observations and texts. But others' books, even the Bible, were not necessarily secure guides to places other travelers had not frequented. Because no prior texts could be relied on, producing certainty in their own narrative demanded recourse to methods other than textual correspondence, particularly with regard to the scriptures. One such was instrumentation, although this was minimal. Another was the adoption of in-the-field inscriptive practices which, later, would produce consistency through corroboration:

We carried with us no instruments, except an ordinary surveyor's and two pocket compasses, a thermometer, telescopes, and measuring tapes; expecting to take only such bearings and measurements as might occur to us upon the road, without going out of our way to seek for them. But as we came to Sinai, and saw how much former travellers had left undescribed; and then crossed the great desert through a region hitherto almost unknown, and found the names and sites of long-forgotten cities; we became convinced that there "yet remained much land to be possessed," and determined to do what we could with our limited means towards supplying the deficiency. Both Mr Smith and myself kept separate journals; each taking pencil-notes upon the spot of everything we wished to record, and writing them out in full usually the same evening; but we never compared our notes. These journals are now in my hands; and from them the following work has been compiled. On thus comparing them for the first time, I have been surprised and gratified at their almost entire coincidence.[75]

In a sense, Robinson's narrative of biblical landscapes began in a text — the book of scriptures — and ended in a text, his scriptural geography, wherein he and Smith established the veracity of the landscape as the basis to the narrative. This is not to see Robinson's narrative as dismissive of science. There was room for method in science and religion since, in his view, science could help one get to religious truth.[76]

More generally, such links were only hinted at, or not made at all. In west Africa, Thomas Bowdich found that "the impression of the Natives that we came 'to spy the country'" was aggravated by the suspicions of the Muslims in the area, and so he resolved to cease note taking and questioning in order to "suppress all curiosity for a considerable time, lest their anxiety to detect us in geographical enquiries . . . might have been gratified."[77] In the face of native sociability as he traveled in Asia Minor, Charles Fellows had to feign illness "in order to get time for writing and the other occupations of a traveller."[78] Later, in a quarantine house in Constantinople, Fellows was glad of the restrictions placed on his movement: "Quarantine & the existence of a regular post marks the verge of our European world, & gives me an opportunity of writing to you. To some people the detention of a Lazzaretto is tedious, but after active travel I must say that I always enjoy the repose of arranging my affairs bodily and mental."[79] Returning from Central Asia, Macdonald Kinneir lost much of his writing to pirates in the Persian Gulf, later reconstructing parts of his printed narrative from memory in the style of his lost field journals.[80] In Borneo, James Brooke's writing was constantly interrupted by "the turmoils of war and the discomforts of savage life."[81]

In Amazonia, Henry Maw's travels meant he could not write by day, but neither could he write at day's end: "The night is so dark, that it was not possible either to make many remarks, or to note them down if made. My notebook is not legible."[82] Lonely on Russia's steppes, Richard Wilbraham found mobility an impediment to writing.[83] No less lonely on the Pampas, Francis Bond Head found writing space and time wherever he could: "sometimes when I was tired, sometimes when I was refreshed, sometimes with a bottle of wine before me, and sometimes with a cow's horn filled with dirty brackish water, and a few were written on board the packet."[84] Returning overland to England from India in 1817, George FitzClarence was too tired to write at all: "The last four days I have travelled with great fatigue; and from the mode of crossing the desert, having no opportunity, it has been out of my power to write regularly."[85] In the South China Sea, Basil Hall's efforts to record and map were overcome by the very geography he had come to explore: "We threaded our way for upwards of a hundred miles amongst islands which lie in immense clusters in every direction. At first we thought of counting them, and even attempted to note their places on the charts which we were making of this coast, but their great number completely baffled these endeavours."[86]

Although evidence of the individual embodied experience of traveling and writing is thus often slight and fragmentary, its collective presence has considerable implications. Writing, traveling, and exploring are productive processes of achievement. While inscription in the field may typically be seen as the first step in the emergence of travel accounts, instructions given beforehand, even before departure, cannot be neglected since they shape what can be written and why. Because both the motive behind exploration and the experience of traveling effected and affected writing in certain ways, the significance of travel accounts does not rest exclusively in the printed record but also in what precedes it. The making of the travel account, and the making of the travel author, involved processes of translating that experience, be it of disorder, bodily privation, loneliness, bewilderment, dismay (at the number of islands and so on), into a narrative, a travel *plot*.[87] Yet precisely because these processes of ordering and authorizing by Murray and author alike were preceded by experiences of disorder in the initial writing, we may ask, with others, "whether the published explorers' records . . . constitute the best versions of the evidence which the explorer actually gathered."[88] Our attention to motive and to the experiences of traveling and writing (as opposed to the experiences involved on return in being and becoming an author), supports the views of others that the very terms "explorer" and "traveler" were not distinct categories, possessed of an essential and fixed meaning.[89] Like the category of author, being and becoming an explorer or traveler depended on how one regulated one's experiences, overcame an adverse nature, suspicious natives, or personal frailty, in order to explore, travel, and write. Adriana Craciun has suggested that the term "explorer" is a later nineteenth-century "back-formation" used to give status to figures after the events of discovery they undertook were judged and legitimated through print.[90] The category of authorship was equally problematic: writing "out there" as the explorer and authoring "back home" were for some influenced by what they had been told to do, why, and through generic convention.

MATTERS OF PRACTICALITY AND PROCEDURE

The idea of regulating the traveler through practical guidance has a long history. Written advice is easier to study than the spoken word. In his *Western Barbary* (1844; app.: 169), John Drummond-Hay, Her Majesty's honorary consul to Barbary, recounts the verbal advice he had imparted nearly a decade before to the African traveler, John Davidson. Davidson was no novice: after completing medical studies at Edinburgh, he traveled throughout the Near East and in America, for which work he was made a fellow of the Royal Society in 1832. But in trying to reach Timbuktu across the Sahara, Davidson was murdered by desert tribesmen in December 1836.[91] Davidson's travels were noted in correspondence to the Royal Geographical Society, and a volume of his travels was published posthumously by his brother.[92] Drummond-Hay's advice had been to the point: Davidson should remain in England "until he was no longer talked of as the African traveller, and during that time . . . improved his Arabic"; change his name on leaving England; avoid well-known ports such as Gibraltar and Tangier; pretend to be a trader with some knowledge of medicine but not advertise himself as a qualified physician; study the local dialect and pick up "if possible, the language of the African tribes through which he would have to pass" so as to secure knowledge of the region before traveling; and, lastly, "make friends with the Arabs who accompany the Kafflas [camel caravans]."[93]

In his counsel, Drummond-Hay addressed those self-same practical questions of disguise in language, dress, and prior declaration of one's intentions that men such as Alexander Burnes faced in Central Asia (to which we turn later). We do not know how seriously Davidson heeded his advice. Other African travelers such as Alexander Gordon Laing who traveled, and

died, in full British uniform during his North Africa travels might have benefited from doing so.[94] But even men like the Swiss-born John Lewis Burckhardt, who endorsed Hay's views in practice in preparing to reach the Niger
by taking "long journeys, on foot, in the heat of the sun, sleeping upon the
ground, and living upon vegetables and water," learning Arabic, residing
in Syria for two years before moving to Cairo, and adopting Arabic name
and dress, could do little in the face of deadly disease. Burckhardt died of
dysentery in Egypt and never undertook the travels for which he had so
diligently trained.[95] Regulating the body so that it performed properly was
a matter of official instruction. It was also a practical issue with profound
epistemological implications.

In the long history of the regulation of travel and of the observant traveler through printed instructive guides, distinctions may be drawn (albeit
not absolute ones) between two traditions of geographical itineraries in
which advice was given on routes and places of note to visit. There were
the more bespoke practical guides to travel whose production in Britain and
elsewhere from the later eighteenth century reflected and drove the rise of
the genteel traveler and discerning tourist. But there were also the more
specialist and prescriptive guides to scientific travel, disciplinary practice,
intellectual conduct, and the regulations of the body in the field.[96]

The house of Murray was involved in such publications in two main
ways. The first was its series of handbooks, begun in 1836, the concern
of which was largely tourist travel. The second, our focus here, was its
involvement, beginning in the mid-nineteenth century, with four related
works concerned with the practical regulation of scientific travel, two of
which were published by Murray: Julian Jackson's *What to Observe; or, The
Traveller's Remembrancer* (1841); Sir John Herschel's *Manual of Scientific Enquiry* (1849), which incorporated guidance on numerous scientific fields;
Hints to Travellers, produced by the RGS in 1854; and Francis Galton's *The
Art of Travel* (1855; app.: 194, for Herschel, and 221, for Galton).

Julian Jackson's *What to Observe* (1841) sought "to point out to the uniniti-
ated Traveller what he should observe, and to remind the one who is well
informed, of many objects which, but for a Remembrancer, might escape
him."[97] To be useful in traveling, and when writing in the field, the memory
had to be jogged. Julian Jackson, one-time secretary, editor, and librarian
to the RGS in the 1830s, was motivated to write the book from a more general concern to improve the intellectual and social standing of geography.[98]
In 1832, the RGS had declared its intention to support the development of
geography in various ways: through premiums and prizes; through instruc-

tive essays on the content and range of geography; through guidance on appropriate instruments; and, not least, through preparation of a traveler's manual of practice that should contain:

> a clear and concise enumeration of the objects to which a Geographer's attention should be especially directed; a statement of the readiest means by which the desired information in each branch may be obtained; a list of the best instruments for determining positions, measuring elevations and distances, observing magnetic phenomena, ascertaining temperature, climate, &c; directions for adjusting the instruments, formulae for registering the observations, and rules for working out the results;—adapted to the use, not of the general Traveller alone, but also of him who, in exploring barbarous countries, may be obliged to carry and often conceal his implements.[99]

In an unpublished memorandum of May 1837, Jackson emphasized the need not only to regulate travel for the purpose of securing geographical knowledge but also to balance the results of exploration with the methodical treatment of geographical knowledge—what he called the "labours of the cabinet." "Far be it from me to wish to depreciate the labours of the traveller," he continued, "but I would insinuate that new facts are but slowly collected, and that while travellers are engaged in the research of these, sedentary Geographers have to perform the Herculean task of duly examining and arranging the facts innumerable already possessed by the Science."[100]

Jackson's *What to Observe* operated in a context where travel and observation as part of geography and science were known to require methodizing. The problems he and the RGS were dealing with were exactly those faced by Leake, Davidson, Wellsted, Burnes, and others. Of course, producing a book about travel and ensuring the safety of the traveler are different matters: there is no direct evidence that Jackson's work was read in the field by these men or in their preparation for doing so. But in its intention, Jackson's book aimed to instill a culture of regulation and procedure about what to look out for and how best to record one's observations—ambitions in which he was far from alone.

The *Manual of Scientific Enquiry*, which Murray published in 1849 under the editorship of the celebrated mathematician John Herschel, appears at first glance to be the epitome of methodological guidance and mid-nineteenth-century scientific authorship. There were substantial chapters by fifteen leading men of science.[101] Herschel acknowledges Jackson's 1841 work.[102] Herschel's was not, however, a work advocating disciplinary precision and methodological rigor. Admiralty commissioners expressed their concern only that "its directions should not require the use of nice apparatus and instruments; they should be generally plain, so that men of good intelligence and fair acquirement may be able to act upon them; yet, in pointing out objects, and methods of observation and record, they might still serve as a guide to officers of high attainment."[103] This emphasis on practical method was directed less at scientists in order that they might define the field in question and more at naval officers who, in this period, effectively were the "professional" men of science. The general intention regarding the key features of "observation and record" was consistent with the training of British naval officers as being themselves regulated instruments, reliable inscription devices suitable for operation in different environments and capable—like the handheld precision devices they used daily—of tolerating hardship in use on Arctic seas or in Africa's deserts while maintaining accuracy of record.[104] Glascock's *Naval Service or Officers' Manual* (1836), for example, offered guidance on the writing of the ship's log and the keeping of account books by first officers and captains in ways that allowed their use in corroborating each other's accounts and on the need for observed facts to correspond with written records of them. Glascock, a close friend of the polar explorer and Murray author George Lyon, knew well the value of proper prose and careful record.[105]

The 1849 *Manual of Scientific Enquiry* was, like Jackson's 1841 work, one expression of the emerging status of science and of prescriptive guides to observational and authorial training. In this respect, it is helpful to examine the manual's essay on geography by William John Hamilton. Hamilton, a leading figure within the RGS, the Geological Society of London, and one-time council member of the British Association for the Advancement of Science, began by arguing that of all the matters to be borne in mind in undertaking observation, the most important was "the necessity of acquiring a habit of writing down in a notebook, either immediately or at the earliest opportunity, the observations made and information obtained. . . . This habit cannot be carried too far."[106] Hamilton stressed that observations should be noted down on the spot and that traveling and exploration was most successful when a culture of regularity was established. To the geographical traveler, notebook and compass "should on all occasions be his constant and inseparable companions": "Let him then acquire the habit of never quitting his ship without his note-book and pencil and his pocket-compass, and although at times it may seem irksome to have to remember and to fetch these materials, the traveller, if he acquires the habit of constantly using them with readiness, will never have reason to regret the delay or the inconvenience which may have temporarily arisen in providing himself with such useful companions."[107]

Regulating the human traveler, naval officers or otherwise, thus demanded inscription, accuracy, and repetition so as to be habit forming in the aspirant author. Inscription and regularity of performance was both a scientific and a moral necessity. Hamilton knew this from personal experience. His own principal publication was *Researches in Asia Minor, Pontus, and Armenia* published by Murray in 1842 (app.: 157). Hamilton had undertaken this work between 1837 and 1839 from a concern for the scientific documentation of a part of the world then not widely known to European scholars, a claim that echoes that made by Leake nearly twenty years before. Personal experience proved a stimulus to more general rules of practical conduct regarding mapping, note taking, and in-the-field writing: "Independently of a very detailed Journal, I succeeded in keeping, with a very few exceptions, a minute Itinerary of every mile of road, noting the exact time of departure, and, with my compass constantly in hand, the direction of the road, as well as every change, sometimes to the number of twenty or twenty-five in an hour, adding remarks suggested by the physical structure of the country."[108]

The practical emphases of Hamilton's essay are evident in this insistence on detail, the separation of travel itinerary from journal writing, and constant recourse to instrumentation. They are evident in his 1849 essay in Herschel's collection and they may have informed that volume more generally: it was Hamilton, not Herschel, who negotiated with Murray over the layout of the Admiralty manual, with Hamilton having control of all the prefatory matter, the form, and the title, and in all dealings with the Admiralty.[109] Herschel's manual proved popular and was reprinted several times. By midcentury, however, the RGS renewed its attention to regulating field observation and instruction for travelers. The RGS's *Hints to Travellers* was published by Murray in 1854, although it had been begun two years earlier as a set of essay instructions. Its moving force was Francis Galton, who was notable for *Tropical South Africa* (1853) and *The Art of Travel*, subtitled *or Shifts and Contrivances available in Wild Countries* (1855; app.: 214 and 221, respectively). Galton's African experiences between 1850 and 1852 led to his membership of the RGS's Committee on Expeditions.[110] It was in that forum, and drawing on his African work, that Galton first aired his concerns about "the want of [guidance on] proper Instruments for travellers."[111] As the above evidence shows, Galton was wrong about earlier guidance as to proper procedure for travelers in the field as were those of his contemporaries who asked him to "draw up a set of general instructions for the use of travellers, to be laid before the Committee as early as convenient."[112] But *Hints to Travellers*, as these instructions became, was an important instructional manual: in later editions Galton even recommended that the discerning traveler should also carry with him or her Herschel's *Manual of Scientific Enquiry*.[113]

Several points follow from this evidence. These works were guides, not guarantees. Even if Tuckey, Burnes, and Moorcroft had read them or something like them, and closely followed their counsel, textual guides could not provide security in the field. They were intended to instill regularity of procedure and of practice, to help explorers establish an authorial regimen of habit and discipline. It is of more than passing interest that these guides were, in general, produced after those decades that witnessed, in terms of Murray's published narratives, the greatest concentration of exploration in the Arctic, Africa, and Central Asia. These were works that sought to establish status for science and disciplinary identity for the constituent sciences, such as geography, then taking shape through institutional procedure and exploration in the field. "Science" and "geography" were just as much terms in the making in the early nineteenth century as "explorer," "traveler," and "author."

The motives of Murray's authors are in most cases discernible even if the motives themselves are not easily reducible to a simple or single circumstance or typology. There was a clear difference in motive between those who explored or traveled on the basis of others' official sanction or government instructions and those whose motive lay in a mixture of personal interest, a desire to test other's texts, or simple curiosity. Rather than think of these individuals simply as Murray's authors, attention to motive and to the practicalities and performance of writing shows them to be "writers" for different reasons. This is not to propose any new taxonomy: the many variations of writing experience in relation to travel type here disclosed would only make for an imprecise Sternian listing of the kind referred to above (p. 37).

The facts of published authorship and the processes of authoring were of course preceded by the experience of writing. Writing in the field, even being in the field, was influenced by the commands of others or by one's state of mind, overattentive locals, mental and physical fatigue, and, even, available hours of daylight and the capacity to secure notes once made or to recall their import if notes were lost. For the many figures between about

1813 and 1847 on whom this chapter has focused, writing an official journal may have reflected a more regulated and openly declared authorial regimen than the writing of those who had to keep such inscription and their true identity secret. But insofar as the firm of Murray was involved, attempts to regulate in print the experience of travel and of observation and writing in the field came sometime after many of the exploratory voyages and travel narratives that the firm published. Several travelers had no thoughts of being authors until they returned home. Several writer-explorers became authors but never came home. Like writing and traveling, authorship was an experience. As we shall see, it was one that, for most people, involved the shaping and reshaping, by Murray and others, of the words of others gleaned from encounters elsewhere.

CHAPTER THREE

Writing the Truth: Claims to Credibility in Exploration and Narrative

Maria Graham was already an established travel writer when, in 1821, she was presented with the opportunity to accompany her husband, a naval officer in command of HMS *Doris*, to South America.[1] The revolutionary movements that, in the first decades of the nineteenth century, had opened up that continent to commercial, cultural, and scientific exchange with the non-Iberian world drew a variety of soldiers, speculators, scientists, and social explorers, each traveling with different purposes in mind.[2] Mary Louise Pratt has identified a gender division in these different modes of travel and travelers, distinguishing between a male "capitalist vanguard" and female "*exploratrices sociales*."[3] This gendered schism was, superficially at least, typified in Graham and her husband and by the factors that motivated their travel: social curiosity for Graham, imperial duty for her husband. Pratt's classification makes clear the gendered dimensions to exploration and travel but elides Graham's multiple and contested identities as artist, author, botanist, and wife and downplays the epistemic concerns that were common to these different modes of travel. Moreover, an unavoidable issue that the capitalist vanguard and the *exploratrice sociale* both faced was what Nigel Leask has called "the problem of credit"—that is, how to evaluate the veracity of those on whom one's own knowledge and travel depended and how, in turn, to demonstrate one's own credibility and the authenticity of one's experiences to different audiences.[4] Being able to trust others (and oneself) in the field and being trusted as a credible author were issues that all of Murray's authors shared, regardless of the purpose for which they traveled. The precise means by which trustworthiness was evaluated and claimed varied between travelers and their social contexts, reflecting the vicissitudes of what we might call the sociology of credibility.

Such questions of credibility and the authenticity of the author were very real for Graham. In her earlier published work on India, she had positioned herself as an "observant stranger," socially and politically attuned, to be sure, and historically and botanically aware but not motivated by any particular regime of scientific investigation.[5] Graham had sought to portray herself as a "rational agent rather than a passive vehicle of passion and prejudice," and so to distance herself from the "excessive sensibility of late eighteenth-century woman's writing."[6] Graham was commended for being an "active and intelligent observer," having a mind "too well furnished, and too discriminating, to be imposed upon . . . by false appearances."[7] The fact that she was an "instructed botanist," a field of natural history in which many middle-class women were then encouraged to participate, lent further credibility to her Indian narratives.[8] Even so, critics identified in her writing a "warm sensibility" in her descriptions of the "beauties of nature"—a loaded compliment that made clear that Graham's descriptive and interpretative abilities could not straightforwardly be divorced from pejorative understandings of her gender (plate 9).[9]

Graham's concern to be judged as authoritative and credible quite apart from the standards of her gender was evident in the preparations she made for her South American journey and in her wish to reposition herself as a "philosophical traveller," not as the observant stranger she had hitherto been.[10] Prior to departure, Graham familiarized herself with what she took to be the most appropriate textual authorities—the initial volumes of Alexander von Humboldt's *Personal Narrative* (1814-29) and Robert Southey's *History of Brazil* (1810-19).[11] Graham's intention to use Humboldt as a guide to correct scientific travel and observation was, however, modified en route. Sailing to Brazil, HMS *Doris* made landfall at Tenerife, and Graham explored that island with Humboldt's account to hand. Although she considered Humboldt to be "a wonderful traveller & wonderfully gifted & qualified," she found his observations "too fine & philosophical."[12] As she confessed to Murray II, "I saw many things . . . that he has thought beneath him [including details of domestic and social organization] & I strained my eyes after many another that has I presumed changed places since he was there."[13] Graham was not alone in using Humboldt's words as a traveling companion, a textual means to direct one's own enquiries. Charles Darwin did likewise on HMS *Beagle*, and in his 1827 book on Columbia (app.: 98), John Hamilton excused the nonscientific nature of his own enquiries (Hamilton's mission was largely diplomatic) by lauding Humboldt's words and procedures even above "the scientific men of Bogotá."[14]

Graham sought to mark herself out as distinctive from, but not inferior to, Humboldt, capable of seeing things that Humboldt had not. Graham reinforced her individuality by offering an interpretation of archaeological artifacts in Tenerife that directly contradicted Humboldt's. While it has been claimed that this contradictory positioning was a deliberate calculation on Graham's part to situate herself "on a plane superior to that of the best-known European traveller of her time," it was, more prosaically perhaps, illustrative of her commitment to sensate empiricism and to the validity of her own interpretations—truth to nature and truth to self (fig. 7).[15]

Graham proceeded in Humboldt's spirit, determined to emulate his careful observation of natural phenomena, but she did not treat his textual account as an unflawed manual of correct travel. What Graham demonstrated was an ability to evaluate the validity of the information on which she drew,

FIGURE 7 Traveling in South America. This frontispiece illustration purports to show the author herself, traveling and exploring in some comfort and style. Source: Maria Graham, *Journal of a Residence in Chile, during the Year 1822* (London: John Murray, 1824). By permission of the Trustees of the National Library of Scotland.

quite apart from the assumed credibility of its source. That this was not a simple rejection of Humboldt's authority is confirmed by the fact that, a decade later, Graham turned to him for support at a time when the validity of her own observations regarding a powerful Chilean earthquake was the subject of heated dispute in the Geological Society of London.[16] Graham additionally sought credibility for herself and her works by her choice of Murray over other publishers.

Graham's first works of travel writing—*Journal of a Residence in India* (1813), *Letters on India* (1814), and *Three Months Passed in the Mountains East of Rome* (1820)—had been published under a partnership agreement between Constable and Company (Edinburgh) and Longman and Company (London).[17] Longman had devolved responsibility to Constable for negotiating with Graham as to her share of the profits, noting that "we shall leave it to you to make the best terms you can with the authoress for her interest in the work."[18] While the terms agreed to appear to have satisfied Constable and Longman, they rankled with Graham who traveled to Edinburgh "to quarrel with . . . Mr Constable" over her contract.[19] While all parties ultimately profited from their investment in Graham's books, which enjoyed healthy sales, the author considered herself a "*patient* of Longman," and she confided to Murray II in 1821 that "I prefer very much your ways to their ways."[20] Graham therefore elected to publish her South American travel narrative with Murray rather than with Constable and Longman. In January 1824, as her books were being readied for the press, Graham wrote to Murray in high anxiety following an unexpected visit from Thomas Longman who had rebuked her for offering her text to Murray:

> Mr Longman has just been here. He began: "Mrs Graham do you know Mr Murray has done very wrong in taking that MS." I told him I was sorry for that since it was my fault. . . . All I say is pray don't dispute or quarrel about me—I should be miserable if I thought that my impatience had led to any thing [sic] of the kind. If as a matter of etiquette you shd [sic] have told him—pray for my sake do so now—or at any rate tell me what I may say. . . . I hope all will be well & that I shall be able to drink a glass of your wine to commemorate the peace.[21]

The peace that Graham sought was achieved through an ad hoc partnership agreement between Murray and Longman who agreed to assume an equal share in the books' production.

Notwithstanding this, Graham was never able fully to suppress the suspicion that Longman retained an animosity toward her and to Murray. When the *London Literary Gazette* issued a scathing critique of the first volume of her account in 1824 — a review that Graham considered "very ill-natured and calculated to do me all the harm possible"—she detected in its excoriating prose the influence of "a certain book seller [i.e., Longman]."[22] Writing to Murray, she confessed her suspicions: "I have little doubt of Messrs Longmans being the instigators of the crying down of Brazil for the sake of their other friends."[23] The "other friends" to whom Graham referred were most probably Johann Baptist von Spix and Carl Friedrich Philipp von Martius, whose two-volume *Travels in Brazil* (1824) had just been issued by Longman.[24] Given Longman's financial stake in the success of Graham's texts, it seems unlikely that the firm would deliberately have sabotaged the project in the way Graham suspected, but this is evidence that Graham felt that her decision to approach Murray had been sufficiently thoughtless to provoke Longman's ire. What this incident illustrates is the financial and reputational value that was placed on works of travel and that credibility could be sought by authors in terms of which publishers took their work. Graham's experiences and writerly strategies were by no means unique, although they are better documented than most.

This chapter explores the means by which Murray's travelers sought to establish the truth claims of the sources on which they drew and about which they wrote. We hope to make clear that standards of credibility and of authenticity were never fixed and inviolate but were defined, at turns, by social context, individuals' expectations, disciplinary conventions, and audience demand. Murray's travelers employed various means — in often contradictory ways — in order to demonstrate a correspondence between what they saw, what they were told, what they wrote, and what they claimed in print. Our focus is on what travelers did during their travels with a view to securing their own creditworthiness. Later, we look at what Murray, the firm's authors, and others did to make authors of explorers, at the efforts made on their return to "authorize" the words of travelers as authors turned their words into print.

TRUST AT A DISTANCE

The principal epistemic problem raised by exploration and travel and its written reports is that of trust, specifically how trust was to be secured at a distance. How could the testimony of "unofficial reporters" and "geographically privileged persons" (as those who traveled and witnessed have been so characterized) be evaluated as to its value and reliability by audiences elsewhere?[25] Maria Graham encapsulated the problem exactly in a letter to Murray II: reporting her pleasure at her recent writing labors, she observed how "Cook's voyages after all are the finest things next to Robin [*sic*] Crusoe & the tale of Troy—his people are real creatures & his places true places— for the rest *voyagers* & travellers for the most part deserve castigation— for of what use is it to tell truth if it looks like a fib?"[26] Knowledge of the world depended importantly on the circulation of individual testimony in the form of what Bruno Latour has called "immutable and combinable mobiles."[27] Written texts, specimens, objects, instrumentally derived data, and visual representations were the tangible proxies of observations made "out there," packaged and made transportable from the field to be recombined, compared, and evaluated in "centres of calculation" (often, although not necessarily always, in metropolitan hubs).[28] This circulation of such testimony and its corroboration in a variety of ways and places was from the Enlightenment hugely significant in promoting what Daniela Bleichmar has termed "long-distance collective empiricism."[29]

The particular problem of credibility that travelers' reports and explorers' claims raised was that they "often lacked witnesses to substantiate their claims" and frequently commented on phenomena that were, at least from a European perspective, "strange and previously unknown."[30] The uncertain quality of the testimony of witnesses who might occupy different social strata, and be in possession of variably sound methodological abilities and investigative expertise, meant that distinguishing fact and fiction in their reports—the "truth" from the "fib" in Graham's terms—was not straightforward.[31] Further uncertainty was introduced when the information reported on by a traveler came not from first-hand witnessing but from second-hand testimony or hearsay. It is for these reasons that the moral and epistemic anxiety that frequently led to conflation of the categories traveler and travel liar has been subject to scholarly attention.[32] Alongside judgments made as to the methodological credibility of travelers' works and words sat assessments of their personal worth. Steven Shapin has shown how, in relation to seventeenth-century science and following John Locke's maxims on trust, social status mattered in the evaluation of an individual's testimony: credibility seemingly mapped neatly onto the "contours of . . . society."[33] Yet while social status was inescapably implicated in the "management of testimony," it does not alone constitute an explanatory or evaluative mechanism.[34] As Palmira Fontes da Costa has shown, methodological rigor, rhetorical appropriateness, and logical consistency often mattered just as much as social status in the evaluation by the Royal Society of individual witnesses' testimony.[35] Credibility was a matter "of *how*, not alone *who*"; gentlemanly or other appropriate social status was simply one indicator among many others of the trustworthiness of testimony.[36]

Consideration of the truth of testimony is further complicated when we take into account the role of genre and literary form in the communication through print of traveler's observations and experiences. It would be inaccurate to assume that travelers' reports were necessarily written, published, and read for their scientific contribution alone or that objectivity (as we might now understand that term) was their obvious goal. It would be wrong, too, to assume that scientific criteria were the only basis on which credibility was judged. The literary conventions that connect the travel narrative and novel are well understood; the travel text as a mode of narration has been shown to share much in terms of rhetorical and discursive construction with the novel—the traveler as protagonist, the itinerary as plot.[37] The travel account as an entertainment was a well-established genre by the eighteenth-century and, often, "popular early travel books intentionally blurred the line between travel 'fact' and travel 'fiction.'"[38] The "suppression of . . . the self" that emerged as a fundamental requirement of scientific travel reporting by the mid-nineteenth century, and did so as part of the growing authority of scientific procedure and of particular disciplinary conventions, was less evident in earlier accounts where the distinction between artistic and scientific work was not so clearly defined.[39] In these accounts, faithfulness of testimony—that is, a presumed truth to nature—did not always involve the "writing out" of subjective experience, and it quite often depended on the author demonstrating the status of active rather than passive observer. Embodied testimony such as that evinced by Head in South America (p. 37 above)—one that exposed such things as sweating effort, profound hunger, and the perils of lice-ridden beds, rancid food, and hostile interlocutors—was often preferred to the detached and remote purview.

In their chronology of publication, the travel texts on which this study is based reveal, over time, a shift away from a more fictive and often subjective discussion of the travel experience to a more distinctly fact-laden and objective depiction. Many of the works, especially those of geographical exploration rooted in official instruction, were intimately associated with the emergence of professionalized scientific objectivity and the privileging of rational empirically derived knowledge over intuition or romantic sensibility. It would be wrong to assume, however, that what we see in the strategies of credit making employed by Murray's travelers is chronologically ordered, a neat transition from subjective sensibility to objective rationality.

In part, this is a consequence of those varying motivations behind Murray's travelers explored in chapter 2. In part, it is simply that the genre conventions of such accounts were necessarily varied. Partly, too, it is a question of the audience to whom Murray's texts were addressed, and the fact that these did not neatly separate into those who read for entertainment and those who read for knowledge. Given that the means by which credibility was assessed varied among genres, between audiences, and over time— being configured differently in the late Enlightenment, through the romantic era, and during the second scientific revolution—we should be cautious about asserting a simple rhetorical move by all of Murray's travelers in ensuring credibility and trustworthiness. Quite apart from the decisions made later by author, editor, and publisher as to appropriate literary presentation, individual authors adopted different strategies within what we may call an ad hoc regime of credibility. This reflected particular judgments as to how best their veracity and that of their observations and judgments might be reckoned truthful to the audiences they sought to address and convince: that they saw it for themselves; were told it by reliable others; or learned of things only from "native authorities."

The regime of credibility that Murray's authors employed is here explored through three overlapping activities and strategies: scholarly citation; authenticity and self-representation (including disguise); and instrumentation. Not all travelers sought to emphasize their credibility in each of these capacities. Those that did so placed different emphasis on specific elements. Nevertheless, these categories are useful in understanding the complex mechanisms that defined travelers' claims to truth and in exposing the importance of credibility at work.

CREDIBILITY THROUGH SCHOLARLY CITATION AND TEXTUAL TRIANGULATION

A confirmation of the accounts of preceding writers, whilst it diminishes the interest of a more recent description, cannot, however, be considered unimportant; correct information is the object in view, and whether that obtained by reference to old, or application to modern authors, is [a] matter of indifference.[40]

In much the same way that Maria Graham had sought information and instruction in Humboldt's *Personal Narrative*, Murray's authors chose to situate themselves within complex textual networks—printed sources and authorities that were brought to bear in the planning, execution, and evaluation of their travels. The texts on which Murray's travelers drew reflected, however, not simply their likely practical value as repositories of geographical knowledge or methodological guidance. These works actively implicated other authors in the positioning of credibility as travelers, witnesses, and authors. As one historian of citatory approaches has noted, "The scholarly authorities whom an author chose to cite must tell us something of their scientific self-fashioning, of their intellectual tastes and imagination, of the sense they had of whom they were in dialogue with as they composed their books."[41] By examining the citatory practices used by certain of Murray's authors, we can reveal not only a sense of the sources they thought valuable but also how they used certain texts as markers of their credibility, either to show that they were following in the footsteps of a respected other or, like Graham, to show that they were offering a necessary corrective to an already-established truth. The practices employed by Murray as publisher—over positioning footnotes, statistical data, and other paratextual apparatus in the firm's travel imprint—is discussed in chapter 6.

Murray's travelers to North Africa and to the Middle East routinely turned to classical authorities to properly situate and contextualize their own geographical contributions. In their *Journal of an Expedition to Explore the Course and Termination of the Niger* (1832; app.: 124), Richard and John Lander offered their readers a summary of the various debates that had, since Herodotus, characterized speculation about the River Niger's direction of flow and its unknown termination. Through reference to the work of Herodotus, Pliny, and Pomponius Mela and to later Islamic geographers, including al-Fida, al-Idrisi, and Leo Africanus, the Landers sought to make explicit a claim as to their scholarship. To do so was not simply a means by which the subsequent contribution of eighteenth- and nineteenth-century travelers "out there" was to be contextualized and their own first-hand contribution defined; it was an important means of demonstrating their own intellectual credentials. Engaging with classical authorities was necessary not only to the planning of a successful expedition but also in demonstrating an intellectual capacity to test ancients' claims to credibility. Authorial and exploratory credibility came not from accepting the speculative conclusions of classical scholars but from being able to evaluate their utility and truthfulness and, where necessary, to show them to be erroneous. The Landers thus hoped to position themselves as part of that class of "well informed and experienced geographers" who could make such authoritative and corrective proclamations.[42] There is every reason why they and others should do so: exploration, perhaps especially in Africa, was then and in later decades marked by a tension between "in-the-field" explorers, for whom trust had to be secured by direct observation or the verbal testimony of others, and "armchair" or "critical geographers," whose truth claims were derived from faith in others' written evidence and from what we may think of as the accumulated evidence of textual enquiry from different language sources.[43]

What is unclear about the Landers' citatory practices is whether, and in what form, they had actually read Herodotus and the other authorities to whom they made reference. While it is conceivable that they drew on eighteenth- and nineteenth-century editions, it is equally probable that they depended on summary publications such as James Rennell's *The Geographical System of Herodotus* (1800) and Hugh Murray's *Historical Account of Discoveries and Travels in Africa* (1817).[44] It is certainly the case that, for a number of Murray's travelers in Africa and the Middle East, Rennell's book was an important source. For Thomas Maurice, assistant librarian at the British Museum and author for Murray of *Observations on the Ruins of Babylon* (1816; app.: 36), Rennell's text was a "valuable work on the Geography of Herodotus" being so comprehensive in its coverage of "the *geography* of this celebrated region of Asia . . . that little can be expected to be added to its instructive details."[45] Rennell's work in this respect was a form of ancient geography wherein the written works of classical and other ancient authors were scrutinized in their own right and in comparison with "modern geography," the results of first-hand exploratory encounter.[46]

Much the same view of Rennell's scholarly worth was held by Archibald Edmonstone, author of the Murray-published *A Journey to Two of the Oases of Upper Egypt* in 1822 (app.: 70). He, too, considered Rennell's text a "valuable" index to Herodotus's thought.[47] That said, both Maurice and Edmonstone also chose to engage in reading classical authorities in full-dress editions. Footnote citations within Edmonstone's narrative are an explicit testament to his scholarship, even while reading as a form of code: "Herodotus, Thalia, Sect. 26. p. 207. Fol. Ed. Wesseling. Amst. 1763"; "Strabo, Xylandri, p. 940"; "Plin. Hist. Nat. lib. 5. cap. 9. Ed. Delph. 4to"; "Ptolemæi Geogr. Montani. Franckfort 1605, folio."[48] The 1763 Wesserling edition of Herodotus and the 1605 Frankfurt edition of Strabo (distinguished for being the first to present the Greek and Latin texts together) were considered by contemporary scholars to be among the best available.[49] Here was evidence of the fact that Edmonstone had read the classics in their full-dress versions, in Latin, and in the best available editions. Having been educated at Eton and at Christ Church, Oxford, such scholarship was a marker of his social status and academic attainment.

Credibility was a pressing concern for Edmonstone in two further distinct respects. The first was that the geographical discovery he claimed to have made—a previously unrecorded desert oasis—needed to be shown to be genuinely novel. The second was that the discovery of that oasis was then being claimed, erroneously as it was later proved, by one of Edmonstone's competitors, the Italian antiquarian Bernardino Drovetti. Drovetti had arrived at the oasis after Edmonstone but claimed precisely the opposite in his printed narrative. Questions of novelty and precedence were central to demonstrating credibility. The claim to precedence came down to questions of competing chronology and to Edmonstone's efforts to identify and expose irregularities in Drovetti's account.[50] The claim to novelty rested on Edmonstone's citation and evaluation of classical and Arab sources and by reference to his own cartographical data: "In order effectually to prove our claim to the discovery . . . I have brought together all I have been able to find respecting these interesting spots; the accounts of ancient and Arabic authors, the reports of travelers, and the various systems of geographers who have treated upon them."[51]

Citation of classical sources was not, of course, the only means by which a traveler's scholarship and preparedness might be demonstrated: that also depended on an understanding and evaluation of "modern authorities."[52] In preparation for the travels that were subsequently related in his *Journal of a Tour in Asia Minor* (1824; app.: 82), William Martin Leake brought together all relevant classical texts and contemporary travel accounts in order to afford comparative judgment of past authors with modern accounts. In his preface, Leake listed more than thirty authorities on whom he had drawn in constructing the map that accompanied the book—part of a self-justifying effort to demonstrate that he was "qualified for the task [of appropriate travel] by literary and scientific attainments."[53] Leake was an army officer and member of numerous scientific and other bodies: the African Association (having been elected in 1813), the Society of Dilettanti (1814), and the Royal Society (1815). Leake could thus demonstrate his credibility even in advance of going abroad by evaluating the relative significance of extant sources: the eighteenth-century English clergyman-traveler Richard Pococke was, in his view, "extremely negligent"; the Frenchman, Paul Lucas, was "very inferior"; but his compatriot, Jean Otter, was "most useful."[54]

By comparing the work of "ancient geographers and historians with . . . modern travellers," Leake sought to establish truth through a form of what, to draw from the related world of terrestrial surveying, we may call "textual triangulation."[55] When textual sources contradicted one another or even when they did not but simply stated ancient claims as established truths, credibility in Leake's view could not be arrived at by reading alone: it necessitated the on-the-spot testimony of an "intelligent traveller."[56] Such a method of arbitration and evaluation was not unique to Leake, of course, and was characteristic of many of Murray's travelers who sought to demonstrate authority from the very fact of travel itself. Leake was representative of a certain kind of scholarly travel—most evident perhaps in relation to the exploration of "antique lands" where human presence had left its traces over time—and an approach to knowledge making that required the combination of various skills and competencies.[57] Credibility came from the simultaneous demonstration of classical scholarship, navigational aptitude, cartographical competence, orthographical knowledge, and gentlemanly status.

Leake was, in turn, a source of travel guidance for others. When historian, geographer, and literary hack Charles MacFarlane undertook his sixteen-month exploration of Turkey beginning in 1827, he was literally guided by Leake's book.[58] As MacFarlane confided in a letter to Murray II, "When in Asia Minor, his [Leake's] valuable little work on that country, was my constant companion—I learned it almost by heart and whenever my road lay on his, I admired with enthusiasm, his short, graphic, correct sketches."[59] MacFarlane had written to Murray in response to the forthcoming publication of Leake's three-volume *Travels in the Morea* (1830).[60] Murray had solicited MacFarlane's views on the text and they were, clearly, positive: "I return you the first Vol. of Col. Leake's work, which I have read through attentively, and with *extreme* delight. My pencil marks, and they are all *admirative*, you will find pretty numerous! Judging from his marvellous accuracy on subjects and matters I am acquainted with, I can give Col. L. credit for *general* correctness—for an accuracy almost mathematical, in everything he touches."[61] Leake's work was, in MacFarlane's view, invaluable "to the scholar, and the antiquary."[62]

Traveling with book in hand, as MacFarlane claimed to have done, was not simply a metaphor: it reflected the necessity to test when traveling the degree of correspondence between written testimony and the evidence of one's own eyes. We should not suppose that the Landers, Edmonstone, Leake, and others necessarily did likewise with editions of Herodotus, Strabo, and Rennell within easy reach. For them and others, traveling afforded moments where, as thoughts turned to later publication, would-be authors recognized the need to seek warrant for their work in others' earlier words. William Bullock realized this as he mused on his Mexico work (app.: 77). He had kept a journal in the field despite "every disadvantage of hurry, fatigue, and superabundant demands upon my attention," but he came to appreciate, too, that his experiences had to be attested to, where it was possible to do so, by reference to secondary sources for which he had "not the smallest reason to doubt the authority."[63]

Others did travel with books at hand. In his *Biblical Researches in Palestine* (app.: 153), the scriptural Edward Robinson sought to establish a correspondence between the Bible and the geography of the Holy Land. He noted how texts guided his travels, listing in detail the books with which he and his companion Eli Smith had traveled, and, on the basis of such books being "field tested," which of them they might recommend to later travelers:

> First of all we had our BIBLES, both in English and in the original tongues; and then RELAND'S *Palæstina*, which next to the Bible is the most important book for travellers in the Holy Land. We had also RAUMER'S *Palästina*, BURCKHARDT'S *Travels in Syria and the Holy Land*, the English compilation from LABORDE'S *Voyage en Arable Petrée*, and the *Modern Traveller* in Arabia, Palestine, and Syria. Were I to make the journey again, considering the difficulty of transporting books, I should hardly add much to the above list, excepting perhaps a compendious History of the Crusades, and the volume of RITTER'S *Erdkunde*, containing Palestine in the second edition.[64]

Such textual baggage allowed Robinson and Smith to make ready checks on the veracity of these authorities, particularly in relation to monumental epigraphy and the fidelity of other travelers in recording it.[65] They found Johann Ludwig Burckhardt "tolerably exact" in this regard but "not so [Richard] Pococke," whose transcriptions showed "hardly a trace of resemblance."[66] Carsten Niebuhr was "[not] much better."[67] Speaking and writing with authority was—as Robinson and Smith showed, with Leake—a question both of citation of extant scholarship and of evaluation and testing through in-the-field observation. Accuracy—the degree of correspondence between the claims made, their representation either in writing or in picture and comparison with the actual feature—was a proxy measure for the fidelity of the correspondence that existed between a credible traveler's account and the audience's belief in the world portrayed. Accuracy was often more than an ideal: it was a question of emotional and aesthetic veracity and of the ability of a traveler to respond subjectively, as well as objectively, in appropriate ways to their experiences in the field.

In preparing his *Geographical, Statistical, and Historical Description of Hindostan* (1820; app.: 60), an updated and expanded version of his earlier

East India Gazetteer (1815; app.: 32), Walter Hamilton made clear the difficulty of evaluating contradictory sources of information and of attributing
authorship to particular geographical facts, while emphasizing the importance of his own eyewitness testimony as a regulating mechanism in this
process:

> The authorities upon which each description is founded are carefully subjoined
> in succession, according to their relative importance, the author being particu-
> larly desirous to give the credit where it is justly due, as well as to establish the
> high character of the sources from whence the original information is derived.
> But no person is to be considered wholly responsible for any article, the materi-
> als being frequently so intimately blended with each other, and with the result
> of the author's own experience during a ten years' residence in India, that it
> would be impossible to define the boundaries of respective properties. In many
> cases the narrative is given as closely as the necessity of condensing the subject
> of many volumes into a small compass would permit; in others it has been
> necessary to compare contradictory and conflicting testimonies, and select that
> which appears to rest on the most solid foundation.[68]

Hamilton's book—one representative of a genre of statistical enquiry
subtly distinct from the travel narrative but sharing with it the requirements
of credibility—was assembled from Hamilton's personal experiences, various printed and manuscript sources deposited at the East India Company
library in London, and culled from the testimony of company servants.
Hamilton made repeated judgments as to those sources he considered the
"best qualified from length of service, residence on the spot, and established
reputation, [in order] to form a correct judgment of their authenticity."[69]
Hamilton's task was one "of selecting, comparing, arranging, and condens-
ing" the sources on which he drew and relying on his ten years' residence in
India as a guide.[70] In first proposing the book to Murray, Hamilton had been
eager to emphasize the scholarship it would necessitate and to show how
the book would be a distinctly more rigorous undertaking than his earlier
East India Gazetteer. What he wished to produce was, he told Murray II, "a
historical, statistical, political &c &c &c account of Hindoostan, as it now
exists, province by province, the same as if England were described shire
by shire, including manners, customs, religion, commerce &c &c."[71] In thus
composing a new regional geography of India as distinct from a gazetteer,
Hamilton would, he estimated, "refer to nearly 100 new authorities, 3/4
th[ou]s[and] manuscript ones, & comprehending most of the eminent men
who have conducted the affairs of India for the last 20 years."[72]

Hamilton's book was a critical but not a commercial success. Its price and format (at £4 14s. 6d. for a two-volume quarto) necessarily limited its reach.[73] Murray and Hamilton were, however, conscious of the commercial perils associated with the book's publication, Hamilton having warned Murray that given "the total indifference felt by the English nation on every subject relating to India," "the expense & risk of such a publication will be considerable."[74] Notwithstanding this commercial context, Murray II retained Hamilton's services as a reader; his ability to evaluate, "with singular industry," the works and words of travelers, particularly in respect to their credibility, making him especially suitable for that task.[75] The manuscripts that Murray passed to Hamilton for his rapid, frequently acerbic, comments either were travel accounts or were related thematically to India.

Hamilton dismissed works of travel, at turns, for being "silly mawkish nonsense" or "mere catchpenny" but found value in those he considered to be scholarly, informative, or authentic.[76] In reviewing Frederick Henniker's account of travels in Egypt and the Middle East, published by Murray in 1823 as *Notes, during a Visit to Egypt* (app.: 76), Hamilton commended Henniker's "lovely familiar style" yet castigated his "attempts at wit," which he considered to be "the crying sin of modern travellers."[77] Although Hamilton concluded that Henniker's account contained "nothing either solid or new," particularly when juxtaposed against Edmonstone's contemporary narrative, he nevertheless felt that it might "entertain a younger class of readers than I belong to."[78] Hamilton did not enjoy the same editorial control that was exercised by Barrow in the shaping of travel texts for publication, but his opinion as to the quality of submitted manuscripts was essential in guiding Murray's decisions as to which to accept and which to reject. It was on the strength of Hamilton's assessment that Murray saw through the publication of, among other works of travel, Alexander Caldcleugh's *Travels in South America* (1825), "being fresh & authentic," and Reginald Heber's "pleasingly written" posthumous *Narrative of a Journey through the Upper Provinces of India* (1828; app.: 85, 104, respectively).[79] Here, too, there is evidence of an accumulative credit agency at work: scholarship borne of first-hand encounter was used to guide others in the field and to provide a means to corroborate others' textual enquiry.

AUTHENTICITY, DISGUISE, AND NATIVE AGENCY

The exegesis of others' texts was alone insufficient in demonstrating credibility. In nearly all cases, traveler-authors depended on personal, first-hand testimony, even where the text in question was a product primarily of sedentary scholarship. This situation was further complicated by the inescapable requirement of travelers to evaluate the reliability of those third parties—agents and brokers, guides and translators, politicians and porters, chefs and chiefs—on whose knowledge, expertise, assistance, or patronage the successful completion of their journeys depended. Such go-betweens had a clearly central, if often neglected, role in Europeans' excursions in the non-European world.[80] The silencing or erasure of the facilitators of Europeans' travel reflected and encouraged a rhetoric that emphasized the individual author's achievement over collaborative effort. With the contribution of third parties commonly removed from view, European travelers were, so it seemed, "invested with magical powers enabling them to float across desert, swamp, forest, mountain and polar ice without the slightest need for support or guidance."[81]

In the case of Murray's travelers, it would be wrong to assume, however, either that the textual elision of go-betweens was systematic or complete or that the tendency to downplay their significance represented a policy on the part of the publishing house. The extent to which the work of go-betweens was acknowledged in Murray's works of travel, and how it was acknowledged, varied between authors and across time. There is no evidence in the correspondence between the house of Murray and their authors over advice given, or requirements stated, on this matter. Yet certain commonalities are evident in the ways in which go-betweens of different races and classes are acknowledged, if not always named. Mostly, indigenous informants, guides, and facilitators—where their existence and role is acknowledged at all—are referred to not by name but by race or tribal affiliation. High-ranking indigenous inhabitants and resident Europeans (or Creole descendants of Europeans) encountered during travels are, by contrast, typically identified by name. The degree to which indigenous intelligence and agency is acknowledged varies considerably between texts but can, in broad terms, be seen to represent a relationship between geographical region, genre of writing, and intended audience: explorers and travelers in North Africa, the Near East, and Central Asia commonly employed native guides and/or dressed themselves as natives to facilitate their travel. Most of Murray's authors in these places make clear the faith they have placed in the testimony of others and, through this, illustrate credit and judgment both in their informants and in themselves for having chosen credible personnel to assist them (fig. 8). Such statements, we suggest, are part of a rhetorical strategy that authors employed to mark their identity as arbiters of truth. Against this, however, there were travelers who were as keen to dismiss the authority of native agency or deny their dependence on it (fig. 9).

For yet others, securing truth via travel depended, if only for a moment,

FIGURE 8 The go-between at work. In this engraving titled *Interview with the Dey*, Salamé stylizes those vital moments of encounter between European explorers and diplomats and local rulers on whose successful outcome safe exploration and travel would often depend. Source: Abraham Salamé, *A Narrative of the Expedition to Algiers in the Year 1816* (London: John Murray, 1819). By permission of the Trustees of the National Library of Scotland.

on "becoming" a native: disguise was held to lend a degree of verisimilitude for the transient explorer even as it raised questions about disclosure of travel facts by deceit. For John Taylor and his companions journeying overland to India at the end of the eighteenth century, disguise was an ever-increasing necessity as they traveled east. Having not yet "quitted the European dress" by the time they entered Syria, they were entirely persuaded of the importance of native dress and disguise by the time they left that country: "Put on the dress of the natives as soon as possible after your landing in Mahomedan country," Taylor counseled, in words that prefigure those of Drummond-Hay to Davidson in North Africa, "or even before if you can procure it: and the dress I would advise to be correspondent to the manner in which you propose to travel, but never to be beyond that of the middle ranks of life."[82] Disguised as they were in native dress and with assumed native names as they crossed parts of the Sahara nearly twenty years

FIGURE 9 Locals assist the traveler. The rigors of travel could be lessened if locals carried both the equipment and the traveler. The engraver, Edward Finden, based with his brother in Southampton Place near London's Euston Square, was employed by the house of Murray to illustrate numerous exploration and travel narratives in the 1820s and 1830s, in addition to illustrating editions of the life of Byron and of the Bible. Source: John Potter Hamilton, *Travels through the Interior Provinces of Colombia* (London: John Murray, 1827). By permission of the Trustees of the National Library of Scotland.

later, Edmonstone and his companions "soon discovered that among Bed-
ouins, the English was a much better traveling character than the Turkish,
and as Yusuf, Ibrahim, and Halil, the names we had assumed, were equally
used by Christians and Mohammedans, we always took care to disclose our
nation."[83]

William Martin Leake may have sought credibility through the citation
of scholarly sources, but the success of his researches was greatly dependent
on a retinue of go-betweens, including "servants . . . Turkish attendants and
postillions," and one unnamed "Tatár courier" whose job it was to "ride
forward to procure lodgings" for the party.[84] The Tatar courier and Turkish attendants were crucial intermediaries between Leake's party and local
inhabitants in securing food, lodging, and directional advice. Throughout
his narrative, however, Leake is reluctant to reveal the sources of his information beyond the scholarly texts he cites. While his narrative hints at information derived from indigenous oral testimony—"We are told that," "we
are informed that" being oft-repeated statements—the sources of such intelligence are never clearly indicated.[85] This is in marked contrast to his careful citation of the classical authorities on whom he drew: "Strabo has told
us"; "we are informed by Herodotus"; "we are informed by Vitruvius."[86]

Leake's choice over the presentation of his sources reflected a number
of distinct concerns. The fact that the inhabitants of the region through
which he traveled were considered by some of his audience in Britain to
be "degraded and semi-barbarous" meant that he was, as one critic noted,
"compelled to restrict himself to a comparison of the routes of former travellers, and the data which they have been enabled to collect."[87] As a fellow
of the Royal Society, Leake's name and affiliation was "a sufficient guarantee" for "the geographer, the historian, the scholar, and the man of science"
of his credibility and his capacity to evaluate the testimony on which he
drew.[88] Given that Leake's intended audience saw his text (and accompanying map) as a corrective to extant scholarship on the area, his citation
of textual sources mattered more than did his occasional reference to supplementary knowledge acquired from indigenous sources. Leake's status,
mode of writing (one free from "surprising incidents and adventures"), and
intended audience shaped his acknowledgment of go-betweens and indigenous agency.[89]

At least Leake did not subject his go-betweens to disdain and dismissal.
Several Murray authors did. The tendency to frame guides, interpreters,
and local agents as impediments to, rather than facilitators of, travel is sufficiently commonplace in works by Murray's authors that we may again

suspect a particular rhetoric at work. That is, such accounts deliberately reinforced the superiority of the individual traveler (and of the European, more generally) and added weight to their written claims by the evident discrimination of native testimony sufficient to reject it. This is evident, for example, in the work of the brothers Henry and Frederick Beechey who, in 1826, conducted a land-based survey of the northern coast of Africa. The results were published by Murray in 1828 as *Proceedings of the Expedition to Explore the Northern Coast of Africa* (app.: 102). In the Beecheys' narrative, indigenous informants are at turns invaluable and infuriating, trustworthy and disreputable: being able to make a distinction between these traits was evidence of the authors' capacity to discriminate and evaluate their sources.

The Beecheys began their survey at Tripoli, where they "consulted with the most intelligent natives," including Sidi Mahommed D'Ghies (functionary to Yusuf Karamanli, the Pasha of Tripoli) who had earlier been of assistance to the George Francis Lyon in his travels in North Africa (app.: 65).[90] Lyon had remarked on D'Ghies's "kind assistance" and "useful information." Given the fact that D'Ghies had "travelled much in Europe" and was "well acquainted" with European customs, he was, in Lyon's view, "perfectly qualified" to offer reliable and culturally relevant information. [91] By dint of his social position within the court of Karamanli, and his experience in Europe, D'Ghies was deemed, by Lyon and the Beechey brothers, a trustworthy and credible go-between. D'Ghies and other "Mahometan friends" at Tripoli advised the Beecheys to adopt native dress "in favour of the European costume."[92] Although the British Consul at Tripoli disagreed with this advice, the brothers felt it "most advisable to adopt the advice . . . which we knew to be formed on an extensive acquaintance with the prejudices, manners, and customs of the Arabs."[93] It helped, of course, that D'Ghies's advice was, as the brothers informed their readers, "quite in union with our own [instincts]."[94]

Shortly before departing from Tripoli, the Beecheys met with Walter Oudney and Hugh Clapperton, then both at the beginning of what was to be an ill-fated expedition whose written account would be published by Murray in 1826 as *Narrative of Travels and Discoveries in Northern and Central Africa*.[95] If the Beechey brothers discussed modes of dress with Oudney and Clapperton, it is unlikely that the latter shared the former's opinion, for Oudney and Clapperton were (as they proudly informed their readers), "the first English travellers in Africa who resisted the persuasion that a disguise was necessary, and who had determined to travel in our real character as Britons and Christians, and to wear, on all occasions, our English dresses."[96] That judgments made as to the credibility of native advice and testimony did not depend on the informant's social status alone was illustrated when, in the late 1820s, D'Ghies's son, Hassuna (who had been educated in France and had served as the Tripolitanian ambassador to Britain in 1821), was falsely implicated in the death of another of Murray's travelers in North Africa, Alexander Gordon Laing, who was to die in full "English dress" on his return from Timbuktu in 1826.[97]

The Beecheys' party consisted of three Europeans, "who acted equally as interpreters and servants," a Turkish infantryman, "five other Bedouin Arabs, and three Arabs to look after the horses."[98] While the Beecheys received welcome advice and assistance from "the better classes of Turks and Arabs" they encountered on their surveying expedition, the brothers often complained about the unreliability of the Arabs in their retinue.[99] They complained of the "exaggerated accounts" of their guides and their tendency to "excuse themselves from any service of difficulty or exertion."[100] In a damning indictment, the brothers offered a robustly pejorative assessment of their Arab go-betweens:

> Truth is so little regarded by an Arab, that when his interest or his comfort will be promoted by a breach of it, he is always prepared with a falsehood; and it is difficult, even for those who are well acquainted with his character, to tell when he is sincere in his assertions. One of two things must necessarily result from this want of proper feeling: they who place too much confidence in Arab sincerity will continually be deceived and imposed upon, or they who distrust it too far will on some occasions be liable to wish that they had been less obstinate in their disbelief.[101]

Rather than dismiss the value of Arab assistance and testimony outright, the Beecheys sought to distinguish—on the grounds of class and social status—those who could be relied on from those who were to be considered untrustworthy. Although the Beecheys considered "the talent of amplification" to be one "generally possessed by Turks and Arabs of all classes," being able to distinguish between truth and falsehood was nevertheless crucial to the successful completion of the brothers' survey work and their self-presentation as credible judges of character. [102]

The fact that their guides were not straightforwardly to be relied on presented the Beecheys, and other travelers like them, with an opportunity to position themselves as makers and arbiters of truth, constantly engaged as they traveled in inquisitorial appraisal. In this respect, the Beecheys were representative of those of Murray's travelers who made explicit in their written accounts the skepticism with which they had treated third-party testimony and the lengths they had gone to in corroborating that information through personal experience or triangulated testimony. As the returning East India Company officer and Murray author John Taylor put it in his 1799 work (app.: 16), "I trust that the caution with which I received the information, and my own want to credulity, will, in adding my testimony, be sufficient to engage the confidence of my readers."[103] For Robert Fortune (app.: 185), at the mercy of what native guides told him about the location and type of plant specimens in the vastness of northern China, "I made a practice of disbelieving every thing they told me, until I had an opportunity of seeing and judging for myself."[104]

For every Murray traveler who complained about their guide's unreliability, obstinacy, or deceit—castigating them for being "of little use" or having "a total disregard for truth and honesty"—there were others for whom their service was invaluable and a matter for praise.[105] For Edward Robinson and Eli Smith, their guides were generous, attentive, and tireless: "With our Tawarah guides, we had every reason to be satisfied. They were good natured obliging fellows, ready and desirous to do for us everything we wished, so far as it was in their power. . . . They walked lightly and gaily by our side; often outstripping the camels for a time, and then as often lagging behind; and they seldom seemed tired at night."[106] Their chief guide, Beshârah, was "active, good natured, and obliging."[107] In so emphasizing Beshârah's qualities, Robinson and Smith were not supposing blind faith in him, but illustrating to their readers their ability as effective judges of character. When Beshârah was obliged to resign his position, Robinson and Smith consulted the written work of other travelers in that region for recommendations as to reliable guides before selecting a replacement, Tuweileb, who had previously guided Edward Rüppell, Alexandre de Laborde, and Alexander Lindsay, the twenty-fifth Earl of Crawford. Tuweileb, like Beshârah before him, was instrumental to the success of Robinson and Smith's journey. Their written narrative reveals the reliance placed by them on Tuweileb's "sagacity and experience"; phrases such as "Tuweileb said" and "according to Tuweileb" pepper their text.[108]

Whether Murray authors considered their go-betweens and native informants deceptive and untrustworthy or reliable and true, the point of importance is about rhetorical claims as to their own credibility, either by showing themselves to have succeeded despite the obstacles presented by unreliable guides or because they had the ability and insight to discriminate between the reliable and the questionable. Authors might also show themselves credible by deliberately not evaluating the reliability of native testimony. For authors who wished to employ modesty and lack of expertise in their strategy of credible self-presentation (an approach discussed more fully in the following chapter), assuming the role of an unbiased reporter of indigenous information was both an epistemic and a stylistic strategy.

Consider in this regard Godfrey Mundy's *Pen and Pencil Sketches* (1832; app.: 125). Mundy, whose narrative described a six hundred-mile tour of inspection of military stations in British India, positioned himself as a guileless witness who "simply described scenes and characters as they appeared, incidents as they occurred, and anecdotes as they were related."[109] Mundy, let us be clear, did not necessarily accept all testimony at face value, but he certainly felt it necessary to present it unmodified and unverified: he relied on "oral authority, and not . . . authenticated history."[110] As he informed his readers, "many of the . . . anecdotes interspersed through this journal were gleaned—after the manner of an indolent, yet inquisitive traveller—more by asking questions than by consulting documents."[111] Mundy, as he saw it, turned "king's evidence against myself": as someone with no geographical or philosophical point to prove, it was more appropriate for him to report, rather than to evaluate, what he saw and was told.[112] As one critic in the *Edinburgh Review* noted, Mundy's work pointed to an important distinction between the scholarly and the casual traveler. The former often sacrificed the novelty and interest of their material by devoting years to the search for "perfect accuracy"; the latter, in preserving "the hue of novelty" through a more rapid narration of their journeys, could nevertheless offer "correct and authentic description."[113] Balance had to be struck between an account that was vivid and timely and one that was unimpeachable but potentially out-of-step with the reading public's interests. As the reviewer further noted, for the reader "in search of amusement," the work of Mundy and travelers like him drew its authority and interest not from its scholarly and considered evaluation of places, people, and geographical facts but from its rollicking pace and adventuresome spirit.[114] As others noted of Mundy, "He stayed long enough in the country to receive a vivid impression of all that meets the eye, and strikes the mind, without remaining until the effect of novelty had ceased."[115]

The antithesis of William Martin Leake's "intelligent traveller," Mundy embodied an ideal of masculine heroism and boisterous bonhomie that commended him to certain audiences.[116] Not for him the careful evaluation and selection of native guides; "a few rupees, and (shall I confess it?) a little gentle corporal persuasion" was all that was required to secure the services of his porters and assistants.[117] That Mundy's appeal as a seeming prototype to George MacDonald Fraser's Harry Flashman was not universal is illustrated by a satirical review published in *Fraser's Magazine for Town and Country*. With tongue firmly in cheek, the magazine's reviewer opined: "Mundy is a man after our own souls—a hearty, adventurous, bold, chivalrous, dare-devilling, tiger-killing, lion-quelling, jackal-murdering fellow, with a firm hand, keen eye, sure aim, exhaustless enthusiasm, and good humour. . . . Of all boon companions, a British soldier is the man for us; one who will laugh, drink, fight for you. . . . You may fancy such a gentle and generous roysterer, standing six feet six out of his shoes, with one eye leering after a pretty girl, and the other sparkling with adventuresome heroism."[118] The qualities that marked Mundy's authenticity for the *Edinburgh Review* were precisely those which rendered him a figure of ridicule for *Fraser's Magazine*.

Mundy's credibility was, in this respect, clearly contingent on the genre of his writing, his strategies of self-presentation, and the opinion of his audience as to appropriate observational and inscriptive behavior. Mundy's self-positioning was resolutely nonscientific. For him (and for the audience to which he wished to appeal), scientific travel carried with it an air of fustiness from which he was keen to distance himself: "I am totally unskilled in botany and geology; by which fortunate default in my education he [the reader] will escape the usual inflictions of scientific travelers (a class, by-the-bye, not included in Sterne's catalogue of touralizing emigrants); such as being delayed half an hour to dissect a daisy, or being planted the best part of a forenoon before a block of stone, to decide whether it be granite or marble, primitive or secondary rock; till the bored peruser becomes almost petrified himself."[119]

Resolutely nonscientific as they may have been, Mundy's travels were, nevertheless, instrumentally regulated. He traveled with a Fahrenheit thermometer (with which he took frequent temperature readings), a watch, and a telescope. On one occasion, Mundy used the thermometer to establish his party's altitude through testing the boiling point of water. In his written account, Mundy again sought to subvert the aura of scientific authority that such instrumental use suggested and to reinforce his status as an amateur observer: "With all the gravity . . . of a Humboldt or a De Saussure—and to the great edification of the gaping natives, who no doubt took us for a couple of scientific emissaries from the Royal Society, instead of two of Sterne's 'idle travellers'—we prepared an assortment of . . . materials for making a fire."[120]

Mundy reveled in the status of Sterne's idle traveler. Leake strove to secure his position as an intelligent observer through book and field testimony. Graham sought to prove her ability as a philosophical traveler. Such differences tell us something significant about the different personal routes to credibility and trustworthiness in dealing with the problem of credit. Credit was attained and secured not through the demonstration of a single "correct" approach to travel but through careful judgment over different practices and strategies of self-presentation. While the example of Mundy makes clear that the use of scientific instruments was not restricted to those travelers who saw themselves (or sought to promote themselves) as intelligent or philosophical, he may have been disguising his own attainment in that regard in order to promote another view of himself. Certainly, in order to travel with authority and to be read as doing so, instrumentally regulated observation was an inescapable requirement of credible scientific work in the field.[121]

CREDIBILITY THROUGH INSTRUMENTATION

For scriptural geographers Edward Robinson and Eli Smith, a traveling library may have been useful but its bulk presented significant practical problems. Questions of heft and transportability mattered less for those of Murray's authors who traveled by sea. A shipboard library—stocked with a "working reference collection of books on navigation, repair, and maintenance, and . . . proper political and religious literature"—was a common provision aboard nineteenth-century ships of exploration.[122] The reading material available to the officers and crew of naval ships and to civilian passengers on commercial vessels reflected practical requirements of navigation, the needs of correct scientific investigation, and concern as to the morals and the morale of the crew.[123]

In his account of the unsuccessful 1818 expedition to the Arctic in search of the Northwest Passage, John Ross offered a detailed description of the equipment and provisioning of his vessel, HMS *Isabella*. Under Ross's direction, a number of books were supplied for "the use of the officers, and quarter-deck petty officers."[124] The ship's library included travelers' narratives, scientific texts, and astronomical and navigational guides, together with "Thirty Bibles and sixty Testaments . . . supplied by the Naval and Military Bible Society."[125] The narratives fell, broadly, into two categories: those such as Alexander Mackenzie's *Voyages from Montreal* (1801) and Samuel Hearne's *A Journey from Prince of Wales's Fort in Hudson's Bay* (1795), which related specifically to the territory the expedition was to navigate, and those, such as Thomas Falkner's *A Description of Patagonia* (1774) and William Dampier's *Voyages and Descriptions* (1699), to be read for pleasure and for the inspiration they might offer over practices of seamanship, observation, and description.[126] Volumes on mineralogy, geology, and botany constituted the ship's scientific collection. The ship's library, part of the exploration's scholarly apparatus, was complemented by approximately sixty scientific instruments recommended by the Royal Society, including seven chronometers, four dipping needles, and eleven thermometers. Almost half of the instruments were supplied by the prominent London maker W. & S. Jones.[127] Part mobile library, part floating laboratory and instrument shop, the *Isabella* bristled with devices designed to fix her position, probe the waters through which she sailed, chart the earth's high-latitude magnetic field, measure the shape of the planet, distinguish ice from land and sea, and reduce to standardized criteria of word and number the aurora borealis, which shimmered silently overhead (plate 10).

The significance of Ross's detailed inventory of his ship's library and instruments as part of claims to credibility becomes clear when we consider how Ross's testimony was evaluated by his contemporaries, not least John Barrow and the expedition's astronomer, Edward Sabine. Ross's attempt to penetrate the Northwest Passage had been prevented by what Ross termed a "chain of mountains" (which Ross named for John Croker, first secretary to the Admiralty) which seemed to block the *Isabella*'s passage through Lancaster Sound.[128] Their route seemingly obstructed, the expedition returned to Britain. Ross's account of events was initially accepted as conclusive fact. But suspicion as to his veracity began to mount when Barrow learned that "only a few officers other than Ross himself claimed to have seen the lands, that Ross had not investigated to ensure it was not an optical illusion, and that the second-in-command, William Edward Parry, doubted Ross's conclusions."[129] Ross considered these suspicions the product of a "formidable conspiracy" between Barrow, Parry, and Sabine, a tension that affected the publication of the expedition's narrative.[130] Relations between Ross and Barrow were irrevocably damaged when, at a dinner hosted by Robert Dundas, first lord of the Admiralty, Ross impertinently questioned Barrow's own veracity. Ross's unpublished memoir recalls the incident:

> After dinner Barrow as usual began to repeat his story of having been at Greenland in a whaler when he was a boy. A significant nod from me and a negative shake of the head attracted the notice of Lady Melville, who quickly said

"Mr Barrow what was the name of the ship you went to Greenland in" Barrow
had unfortunately no name ready replied "I, I do—not—remember the name
just now."—"What was the Captain's name" says her ladyship. Barrow had also
no Captain's name ready and explained "I, I, don't remember just now but it
was out of Whitehaven"—"Whitehaven!" says I "that is a small dry harbour
no whaler ever sailed out of that!" Barrow being now completely caught out
in a Bouncer—coloured up to the eyes, while the whole company burst out in
laughter but from that moment Barrow was my bitter enemy.[131]

Ross's mockery of Barrow's dinner-party anecdote was ill-timed given
that Ross had just started working up his journal for publication under
Barrow's oversight. Ross later suspected Barrow of deliberately sabotaging
elements of his narrative and, thereby, damaging Ross's own credibility.[132]
While Barrow claimed that the intention of his textual mediation was to
make the narrative "read better," Ross later concluded that his [Barrow's]
editorial interventions had been deliberately calculated so that Barrow
might "hold them up to ridicule in the Quarterly Review."[133] When Ross
objected to a particular intervention on Barrow's part, which he considered
ungrammatical, the two exchanged angry words in Murray's office, and
Barrow (as Ross later recounted it) "turned on his heel and said 'I'll have
nothing more to do with you' and walked out."[134]

Ross's expedition had been well provided with scientific instruments because of the importance of the science they hoped to reveal: the development
of what would in midcentury become the science of oceanography then
focused on the study of sounding, marine zoology, the chemistry of sea-
water, and what contemporaries called "tidology" alongside hydrographic
work and the surveying of islands.[135] Ross, moreover, had advanced claims
about the restrictions to his ship's movements by the supposed mountain
chain in the face of others' opinions on board and so had cast doubts about
his instrumental competence as captain to command men and ship—to say
nothing of trusting his own observations over and above the visual corroboration of others. Clearly anxious as to Barrow's suspicions, and keen
to counter the negative criticism that his narrative might engender, Ross's
decision to offer a full-scale account of the literary and scientific equipment of the expedition was, we may surmise, part of a deliberate (although
inadequate) strategy to show careful planning, sound scholarship, and the
production of relevant knowledge while at sea. Once home, Ross drew on
others' help, including the Scottish geologist John MacCulloch who assisted
in providing geological material for one appendix. MacCulloch received
£100 for his role; either a generous fee for limited work or Ross used him
in ways that are not disclosed in the evidence.[136]

Throughout the publication process, Ross maintained "constant communication with Murray" and his shaping of the narrative was undoubtedly influenced by the fact that he, Ross, knew that his credibility and veracity were
at question.[137] At one meeting in Murray's office, Ross learned from William
Gifford, the editor of the *Quarterly Review*, that his narrative "was to be
reviewed by Barrow."[138] Barrow's subsequent review was "scathingly negative" and Edward Sabine objected to the fact that his magnetic observations
were presented in the narrative in such a way as to "give the impression that
the work had been done by Ross himself."[139] Shortly after the publication of
Ross's narrative, Sabine issued a critical pamphlet, *Remarks on the Account of
the Late Voyage of Discovery to Baffin's Bay* (1819), in which he accused Ross
of plagiarism.[140] Sabine's pamphlet included testimonials from members of
the Royal Society as to his scientific credentials and his status as a "proper
person to conduct . . . experiments."[141] Sabine's intention was to mark himself out as better trained and more able as both user and interpreter of scientific instruments than Ross, and so to distance himself from the content
and tone of Ross's 1818 work. Ross swiftly prepared a reply (a "complete . . .
refutation of all the complaints made against me," as he saw it), which was
published by Murray. Barrow, in his turn, sought to repress its publication,
but five hundred copies of the pamphlet—*Explanation of Captain Sabine's
Remarks* (1819)—were printed.[142] In his memoir, Ross later recalled that his
pamphlet had failed to reach a wide public because "Sabine's brother in
law . . . purchased the whole . . . and committed them to the flames!"[143] The
cycle of claim and counter claim concluded with an official hearing at the
Admiralty in 1819 at which Ross sought to defend himself against Sabine's
accusations. Although officially cleared of the charges, suspicion among the
public and Ross's naval contemporaries as to his veracity persisted. Ross's
efforts to manage his reputation in part through demonstrations of scholarship and instrumental ability while at sea were, in part at least, limited and
ineffectual. The Croker Mountains were subsequently proved to have been
a mirage, when William Parry (who had been dispatched there by Barrow)
successfully navigated Lancaster Sound in 1820.

Quite what took place daily on board Arctic-bound ships over proper
instrumental use and who had what authority over the instruments and in
interpreting their results is hard to know in detail, especially when, as in
the case of John Ross, authors had the capacity to alter accounts of their
activities. One of the emphases of that body of printed guides to scientific
travelers from the later 1840s was for the attainment of appropriate skills
in instrument use and the regular taking of instrumentally derived measurements as a mean of providing a record of travel and standard measure-

ments that would be understood by later travelers and distant readers alike. In the field, the experience of many of Murray's authors was often one of frustration, missed opportunities, and effort expended for little reward. Even getting hold of proper devices could be a problem. For the itinerant traveler Alexander Caldcleugh in South America, keen to demonstrate his scientific abilities, the difficulty of obtaining the correct instruments was a cause of frequent disappointment. As Caldcleugh and his party prepared to traverse the Cordilleras from Argentina into Chile in 1821, he lamented his "almost entire deficiency of philosophical instruments," particularly those that would allow him to ascertain altitude and thus contribute to the geographical understanding of the Cordilleras.[144] As he noted with regret, "The barometers, which had been ordered from England, had not arrived previous to my departure from Rio de Janeiro; and could such instruments have been procured in Buenos Ayres, it was hopeless to imagine that with their delicacy they could have reached unbroken the Cordillera of the Andes. A simple instrument for ascertaining the dip and direction of strata, and two small thermometers, comprised my entire stock."[145] But if having and using scientific or "philosophical" instruments was for some authors a mark of their scientific credibility and was, for official voyages, an absolute necessity, making do with limited resources became, for Caldcleugh and others among Murray's more "amateur" travelers, an opportunity to demonstrate resourcefulness and to focus on the kinds of observations and practices, such as botanical and geological collecting and naming, that could be made without recourse to instrumentation. As it turned out, Caldcleugh was subsequently elected to the fellowship of the Geological Society (in 1822) and Linnean Society (1823) precisely for his contributions in these areas.

Credibility was not straightforwardly contingent on instrumentation, although the desire to supplement and extend individual testimony with instrumentally derived data was pressing for many travelers at work from the 1820s because of associated changes in the status of natural philosophical enquiry. Even on those occasions when travelers were provisioned with appropriate and well-functioning instruments, limitations of bodily ability—whether as a consequence of illness, forgetfulness, or failure to adhere to a regimen of regularity—necessarily limited their usefulness. Struck down by fever, Alexander Laing was incapable of winding his chronometer, whose working was vital in calculating his location and the distance traveled through dead reckoning; as a result, he lost his ability thereafter to make statements about his geographical position with any confidence. He recorded his frustration at losing time (in several senses) and at being unable to maintain his otherwise regular meteorological observations: "I took up the date of the month from my interpreter, who informed me of the number of days I had been insensible, and on my return to Sierra Leone, I found the reckoning correct; but, during this illness, my meteorological observations ceased, and it was with a grief bordering on distraction that I thought of my chronometer, which, as nobody could wind but myself, had unavoidably gone down."[146] Laing's chronometer, manufactured by the French horologist Pierre-Louis Berthoud (clockmaker to the Bureau des Longitudes, the French equivalent of the British Board of Longitude), had come into his possession in Sierra Leone following the wreck of a French frigate, the *Meduse*, on which it had previously been carried.[147] Laing's use of the chronometer as a tool to determine longitude was dependent on his interpretation of its reliability, especially of the number of seconds it gained or lost in a twenty-four-hour period. These data were presented to Laing's readers in the appendix to his narrative but could not compensate for the lapse in health that had led to the introduction of a nonsystematic error.

The instrumental problems that beset Laing's first expedition in Africa were nothing compared to those experienced during his return in 1825. Extremes of temperature, wind-blown sand, and the juddering effects of prolonged rough camel trekking had a deleterious effect on his delicate equipment. Laing's biographer records quite how extensive the losses and malfunctions were reported to be: "Laing reports the destruction of all his instruments from the heat of the weather, and the jolting of the camels; his barometers broken; his hygrometers rendered useless from the evaporation of the ether; the tubes of most of his thermometers snapt by the warping of the ivory; the glass of the artificial horizon so dimmed by the friction of sand which insinuated itself everywhere, as to render an observation difficult and troublesome" (plate 11).[148]

The rigors of travel meant that one could not replicate in the field the laboratory procedures necessary to make authoritative claims in certain forms of science: working in an unregulated nature simply meant that things broke, slowed down, did not keep to time. [149] That Laing, and others, had *tried*, if not succeeded, was nevertheless central to their self-positioning as credible, scientifically minded observers. The chronometer was faulty, of course, but the instrument that really failed here was the sentient author: overcome by the facts of travel, he had allowed himself to be run down.[150] Stories of failure—of instruments and of bodies—were nevertheless used to underline the perilous physical demands of travel and the dedication displayed by travelers to the "heroic" cause of science.

Several of Murray's travelers positioned themselves as what we might call "suffering scientists," choosing to emphasize in their writing their bodily exertions and privations in the service of data collection.[151] One such was the geologist Joseph Beete Jukes who, on the strength of a recommendation from William Whewell, president of the Geological Society of London, had been appointed geological surveyor of Newfoundland. In his role as professional geologist, Jukes traveled with various instruments including "a prismatic compass, box sextant, mountain barometer and thermometer."[152] He permitted his assistants to carry his heavy theodolite, but he "did not intrust [sic] the barometer to anyone," assuming personal responsibility for that delicate instrument.[153] The nature of the landscape Jukes sought to survey presented a challenge to the standard geological method of making transects and he frequently endured physical discomfort in order to complete his scientific observations: "The wind was blowing, and it was piercingly cold on the top of the hill, but we stayed there an hour while I took a round of angles and bearings with the prismatic compass and box sextant, and left, completely chilled."[154] Despite Jukes's careful traversing of uneven ground, he had arrived at the summit of that hill after a difficult ascent to find his mountain barometer "broken and the mercury gone."[155] In the absence of this instrument, Jukes could only estimate the hill's height and so lost the ability to turn physical exertion into a commensurable data point. Failed instruments did not necessarily mean, however, a failure to secure credibility. Credibility was not a matter of things always going right but of doing the right thing when things went wrong. Even then, honesty did not guarantee literary success: Jukes' work was a loss maker.[156]

Our examination of Murray traveler-authors in the first four decades or so of the nineteenth century suggests that they occupied, in broad terms, one of two principal categories: the half-pay military officer seeking adventure or financial return in the period following the Napoleonic Wars and the scientific or antiquarian traveler (commonly male with, like Maria Graham, some noteworthy exceptions), motivated by the intellectual concerns of emerging disciplines and supported in their endeavors, more often than not, by one or more scholarly society. It is imprudent, however, to draw any artificial distinction between a "scientific vanguard" and "exploratrices sociales," between the amateur and the professional, the self-deprecating Mundy and the haughty Beecheys. Yet it is evident that the professionalization of science through the first half of the nineteenth century was reflected in the self-positioning of certain of Murray's authors in the kinds of travel they undertook and in the strategies employed to demonstrate their credibility and that of their observations.

For Murray's explorers and travelers, credibility was a matter of demonstrating in appropriate ways the epistemological and moral warrant that, if properly wrought, would facilitate the circulation of their testimony and secure their reputational status as tellers of truth. The specific means by which individual explorer-authors sought to demonstrate the appropriateness of their methods and the verisimilitude of their accounts—their ad hoc regimes of credibility—varied, however, on the basis of genre, geographical region, social status, and intended audience. As a consequence, at least as far as the evidence for explorer-authors in the late eighteenth and early nineteenth centuries is concerned, there was never a single, always-recognized, marker laid down by which to distinguish a traveler's credibility. The strategies employed by Murray's authors were often explicitly contradictory: evoking, on the one hand, serious scholarship and academic dedication and, on the other, physical endeavor and a sometimes playful disregard for questions of science, even as they sought to justify why their claims ought to be taken seriously in remarks about overcoming nature's rigors or the failure of instruments, human and scientific alike.

Murray's role in the production and maintenance of credibility was to ensure that the right authors met with the right audiences. While those audiences were the ultimate arbiters of an author's credibility, the house of Murray depended in its professional judgments on the opinion and advice of its readers as to the stylistic and epistemic worth of the texts under consideration. Individual travelers were, in their turn, required to act as arbiters of the knowledge and testimony on which their travels depended. Credibility was, then, earned in the field and in later writing and attributed through a chain of judgments made as to veracity and authority. Although the emerging requirements of science during the first few decades of the nineteenth century placed a gradually greater emphasis on objectivity and the writing out of the self in works of travel, this was neither rapid nor total. An assumed or sought-for correspondence between the world as experienced and the world as written could, for certain authors, result just as well from subjective and embodied testimony, if correctly pitched and delivered with a suitable stylistic flourish, as it could from an instrumental approach to observation, rigorously followed. As we shall show in the following chapter, Murray's editorial guidance, the views of the firm's readers, and those of the author were also crucial in questions of credibility. But even before the author and his or her words came home, they had already been subject to, and were the subject of, different routes to credibility.

CHAPTER FOUR

Explorers Become Authors: Authorship and Authorization

The implications of George Francis Lyon's remark that his 1824 *Private Journal* (app.: 83) had been written "solely for the amusement of my own fire-side, and without the most distant idea that it would ever see the light in any other shape than of its original manuscript" would have been clear to his readers: Lyon was author by circumstance, not by profession.[1] For Lyon and others, the category of author could carry with it the whiff of vainglorious intent, of a purposiveness that might limit credibility among a public uncertain about the "craft and mystery of book-making" and of the self-aggrandizing author.[2] Lyon chose to present himself as at once guileless writer and reliable witness—an authorial positioning that contrasted remarkably with a personal tendency to be very unretiring (see fig. 10), a "moustachioed extrovert" who described his main interests in life as "balls, riding, dining and making a fool of myself."[3]

What the example of Lyon highlights is quite how uncertain and troubling the notion of authorship could be for Murray's travelers and, for modern researchers, how important it is to understand how and why this was so. We now explore the connections between travel, writing, authorship, and authorization in order to cast light on how Murray's travelers negotiated and managed their writerly identities on their return. In examining strategies of narrative, testimony, and truth telling, we look at the idea of the "modest author" and show why protestations of modesty and humility were not simply a fashionable default but were integral to the construction of an author's authority. We will show how the category of author was sometimes mutual and collaborative, representative of what one scholar has

FIGURE 10 The immodest author? George Francis Lyon here chooses to use disguise as a marker of social status and of personal display. *Captain G. F. Lyon, R. N. in his Tripoline Costume*, by Richard J. Lane. Lithograph, ca. 1825. By permission of the National Portrait Gallery (NPG D5145).

called "aggregate institutional authorship."[4] It is clear, too, that the nature of authorship varied between amateurs and professionals, a factor in determining the ways in which credibility and authority were secured in print. We look, too, at the ways in which the house of Murray and its readers and editors acted to fashion the words of explorers into published narratives.

Several Murray authors were also critical readers for the firm. As Walter Hamilton labored through the compilation of his *East India Gazetteer* (1815; revised in 1828) and his *Geographical, Statistical, and Historical Description of Hindoostan* (1820), both published by Murray (app.: 32, 60), the publisher drew on Hamilton's expertise to judge the submissions of others. Numerous works of travel came his way. Proposed works on Italy were rejected as "too familiar" in content and poorly written; accounts of the Peninsula Campaign were deemed excellent; Mexican travelogues dismissed; a work on Cuba, dedicated to Croker even before its acceptance by the publisher, was passed to Croker on Hamilton's say so, who additionally suggested that Barrow "look into it"; and an account of Ceylon, which "I think a very good one (certainly the best I have yet seen)," was rejected nonetheless: "You know the subject is one that does not at all interest the great mass of the English nation."[5] Writing to Murray III in 1846 over the possibility of another work by Pascoe Grenfell Hill, Murray's editorial assistant, Henry Milton, was blunt in his assessment: "There is nothing either in the matter or the manner of this M.S. to render it at all worthy of your attention."[6]

Murray III occasionally responded to intending authors directly. Writing in October 1847, he counseled the diplomat Charles MacFarlane over his proposed volume on Turkey at a time when changes to civil government there had exposed its feudal governance to Western condemnation and alerted British politicians to the precariousness of West-East relations: "The incidents of a journey in Turkey have lost their novelty—you will relive [sic] your vessel by throwing them overboard, if not you will incur the risk of being swamped by your cargo. The stringent abridgements which I recommend would amount to *one half*—of the M. S.—but that is a matter of conjecture."[7] Undaunted, MacFarlane brought forward a shorter manuscript that Murray published in 1850 (app.: 203). For perhaps the most famous of Africa's explorers, David Livingstone, Murray's construction of Livingstone the Author out of accounts written by Livingstone the Explorer involved fashioning a biographical account to preface the work and placating the author who felt that his evolution into public celebrity from African explorer would not be helped by John Milton's redactions of his prose (fig. 11).[8] Livingstone appealed to others to justify his views on his own words: "I have the opinions of Prof. Owen, Sir Roderick Murchison, Mr Elwin & Mr Binney that it is more likely to be popular and saleable than if diluted or emasculated as this man has impudently presumed to do. It is therefore both for your advantage & mind that I reject *in toto* every change introduced in red ink."[9] Authorship, as these and other writer-explorers were to discover, involved more than getting words written.

THE MODEST AUTHOR

Prefatory declarations of modesty and reluctance in authorship are sufficiently commonplace in works of nineteenth-century travel writing that they risk being dismissed simply as "highly conventionalized" defaults, rather than being acknowledged as crafted elements of rhetorical strategy.[10] Scholarly attention to the role and evolution of the textual preface has shown its

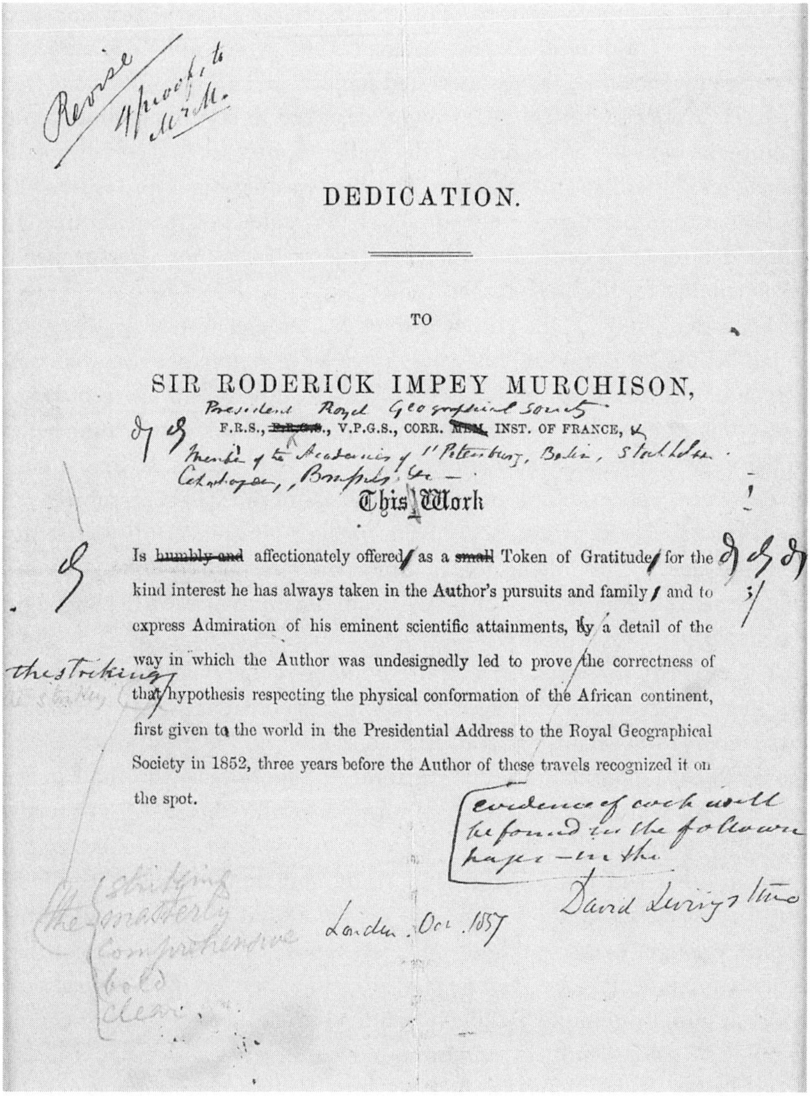

DEDICATION.

TO

President Royal Geographical Society

SIR RODERICK IMPEY MURCHISON,

F.R.S., ~~D.C.L.~~, V.P.G.S., CORR. ~~MEM.~~ INST. OF FRANCE,

Member of the Academies of St Petersburg, Berlin, Stockholm, Copenhagen, Brussels &c —

This Work

Is ~~humbly and~~ affectionately offered as a ~~small~~ Token of Gratitude for the

kind interest he has always taken in the Author's pursuits and family and to

express Admiration of his eminent scientific attainments, by a detail of the

the striking

way in which the Author was undesignedly led to prove the correctness of

that hypothesis respecting the physical conformation of the African continent,

first given to the world in the Presidential Address to the Royal Geographical

Society in 1852, three years before the Author of these travels recognized it on

the spot.

evidence of which will be found in the following pages — in the

David Livingstone

London Oct 1857

(striking
masterly
comprehensive
bold
clear

FIGURE 11 The author establishes his authority by association. The proof copy of the dedication page of his *Missionary Travels and Researches in South Africa* (1857) shows the author, David Livingstone, adding to the cachet of his work by enhancing the standing of the person to whom the work is dedicated, Sir Roderick Impey Murchison, president of the Royal Geographical Society. Source: National Library of Scotland, MS 42341. By permission of the Trustees of the National Library of Scotland.

origins to lie in classical oratory.[11] For Roman philosopher-rhetoricians such as Cicero and Quintilian, effective oratorical self-presentation depended on a "deferential attitude," a "disinterested stance," and a "low, unornamented style."[12] Correct rhetorical deportment, designed to secure credibility and capture the benevolent attention of the audience, necessitated certain topoi: reluctance, humility, and modesty. The first was designed to position the speaker as one persuaded only to talk at the request of the dedicatee. In talk and in text, this conferred two advantages: "The author's [or speaker's] responsibility for his work and its imperfections is shared by his employer [or dedicatee], and . . . the author is freed from the charge of presumption in publishing [or speaking] his work, since he is merely obeying the commands of his friend or employer."[13] A second function of the reluctance topos was to demonstrate that the motivation to speak came from belief in the utility of the subject matter, rather than the orator's desire for public exposure. This emphasized the "worthiness of the speaker's subject . . . [and] the relative worthlessness of the speaker himself." Utility, together with the status of the dedicatee, justified the presentation of content and subordinated the speaker to the fact. Effective communication depended on the performance of humility.

The modesty topos was one whose influence came from demonstrating an "inverse relationship between perceived and actual authority." Modesty was employed to secure the audience's attention on the grounds that "the more effective a speaker's self-abnegation, the more seriously the listener will take his words on a subject."[14] In demonstrating the absence of stylistic ability and rhetorical flourish, modesty was "a device for deflecting the suspicion that verbal art distorts 'the truth' and similitudes . . . veil it."[15] But modest claims could become precisely the kind of stylistic apparatus that they sought to disavow: the "modesty topos . . . becomes a device that actually prepares the audience or reader for just the kind of rhetorical skill which it initially denies."[16] While it is difficult to determine whether claims to modesty were genuinely held or employed merely rhetorically, it is undeniable that declarations of reluctance, modesty, and humility assumed a central importance in the rhetorical construction of travel literature. What was Lyon if not a reluctant and modest author, one persuaded to publish his stylistically unsophisticated and intellectually limited account on the advice of his superiors and in the belief of the utility of his subject matter? Whether Lyon's modest and reluctant protestations were genuine is uncertain, but it is clear that he and Murray felt them to be appropriate, effective, and necessary.

The notion of authorship has been subject to considerable attention and the category author revealed to be not fixed but actively contested and constructed.[17] Michel Foucault theorizes the means by which "the writing subject" is transformed into an author and how normative notions of authorship are produced and sustained.[18] Foucault has considered how writers come to be designated as authors, and how authors' textual inscriptions come to be regarded as their "work," a function, for Foucault, of church and state censorship (which could hold writers to account for their work) and of the emergence of formal systems of copyright (which sought to attribute texts to their writers, and intellectual property to their creators). The author in Foucault's model is not simply identical with the writer but is always discursively and culturally defined. The "author function" serves to codify, commodify, and classify the writer's inscriptions as intellectual property and to embody claims to truth and validity within certain discursive arenas.[19] While Foucault's perspective on authorship has been subject to scrutiny and revision, the general notion of the author function as a legal and civil construction has important implications for assessing the means by which Murray's texts were produced and consumed.[20]

Authorial creativity rests, at turns, with the writer of a text, its publisher-printer, and its translator. The uncertain connection between authorship and authority in this respect presented particular difficulties for what we might now call the scientific writer. For seventeenth- and eighteenth-century natural philosophers guided by the social desiderata of gentlemanly conduct, authorship was often approached with ambivalence and reluctance.[21] Polite social codes might encourage intellectual openness, but gentlemen experimenters often viewed authorship with trepidation, concerned that the "mercenary practices of the book trade" would divest them of control of their work, misrepresent it, and so reduce their credibility.[22] Relinquishing control of written work to a publisher who might then engage in undesirable piratical practices such as unauthorized reprinting, abridgment, and translation was also deeply troubling. For a gentleman, assuming the role of author meant, in social and intellectual terms, "losing one's self."[23]

While organizations such as the Royal Society would come to assume a role in warranting, attributing, and protecting authorial recognition in scientific writing, gentleman-scientists themselves employed strategies of self-representation, namely, reluctance and modesty, to satisfy the expectations of their social position and demonstrate their credibility. Scientific texts were accompanied by prefatorial declarations that made clear that publication arose only as a result of the entreaties of others, that the material presented, even if partial, would serve a public or intellectual benefit, and that the writer was not well practiced in terms of literary ability. Underpinning such strategies was the assumption that "those who presented themselves naturally in the person of an author were seekers after fame and celebrity, individuals of compromised integrity who sought personal identification with the claims and systems presented under their names because, so to speak, *it was their trade*."[24] There was a desire among scientific writers to frame their texts as revelations of an "established or always redemonstrable truth," rather than as offering a personal perspective on it.[25] The role of the scientific writer was to reveal truth rather than to construct it. Such strategies were designed specifically to deny (or obscure) nonaltruistic motivations and to secure authority through demonstrating appropriate decorum. Exhibitions of humility, both socially and in writing, would prove the author to be "a disinterested observer and his accounts as unclouded and undistorted mirrors of nature."[26]

TRUTH AND PLAIN STYLE

Fundamental to the convincing portrayal of modest and reluctant authorship was an appropriate stylistic approach to writing: one that emphasized plainness.[27] Francis Bacon's *Of the Proficience and Advancement of Learning* (1605) had long since codified what empiricism demanded of its practitioners: simple truths, plainly stated. For Bacon, "ornaments of speech, similitudes, treasury of eloquence, and such like emptinesses" were incompatible with credibility and should be "utterly dismissed."[28] Plain style made a particular claim to—and was implicated in the construction of—ideas of objectivity. Naked talk and naked text, seemingly free from rhetorical sophistication and the stylistic display of the author, would allow "the facts [to] shine through."[29] As a scientific necessity and as a convincing rhetorical device, plain style and modest self-presentation were founded on (and codified in) the production of credibility: "Ostensible humility acted as a claim to empirical truth."[30] The long-run association between facts and credibility was for some Murray authors a function of bodily and writerly regulation, not merely in terms of prescribed practices of observation, experimentation, and demonstration but with respect to defined modes of speech and of writing. The establishment of authority and the warranting of claims to authorial competence depended on a particular performance of authorship, one that emphasized as much what the writer *lacked* (overt rhetorical aptitude, financial motivation, ego) as what the writer *possessed* (brevity, dedication to fact, humility). Securing trust depended, in a sense, on seeming to be "someone dull enough to tell the truth."[31] In reality, of course, plain style was itself a form of rhetoric, one whose complexity had been recognized by Cicero: "Plainness of style may seem easily imitable in theory; in practice, nothing could be more difficult."[32]

The extent to which Murray's travelers were content to submit to such a mediation of authorial identity varied. It did so in relation to the emergence during the late eighteenth century of the professional travel author, an individual for whom writing about travel was not simply an accident of circumstance but a choice of livelihood.[33] It varied during the early nineteenth century on account of the imperatives required of the professionalization and institutionalization of science and "scientists," the significance placed on academic credibility, and the demonstration of appropriate method in the making of geographical truth. And it varied as a consequence of contemporary literary and cultural sensitivities over sensibility, particularly a requirement that traveler-writers should not be simple conduits for facts but that they should embody personal experience. Certain of Murray's travelers were content for their texts and their authorial personae to be mediated and managed in whatever way the publishing house felt most appropriate. Sending Murray III part of his journals kept while traveling in the Near East, the Glasgow merchant John Kinnear (who had been persuaded to contact the publisher only by his friends' recommendation) was quite prepared to let Murray be the arbiter as to their worth: "The MS is little more than a transcript of my letters to Mrs Kinnear, with some additions from notes taken on the journey. I am not so far gone as to imagine that it possesses any literary merit whatever, . . . but whatever you may think of my appearing before the public, I shall feel much obliged by your candid opinion."[34] Kinnear's self-effacing attitude may not have been genuine, of course, but it stood in marked contrast to the attitude of men like Woodbine Parish who, with others, took the staging of their work and the positioning of their authorship as matters of profound concern. The model of the "immodest author," Woodbine Parish bombarded Murray II and Murray III for nearly forty years between the late 1830s and the mid-1870s with correspondence over his South American book, its commendation by governments in that continent, new materials for inclusion in later editions, notices of reviews, and the like.[35] For different kinds of travel text, and for different authors and audiences, different expectations existed of authorship. Like credibility, authorship—and the authority that was presumed to go with it—was secured in different and occasionally contradictory ways: through humility or decisiveness, through restrained and careful prose or rough and adventuresome writing, through detailed, measured, and often plodding science, or in immediate and subjective responses.

ANONYMITY AND AUTHORIAL CONSTRUCTION IN MURRAY'S EIGHTEENTH-CENTURY TRAVEL TEXTS

Most of Murray's accounts of non-European travel between 1773 and 1800 were issued without authorial attribution. In the half-century that followed, only one—Julia Maitland's *Letters from Madras* (1843; app.: 161)—is so attributed.[36] The shift from anonymous to onymous publication reflected, in part, the emergence of the professional travel writer and changing sensibilities among audiences for works of travel. It would be wrong to assume, however, that this transition was replicated as rapidly or as fully elsewhere in British literary publishing. Anonymity remained a default for the majority of novels and periodical literature in the first half of the nineteenth century.[37] The factors that motivated the anonymous presentation of travel texts varied considerably, whether to obscure gender or social class, to facilitate the presentation of scurrilous or controversial ideas, or to lend credence to the pretence of reluctant authorship.[38]

The anonymous publication of William Macintosh's *Travels* (1782) allowed the author (and, by implication, Murray I) freedom to criticize the East India Company's administration of British India. Anonymity was used to a different end in a short pamphlet, *An Account of the Loss of His Majesty's Ship* Deal Castle, issued by Murray in 1787 (app.: 7). The pamphlet (which has since been attributed to one Robert Young) concerned the destruction of much of the British naval fleet in the Antilles during the Great Hurricane of 1780. Young's was an eyewitness account of this infamous and well-publicized event and part of a longstanding genre of travel writing, the shipwreck narrative.[39] Part of the then popular interest in shipwreck narratives came from what they revealed about human nature and the operation of ordinary everyday social norms in extraordinary circumstances. Such narratives were deliberately intended to evoke sympathy. The extent to which faith could be placed in their authenticity was, therefore, an important concern for authors and publishers alike. In the preface, Murray I deployed several strategies to demonstrate the credibility of his anonymous author. The fact that the narrative was being published seven years *after* the events it described showed that its author was not seeking to exploit a recent tragedy for commercial gain: the text was being presented merely to

gratify "the mind of the curious or humane."[40] The author was an eyewitness who had committed his experiences to his diary "immediately after the
disaster, whilst it . . . remained fresh in his remembrance."[41] The narrative
was thrilling, but not exaggeratedly so. Positioning Young as an anonymous
author was a strategy that lent additional credence to the claim that he was
a writer as a consequence of circumstance, rather than of literary ambition,
and that he was not motivated by vanity or the pursuit of celebrity. The
identity of the author need not always be known for the work to be established as authoritative.

A different strategy was pursued the following year when Murray issued
a short twenty-four-page maritime account written by George Robertson,
a member of the Society of East India Commanders. Although Robertson's
name did not appear on the title page of his quarto volume—*A Short Account
of a Passage from China* (1788; app.: 8)—it was appended to the epistolary
preface.[42] The absence of Robertson's name from the title page was not a
strategy of anonymous presentation but, rather, reflected what others have
seen as the privilege given to the volume's content and subject matter.[43]
Robertson's introductory comments positioned him as a reluctant author,
one who wrote, mapped, and sketched for "private information and use,
[and] without any design of publication" but who had felt compelled to
publish by the entreaties of "several experienced Gentlemen" of the Society
of East India Commanders.[44]

Robertson approached the position of public authorship with apparent
reluctance, but he proclaimed with confidence the rigor and detail of the
observations that formed the basis to the accompanying three navigational
charts. Robertson's maps (engraved by Thomas Harmar, who, under the
direction of the hydrographer Alexander Dalrymple, had previously executed several East India Company charts) were presented in a separate
folio, *Charts of the China Navigation* (1788).[45] Robertson's narrative account
served, in effect, as an extended prefatory statement to the navigational
charts. Robertson positioned himself in his text as a careful and dedicated
scholar (having "examined many journals" prior to his departure in order
to evaluate the current state of geographical intelligence, and as a depend-
able eyewitness observer).[46] Robertson was painstaking in detailing the
sources of his information, in describing his own observational approach,
or in evaluating the testimony of others (particular faith being placed in
the testimony of other East India Company navigators and captains). For
Robertson, authorial credibility depended on declaring himself but only as
something of an afterthought, and, more important, by his correct dem-

onstration of expertise and evaluation in the field. Crucial in this respect was his recourse to the marine chronometer, a device "of the greatest use imaginable."[47]

In 1791, Robertson published with Murray an updated maritime map, *A Chart of the China Sea*: his accompanying memoir making clear the significance of the chronometer and explaining to his readers his working method (fig. 12): "The following Chart is almost entirely constructed from corresponding observations made with that invaluable machine [the marine chronometer], corroborated by the astronomical and lunar observations of many of the commanders and officers in the service of the Honourable East India Company; by whose obliging communication and assistance I have been enabled to submit to the Public a select series of information, compared with and collected from a variety of valuable original observations and journals, which otherwise would in all probability have remained generally unknown to the world."[48] Because Robertson was keen to avoid any self-aggrandizement in his publications, he secured his audience's attention by positioning himself as subservient to geographical truth: "No other merit is claimed from the present publication than what arises from a strict regard to truth; having throughout laid down no point till ascertained to the nearest truth by a careful and repeated comparison with different observations."[49] Robertson was sufficiently confident in regard to his instrumental observations and calculations for him to donate copies of his charts to the Royal Society in London and to the Royal Society of Edinburgh (he was elected an ordinary fellow of the latter society in 1792, proposed by the mathematician John Playfair and geologist James Hall).[50]

While Young's shipwreck narrative and Robertson's marine charts addressed a broadly similar audience—those involved in East India trade—they communicated authenticity and positioned their authorship in precisely different ways. For Young, subjectivity and bodily experience allowed his text to be an affective account. For Robertson, objectivity and instrumentally regulated observation mattered to his authorship. Anonymity lent credence to Young's claim of reluctant authorship. This strategy was less suited for Robertson because it would have been difficult to achieve among the intimate social circles of the Society of East India Commanders of whom he was one and because allowing his name to be evident strengthened his geographical claims. Robertson's reputation as an experienced seaman—he was a veteran of numerous journeys between Britain and the Far East—served as further warrant to his authorial status.

For Young and Robertson both, the claim to be writers by circumstance

MEMOIR

OF A

CHART OF THE CHINA SEA;

INCLUDING THE

PHILIPPINE, MOLLUCCA, AND BANDA ISLANDS,

WITH PART OF

The Coaſt of New Holland and New Guinea.

DEDICATED, BY PERMISSION,

To the Society of Managing Owners of the Ships
in the Eaſt India Company's Service.

———

BY

GEORGE ROBERTSON.

═══════

PRINTED FOR THE AUTHOR,

AND SOLD BY J. MURRAY, FLEET-STREET.

———

M,DCC,XCI.

FIGURE 12 The work of late Enlightenment exploration. The memoir was commonly used to supplement the map in late Enlightenment publishing by giving an account of the sources used, and their credibility, in the making of the map. Source: George Robertson, *Memoir of a Chart of the China Sea* (London: John Murray, 1791). By permission of the Trustees of the National Library of Scotland.

(rather than by profession) was instrumental to their self-presentation. Many of Murray's eighteenth- and nineteenth-century travelers satisfied that requirement (or could appear to do so). Yet the emergence of the professional author as a specific class of writer meant that the guise of reluctant authorship became more difficult to communicate with sincerity. One illustration of this is the case of Robert Charles Dallas, whose anonymous travelogue, *A Short Journey in the West Indies*, was published by Murray I in 1790 (app.: 10). Dallas's text was based on a single voyage to the Caribbean following his inheritance of a sugar plantation in Jamaica.[51] Dallas clearly entertained literary ambitions: this is apparent in the text's rococo style; its opening sentences, describing an aborted altercation with an American sloop, are typical of what follows: "A battle—yes, a battle, Eugenio—I had nearly witnessed, and been engaged in, the horrors of a sea-fight.—I was begirt like a warrior; a cartouch-box [*sic*] on my right side, a cutlass on my left, and a heavy musket in my hand."[52] Dallas's is not, then, the plain style of Young and Robertson, but something altogether different—sensational, personal, subjective. Much of Dallas's account, a series of anecdotes addressed to his eponymous friend, Eugenio, examined the nature and evils of Caribbean slavery, a "disgrace to nature and reason" in his view.[53]

At one point in the text, Dallas assumes the literary voice of a "sooty son of Afric"—a device that allowed him to gather "into one point of view various incidents, hideous in their nature" and to present them as though related from a first-hand witness.[54] Dallas's purpose in so doing was to "shew you [the reader] the picture of slavery in its monstrous size."[55] Dallas was keen to assure readers that the various and collected incidents were "nevertheless true, some to my own knowledge, others I have from credible information."[56] The extent to which Dallas was successful in communicating to his readers the plight of Caribbean slaves varied. The *Monthly Review*, while recommending the book's "true and very amusing sketches of West Indian manners," thought its real purpose was "to exaggerate the hardships of Negro-slavery."[57] The *Review* suspected that Dallas's intention had been to court the public's interest in slavery and thus to gain financially through his narrative.[58] Dallas's decision to offer his readers an account of slavery's "monstrous" size was identified as an impediment to his credibility. As the *Review* framed it, "Would not the cause of truth have been better aided by exhibiting the picture in its *natural* size? . . . A person who describes his travels, should adhere to truth: but can we repose confidence in a writer who professes to have raked all the horrid stories he has heard of the barbarous treatment of the Negroes, into one collected point of view, purposely to form a 'picture of slavery in it's [*sic*] monstrous size?'"[59] Transgressing the expectation of plain style, Dallas had, in the eyes of the *Monthly Review*, cast doubt on his standing as an author. Dallas's *Short Journey* was his only excursion into travel literature. He thereafter became a prolific author in other genres and was subsequently instrumental in introducing Byron to Murray II and in facilitating the publication in 1812 of *Childe Harold's Pilgrimage* (which he also edited).[60]

Positioned either as anonymous author or reluctant writer (or as both), many of Murray I's eighteenth-century travelers typically made such claims to authority based on first-hand experience and eyewitness testimony, seeking in so doing to link what was witnessed to what was written. This connection was less clear in the case of the 1791 *A Narrative of the Loss of the Grosvenor East Indiaman* (app.: 12), a shipwreck account that had been "compiled from the examination of John Hynes, one of the unfortunate survivors" by the artist George Carter.[61] The wreck, which had occurred in 1782, had become a cause célèbre: the uncertain fate of the ship's female passengers being a particular source of sensational speculation and reporting in the London press.[62] Popular interest in the story arose in large part from the fact that the survivors had endured a long overland journey from the site of the wreck at the Cape of Good Hope to reach English and Dutch settlements. During the journey, the party of survivors had become separated and, as the speculation of English and Dutch settlers had it, the female members had been absorbed into (or kidnapped by) the local Xhosa community. As these stories began to reach the London press, both shipwreck and kidnap were reported in sensational terms.[63]

The publication by Carter and Murray of their account, based on the eyewitness testimony of a crewmember, was still a potentially profitable decision, even nine years after the wreck. Carter's text offered, without introductory preamble or contextualizing preface, a third-person account of the wreck interspersed with direct quotations, the testimony of John Hynes, a member of the *Grosvenor*'s crew. The text was not simply a transcription of Hynes's eyewitness testimony, as is clear from Carter's "authorial interjections, additions and digressions."[64] In Carter's role as editor (he described himself thus in the text), he contextualized Hynes's account by incorporating "a variety of anthropological, historical and geographical descriptions" drawn largely from two contemporary travelers to the Cape: William Patterson and François le Vaillant.[65]

Authorship of *Narrative of the Loss* was a collective endeavor, involving Hynes as the authoritative source and Carter as the mediator and evaluator of that testimony. For several critics, Carter was thus guilty of book-making (as opposed to "bookmaking" in the more general sense of simply making book), of assembling existing material and attempting to pad it out to form a book-length account: "We must observe, that it contains little, if any thing [*sic*], which may be called *new;* and that it is, on the whole, but a meagre performance, eked out with a few extracts from *M. Vaillant*'s Travels, and two or three borrowed prints."[66] The fact that the volume's authorship rested not with the traveler whose testimony it communicated but with a sedentary other was a source of critical concern to critics. Carter's method of authorship-by-assemblage and the fact that he had chosen neither to include a preface nor to explain his decision to pursue authorship all meant that he had abrogated a fundamental literary responsibility: that of modest, reluctant authorship.

For the house of Murray, the decade that separated Macintosh's *Travels* and Carter's *Narrative of the Loss* was one in which the firm gained experience in managing the authorship and publication of travel texts. Critical response to several of the published works of overseas travel show that Murray I and II, and their respective authors, got questions of authorship wrong as often as they got them right. Not until Murray II assumed sole control of the firm in the early nineteenth century did the firm achieve regular critical success as an authority on geographical and exploration publishing. In building on collaborative relationships with institutional authorities such as the Admiralty, the production of travel texts became from that period "more regulated, more collective, and more dependent on co-present scribal and oral networks than was commercial print culture."[67] Murray's travel texts were necessarily always collaborative undertakings, emerging from negotiation "among travelers, between travelers and editors, between editors and printers."[68]

"ANXIETY AFTER TRUTH": AUTHORSHIP AND AUTHORITY IN MURRAY'S NINETEENTH-CENTURY ACCOUNTS OF WEST AFRICA

Trust in Authors: The Narrative of Robert Adams *and* Loss of the American Brig *Commerce*

Where the shipwreck narratives issued by Murray I in the late eighteenth century made little claim to interest or utility beyond the sensations they provoked by the sufferings they described, two such works published by Murray II in 1816 and 1817—the first of this type published by the house after more than quarter of a century—had an altogether different *raison d'être*. *The Narrative of Robert Adams* (1816), and James Riley's *Loss of the American Brig* Commerce (1817; app.: 34 and 42, respectively) both met the conventions expected of shipwrecked and suffering travelers—"captivity narratives" as they have been termed—but the motivation for their publication depended equally on their topicality and potentially revelatory geographical contents.[69] Both books dealt with the geography of West Africa and, crucially, with related geographical puzzles then the focus of scientific and commercial debate: the River Niger and Timbuktu.

Murray's first foray into this field had come in 1815 with publication of Mungo Park's posthumous *Journal of a Mission* (app.: 33). Park's death on the Niger in 1806 amid uncertain circumstances, which his *Journal* did little to clarify, and continuing British interest in the commercial opportunities of West African trade meant that the Niger, and West African exploration more generally, was of great interest for scientists and traders alike.[70] The opportunity to supplement a significant exploration narrative with additional accounts based on eyewitness testimony was, for Murray, an alluring prospect. The genesis of *The Narrative of Robert Adams* was almost as improbable as the events of travel and bodily suffering in Africa it sought to relate. Its production illustrates more generally how fluid the distinction between author and editor could be.

The text's titular author was an "illiterate African American sailor," who, in 1815, had been found living destitute on the streets of London. He was subsequently introduced to Simon Cock, secretary to the Committee of Merchants Trading to Africa (hereafter, the Africa Committee).[71] Having assumed the name Robert Adams—his original name is thought to have been Benjamin Rose—the sailor recounted to Cock a thrilling tale of his shipwreck on the Barbary Coast in 1810, his capture by local peoples, and subsequent sale as a slave to an Arab merchant.[72] Adams had been taken to Timbuktu and resided there for four months before escaping and making his way to London via Cadiz. The recent departure of an African-Association-backed expedition under the direction of John Peddie to explore the Niger meant that, for Cock, "the interest of the story was very much heightened at that particular moment."[73] Adams was brought to the offices of the Africa Committee where he related a "series of adventures and sufferings" to excited but skeptical audiences whose number included Joseph Banks and John Barrow.[74] Initially, Adams's account of Timbuktu was regarded with skepticism because it differed from "the notions generally entertained," and so excited among his audience "a suspicion that his story was an invention."[75] In Adams's rendition, Timbuktu was less grand and magnificent than was supposed. Cock was convinced, however, of Adams's fundamental credibility: from his "plain and unpretending answers" to Cock's enquiries, Adams conveyed "a strong impression in favour of his veracity."[76] Cock's strategy in questioning Adams had been to "draw from the sailor, not a continuous and straightforward story, but answers to detached and often unconnected questions."[77] These questions were repeated at a subsequent interview in order to test the degree of correspondence between Adams's present and earlier recollections. Cock found that Adams's answers had, on reinterview, been "repeated almost in the same terms," a fact that strengthened Cock in his view that "the man's veracity was to be depended upon."[78]

These recollections, assembled and transcribed by Cock, formed the basis of *The Narrative of Robert Adams*. As Cock noted in the preface, "There was not one [gentleman] who was not struck with the artlessness and good sense of Adams's replies, or who did not feel persuaded that he was relating simply the facts which he had seen, to the best of his recollection and belief."[79] In detailing the systematic approach to Adams's interrogation, Cock's intention was to demonstrate that Adams "possessed an accuracy of observation and memory that was quite astonishing" and that his account had been scrutinized, and attested to, by "every gentleman who might wish it."[80] Cock sought to establish Adams as a credible witness (and, of course, reassure readers in their view of him, Cock, as similarly reliable). Together with footnotes, commentary, and appendixes (paratextual "devices of authentication" as they have been termed), Cock's editorial additions expanded Adams's fifty-odd-page account into a two-hundred-page book.[81] But Cock's intervention also had the effect of denying Adams "all opportunity for self-assertion": even as the text gave him a voice, it limited what he could say.[82] Cock's efforts to establish Adams's credibility were sufficient to persuade Murray to undertake the book's publication. It was placed by Murray "on a footing with the two volumes of [Mungo] Park . . . [and] accordingly printed in a uniform manner with those volumes," that is, in quarto format at £1 5s.[83]

The response of the British press to Adams's credibility was varied and involved careful assessment. For the *Edinburgh Review*, the fact that Adams's account of Timbuktu was of a city less grand and magnificent that was otherwise assumed functioned "in favour of his accuracy": it demonstrated a reluctance to adapt his account to conform to existing expectations and reflected the common occurrence that "authentic information" tended to "diminish the wonders related and credited [about distant and exotic

locales] during the period of ignorance and vague reports."[84] Others lauded
Adams for not book-making: "though *he did not make it,* he is pretty nearly
as much the author as some gentlemen are of certain books, through which
they have acquired the reputation of *literary men*."[85] The success of Cock's
endeavors in making Adams believable as author and genuine eyewitness
was commented on by the *Monthly Review*: "Great pains have been taken to
sift the testimony . . . and the proofs of Adams's veracity are established on
grounds which, to our minds, are very satisfactory."[86]

The French and the American press thought otherwise. The publication
in 1817 in Boston of an American edition of Adams's narrative was met with
disapproval and censure. For one reviewer, Adams's account was "a gross
attempt to impose on the credulity of the publick," the work, being so obviously replete with lies and mistruths that, were it not for the fact that it had
"gained universal belief in England," it was unworthy of serious examination.[87] To the French cartographer Edme Jomard, whose authority came
from having edited the multivolume *Description de l'Égypte*, Adams—being
both illiterate and of mixed racial heritage—was "morally and physically
inappropriate to act as an explorer and scientific observer."[88] While for certain British critics the matter of Adams's race evoked romantic associations
with Othello (a point made by the *British Lady's Magazine*), and for American critics may have had a part in their generally condemnatory tone, it
was for Jomard altogether an impediment to Adams's reasoning, reliability,
and authenticity.[89] To him, Adams's race signified unreason and served to
undermine his ability "to comment with authority and objectivity on the
African places he claimed to have visited."[90] Jomard's views were motivated
by his attitude more generally as to the authority of French exploration:
he would go on to support René Caillié in his journey to Timbuktu in the
early 1820s, in advance of the British officer and Murray author Alexander
Laing, who died on the return leg. No self-regarding French authority could
countenance his compatriot's endeavors to have been bested by an illiterate
American half-caste.

Murray II perhaps had Adams's account in mind when he turned in 1817
to publish another American shipwreck narrative, James Riley's *Loss of the
American Brig* Commerce. Riley's authorial competence was shaped rather
differently. Comparing the two texts, we gain insight into Murray's views
over the nature of authorial identity and of the ways in which authorship
and authority were related. In his preface, "To The Reader," Riley, a ship's
captain from Connecticut, epitomized authorial modesty: "The following
Narrative of my misfortunes and sufferings, and my consequent travels

and observations in Africa, is submitted to the perusal of a candid and an
enlightened public with much diffidence, particularly as I write without
having had the advantages derived from an academic education, and being quite unskilled in the art of composing for the press. My aim has been
merely to record in plain and unvarnished language, scenes in which I was
a principal actor, of real and heart-appalling distress."[91] Riley brought forward a sensational account of physical endurance and the subversion of assumed racial hierarchy. Following shipwreck on the Barbary Coast in 1815,
Riley and his crew endured the twin privations of a hostile environment
and aggressive inhabitants who took them captive and subsequently sold
them as slaves to Arab merchants. Popular interest in Riley's text centered
on its account of bodily suffering and in the inversion of racial hegemony:
white Americans being enslaved by black Africans serving as a prompt to
moral reflection on the ignominy of slavery.[92] Appended to Riley's narrative was an account of Timbuktu, which Riley had obtained from an Arab
merchant, Sidi Hamet, who had purchased Riley as a slave from his original
captors.[93] On being released from Hamet's ownership by William Willshire,
the British vice-consul at Mogadore, Riley had the opportunity to gather additional geographical and historical information respecting Timbuktu and
the River Niger. From Hamet's testimony, together with "the best authorities," Riley had a map prepared by the New York State geographer, John H.
Eddy, which communicated Riley's belief that the Niger "discharges its waters with those of the Congo into the gulph of Guinea."[94] Consequently,
Riley's text and map were important contributions to contemporary debate
over the source and course of the Niger and the location and character of
Timbuktu.[95]

Riley's narrative was printed and published for the author by T. & W.
Mercein in New York in 1817 as *An Authentic Narrative of the Loss of the American Brig* Commerce. Riley made clear the process by which his experiences
in Africa were committed to the page and so sought to confirm the veracity
of his account: "The Narrative, up to the time of my redemption, was written entirely from memory, unaided by notes or any journal, but I committed
the principal facts to writing in Mogadore, when every circumstance was
fresh in my memory, (which is naturally a retentive one,) and I then compared my own recollections with those of my ransomed companions: this
was done with a view of showing to my friends the unparalleled sufferings
I had endured, and not for the particular purpose of making them public
by means of the press."[96] Riley's claim to plain style and lack of authorial
ambition was accompanied by acknowledgment of editorial assistance, his

"crude notes, journals and log-books" having been prepared by the New York lawyer-author Anthony Bleecker.[97] As Riley noted, Bleecker had "revised the whole of my written manuscript, and suggested some very important explanations. I have been governed, in my corrections, by his advice throughout."[98] In addition to Bleecker's editorial intervention, Riley's narrative had been scrutinized "in point of diction and grammar" by his "very intimate friend," Josiah Shippey Jr.[99]

Authorship of Riley's text was, then, a collective and collaborative endeavor. Riley worked to assemble his own testimony and to verify it by comparison with that of his crew. Riley was prepared and willing to "sacrifice truth for plausibility" and, through exclusion and elision, to secure credibility.[100] Riley felt it unnecessary to portray the text's authorship as an individual undertaking. The fact that he sought assistance in the book's production lent credence to his claim to be unskilled in literary production (a writer by circumstance) and demonstrated his commitment to accuracy and authenticity. Whatever its route to authorship and the merits of its collective origin, its success was remarkable: "Over a million copies were distributed in a short period of years," and "no book published in the United States during the first half of the nineteenth century attained so extensive a circulation in so short a time."[101] Given the text's powerful antislavery message, its influence on the abolitionist cause was significant and helped shape Abraham Lincoln's antislavery ideals.[102]

The publication in New York of Riley's narrative coincided with the establishment of a reciprocal publishing arrangement between Murray and the New York firm of Kirk & Co., whereby the publishers would exchange and issue works likely to appeal to one another's readers. Riley's was the first text proposed to Murray, and, at the request of Kirk and Thomas Mercein (whose firm had printed *Authentic Narrative*), Riley wrote to Murray outlining his credentials and the potential popular interest that his book might provoke: "This work tho' written by a Seaman, I am confident will be read with avidity & interest by every class of readers throughout the civilized world, & in Great Britain cannot fail to prove a source of uncommon profit to its publisher & proprietors. . . . The matter is all new, highly interesting & sufficient to form a handsome Book either in quarto or octavo."[103] Given that Murray's *Quarterly Review* had commented on the New York edition in glowing terms, highlighting particularly Riley's "intelligence and unquestionable veracity," Riley needed to do little to convince Murray.[104] Although Riley's credibility rested on his claim not to be a book-maker—someone with a potentially vainglorious motivation and an eye to profit—it is clear that Riley was savvy with respect to the financial rewards that might attend publication with Murray. Riley offered Murray "the Copy right for my Book" for sale and distribution "throughout the British dominions."[105] He requested, in return, "a share of the profits, (not to exceed but one half) arising from the publication of the work" for, as he magnanimously stated, "the use of the Widows & children of my unfortunate shipmates who are left destitute."[106]

Murray II undertook publication of the text on Riley's terms, issuing the volume as a quarto at £1 16s. Murray made two important changes in his version of Riley's text. The title was changed from *An Authentic Narrative of the Loss of the American Brig* Commerce to the more prosaic *Loss of the American Brig* Commerce. The second was that Riley's preface was altered to remove certain material concerning illustrations and to disguise the editorial assistance afforded Riley by Bleecker and Shippey. This decision demonstrates a difference of opinion between Riley and Murray II as to how his authorship was most appropriately presented. Murray took the view that to make evident any editorial intervention might damage its claims to authenticity since, in this instance, authority rested on authorship as a seemingly individual and solitary endeavor. Riley was of the opposite view: to declare dependence on others lent credence to his authorial positioning as unskilled, a literary naïf. These issues were the subject of discussion in the press.

For the *British Critic*, Riley's text was "one of the most curious and interesting narratives which has ever issued from the press."[107] It displayed literary style and conveyed credibility: "The style is natural and unaffected, and the facts are both illustrative of other travels, and are themselves verified by them. We have no hesitation in placing our confidence in the author, and in believing his narrative to be both faithful and accurate. It is one of the most valuable books of travels which has lately been published in this country, and as such we have no hesitation in recommending it to general notice."[108] The *Monthly Review*, which concentrated in its commentary on the information the book offered about the Niger and Timbuktu, found Riley's narrative generally credible: "It contains nothing to impeach its own veracity, but much to corroborate it; and it is externally supported by the printed correspondence of those individuals who procured the author's redemption [from enslavement]."[109] Others were less persuaded:

> Captain Riley has certainly not written in a manner adapted to prepossess his
> readers in favour of his veracity. He has been, to say the least, indiscreet in

some of his statements, and has exercised a very unsound judgement in the selection of his facts. His manner of narrating his adventures, is entirely destitute of simplicity; he always manifests a disposition to make the most of every thing [*sic*] which occurs to him, and without being able to fix on any particular ground of invalidation, we feel altogether, that he does not possess the art of securing our implicit confidence.[110]

Although the sensational qualities of Riley's text were a source of uncertainty for some reviewers, they were in large measure precisely what served as a prompt to popular interest in the volume. In May 1817, Murray II received a letter from the Edinburgh publisher William Blackwood, commenting on his and his wife's appreciation of Riley's account: "What a treat you have given to Mrs. B. and me in 'Riley'! I never read anything so affecting and interesting. We cried over it yesterday like children. Surprising and almost incredible as the events are, yet there is a verity and touching simplicity, with a natural eloquence of language, which have perhaps never been surpassed."[111]

Despite Riley's news in September 1817 that the second American edition of his *Authentic Narrative* had sold out its five thousand–copy run and that he was at work on a revised and abridged duodecimo edition (targeted to "country readers at a cheaper rate"), Murray chose not to reset the text as an octavo.[112] By then, Murray was taken up with African travels of a different sort, working with Barrow to undertake publication of the late James Tuckey's *Narrative of an Expedition to Explore the River Zaire* (chap. 2). The example of Riley's "authentic narrative" illustrates how authors had to work to have their narratives treated as authentic: negotiations between explorer-writers and their interlocutors could dramatically affect the ways accounts of travel were shaped for the press and how authorship was arrived at. As one historian of the genre has noted, the explorer-writer was always a "hybrid figure," occupying a "distinct subject position . . . outside our prevailing evolutionary models of print and authorship."[113] The observation that travel writing is a hybrid genre, where both text and author resist simple classification and delineation, might well have been made with Riley's book in mind.[114]

"Instruments in the Hands of Others": Molding Texts, Shaping Reputations

Throughout the 1820s, the territorial, commercial, and scientific ambitions of British overseas geographical endeavor focused on two fronts, the Arctic and North and West Africa. While these frigid and tropical environments presented distinct physical challenges to the travelers and explorers concerned, the journey their written works took on return from in-the-field notes to printed fact followed a similar if difficult passage. The way in which these travelers' texts were managed through censorship, redaction, and writerly manipulation reflected the fact that the travelers were typically perceived by their institutional benefactors "as instruments in the hands of others," mobile representatives of sedentary British authorities, animated and sentient tools charged by a metropolitan elite to create geographical truth.[115] Travelers such as Hugh Clapperton, Dixon Denham, and Richard and John Lander allowed the Admiralty, Colonial Office, and the African Association to operate at a geographical distance. Because their travel was in essence institutional science and commerce on the move, the knowledge created was to an extent seen as property, a literary raw material that, if necessary, needed qualification and refinement prior to public use.

If one Murray traveler alone represents the geographical scope of Britain's (and Barrow's) imperial and scientific ambition in both these contexts, it is George Francis Lyon. Prior to his 1821 Arctic expedition with Parry aboard the *Hecla*, Lyon had completed a series of travels in North Africa, a journey that formed the basis of *A Narrative of Travels in Northern Africa* (1821), the first of the four travel texts he published with Murray (app.: 65, 83, 89, 106). With an eye to career advancement, Lyon had volunteered his services to assist the British vice-consul at Fezzan, Joseph Ritchie, who had been recruited to investigate the geography of North and West Africa with particular reference to Timbuktu and the Niger. Lyon and Ritchie and their party traveled in disguise and under assumed names, adopting "the dress and appearance of Moslems" as a device to smooth their passage.[116] Illness and physical fatigue led to Ritchie's death in 1819, and the expedition failed in its goal of reaching the Bornu Empire. Following Ritchie's death, responsibility for communicating the expedition's findings defaulted to Lyon, a task made more difficult by the fact that Ritchie had failed to maintain his journal with any regularity. Ritchie's "unfortunate procrastination," as Lyon described his superior's inscriptive nonperformance, meant that "much valuable information . . . [had] been lost to the world."[117]

Nevertheless, what did survive, alongside Lyon's own words, was sufficiently rich in its store of first-hand geographical, cultural, and scientific observations for Murray to publish it as a £3-3s. quarto, including several aquatint plates based on Lyon's own watercolor sketches (plate 12). This was Lyon's first venture into print. The modest deference he would demonstrate later in his account of the *Hecla* expedition (app.: 83 and see p. 100 above) was apparent in his text's preface.

Like Riley before him, Lyon made it clear that he had deliberately suppressed elements of his narrative that might otherwise arouse doubt:

> The situation of an author, when he presents himself to the scrutinizing observation of the public, must ever be one of the greatest doubt and anxiety; but as the following pages are intended only to detail facts in the plainest manner, without attempt at embellishment of any kind, it is hoped that they will not only meet with indulgence from the general reader, but escape, without very severe comment, from the examination of the critic. All that can be said in their recommendation is, that they adhere strictly to truth, and that not a single incident described by the author is in the slightest degree exaggerated; on the contrary, he has not only abridged but, in some instances, entirely omitted to mention circumstances which occurred to him, fearing either to excite doubt in the minds of his readers, or by too long details to trespass on their patience.[118]

Yet, there is a sense in which this rhetorical gesture and positioning was formulaic. The words and tone Lyon adopts in written defense of his African account bear great similarity to those used by Mungo Park to establish his own modest authorial position in his 1799 *Travels in the Interior Districts of Africa* (app.: 37), and Lyon's elision of personal circumstances is very close to Park's admission in private (to Walter Scott) that he, Park, had omitted personal anecdotes from his exploration narrative for fear that the public might doubt his word.[119] Lyon's later positioning of himself with respect to Parry's *Hecla* expedition may thus in part have been something expected of him and something he had consciously created in his African authorship. Further, Lyon placed his own recollections as inferior to those that Ritchie would have produced had he survived: "He [Lyon] offers his little work only as an humble substitute for one which would have been far better arranged."[120] Lyon deployed Ritchie's reputation as a mechanism to demonstrate his own modesty and may have deliberately drawn on Park's words to this end; he certainly used Ritchie's death as a prompt to public sympathy. In Lyon's and Murray's hands, the posthumous Ritchie (and Park) become central to the authorization of Lyon's text.

There was some uncertainty about what Lyon's book should be titled, although the archival record is silent as to why. Early newspaper advertisements listed it as a forthcoming Murray work under the title *A Narrative of Travels from Tripoli to Mourzouck, the Capital of Fezzan*, but this was changed for publication to *A Narrative of Travels in Northern Africa*. Given that Lyon was later active in discussions with Murray about the title of his travels in Mexico, it is possible that a similar conversation took place regarding his African narrative.[121] In any event, Lyon's hopes for sympathetic reviewers were not wholly realized. The *Quarterly Review* offered a backhanded compliment: "Though the 'Narrative' before us has no claim to merit as a literary composition, no pretensions whatever to abstract science, antiquarian research, or discoveries in natural history; though it frequently returns on itself, and sets all arrangement at defiance; yet, with all these defects, we are not afraid to hazard an opinion, that it will not only be read, but be found to afford both 'entertainment' and 'instruction.'"[122] The book's plates, which Lyon claimed to have drawn from nature (see plates 11 and 12, and fig. 6), were subjected to particular scrutiny by the *London Literary Gazette*, being thought "far too fine to be accurate."[123] The *Gazette*'s reviewer thought Lyon's *Narrative* "quite in the extreme of modern book-manufacturing."[124] Both the *Monthly Review* and the *British Review* were of the opinion that Lyon had struck just the right note in terms of sensibility and literary positioning. The superficial weaknesses of Lyon's book were, in fact, its principal strengths: "His detail is in truth simple and unpretending: it contains not one trace of professed authorship: it claims no merit in a literary point of view; and it abounds neither with scientific disquisition nor with antiquarian discovery."[125] For the *British Review*, "Without the slightest affectation of fine writing, Capt. Lyon has succeeded in giving considerable interest to his researches. Indeed we have always considered unpretending simplicity of narration to be one of the principal recommendations of a book of travels."[126]

The critical response to Lyon's narrative indicates that there never was a singular and always appropriate way in which to position authorship, define textual content, and specify literary style—in the minds of reviewers at least. Readers could be led toward certain notions of the author: by style, presumed completeness, authoritativeness of tone, and declarations of prior competence. It is clear, too, how notions of authorship could be made before the printed word reached its different audiences.

Lyon had been eager to return to Africa to continue his investigations, but his immodest efforts to secure promotion sufficiently antagonized Barrow that he was instead sent to accompany Parry to the Arctic: if nothing else, "Barrow had a sense of humour."[127] The expeditionary team sanctioned by Barrow for the second attempt to penetrate the Sahara, and to gather first-hand information regarding the Niger and Timbuktu, included two Scottish naval officers, Walter Oudney and Hugh Clapperton. Oudney and Clapperton were joined by a third man, chosen not by Barrow but appointed directly by the Colonial Office: the "capable but arrogant" English army officer, Dixon Denham.[128] Denham's differing views as to the likely source and course of the River Niger and persistent uncertainty regarding the expedition's leadership meant that relations between the members of the party were from the start actively hostile.

Since much has been written on the expedition's activities (and Denham's campaign of calumny against Clapperton, which included false allegations of homosexuality), the precise contours of their relationship and of the expedition are not repeated here.[129] The details of the expedition's narrative, however, have not been studied before. Despite their quarrels, the party was successful in reaching Kuka, capital of the Bornu Empire (which Ritchie and Lyon had failed to attain in their earlier expedition), and in sighting Lake Chad, being the first Europeans to do so. At Kuka, Denham parted company with Clapperton and Oudney: he to visit the Hausa Kingdoms, they to explore the southern shores of Lake Chad and to continue the search for the Niger's termination. Early in 1824, Oudney (long since ill) died at Katagum in the Sudan, but Clapperton pressed on to Sokoto, capital of the Fulani Empire, where by order of Sultan Bello he was required to stop, though the Niger was only five days away. Despite being prevented from reaching the Niger, Clapperton had gained much valuable information about the river's likely source, course, and end point. Clapperton was reunited with Denham at Kuka and both men returned separately to Britain via Tripoli. While Denham at once began work on a written account of the expedition, Clapperton petitioned the Colonial Office to fund a return expedition to Africa in order to test his supposition that the Niger terminated in a delta on the coast at Guinea (as James Riley's earlier testimony had indicated and Mungo Park had come to believe, prior to his death in 1806): this view, later proven to be correct, contradicted prevailing orthodoxy and, crucially, was at odds with Barrow's own.[130]

Given widespread popular and commercial interest in the expedition's subject, Denham, ever the opportunist, proposed to Murray II that £1,500 would be an appropriate sum for the copyright. Murray rebuffed Denham, noting that that figure was "more than ever was given for any Volume of Voyages or Travels" and proposed, instead, the sum of £1,200—£700 to Denham, £400 to Clapperton, £100 to Oudney's estate.[131] Clapperton's return to Africa in 1825 necessitated that his journal and documents were left with Barrow, "with a request that he would see them through the press."[132]

What this meant, in effect, is that they would be mediated by Barrow and Denham, Clapperton's antagonist. And so, as he prepared the account of the expedition, Denham "suppressed as much as possible all mention of his companions, and took the credit for some of their discoveries."[133] In the preface to *Narrative of Travels and Discoveries in Northern and Central Africa* (app.: 93), Denham—echoing Barrow's own assessment—was critical of Oudney's surviving texts, indicating that they were not "deemed fit for publication *in extensor*, from their imperfect state." They contained, moreover, "very little beyond what will be found in my own journals."[134] In effect, Denham used his dead traveling companion as an indication of his own superiority (in contradistinction to the strategy adopted by Lyon in respect of Ritchie). Denham did, however, seek his readers' approval through his modest claim to lack of literary skill: "I have to trust to its [the book's] novelty, for its recommendation to the public, rather than to any powers of writing, which I pretend not to possess."[135]

Clapperton's journal (which detailed the journey from Kuka to Sokoto) was prefaced by an explanatory introduction from Barrow, who sought to make clear that he had exercised only a light editorial touch. This assurance was necessary since there were members of the public who suspected the repression of Clapperton's views as to the River Niger's termination.[136] As Barrow remarked in his editorial preface, "I have carefully abstained from altering a sentiment, or even an expression, and rarely had occasion to add, omit, or change, a single word."[137] Elsewhere, however, Barrow confided to Murray that he had, in fact, been rather more vigorous: "dishing and trimming him [Clapperton] as much as I dare."[138] Barrow did, however, make the exercise of his editorial control more explicit in his preface in respect of the disarray of Oudney's various written materials—"scraps of vocabularies, rude sketches of the human face, detached and incomplete registers of the state of temperature" as he termed them—which he deemed to be "wholly uninteresting" and so could be excluded from the printed account.[139] Clapperton never saw the final printed materials in draft, dying in Africa before he could do so. In addition, Murray and Barrow together "resisted Denham's persistent efforts to omit references to any achievements other than his own."[140] Given the antagonism between Denham and Clapperton, and Clapperton's discord with Barrow over the Niger's termination, the volume was in no straightforward sense a balanced reflection of the expedition members' views or of what they had achieved. As Barrow wrote in his own review in the *Quarterly*, "The information obtained by Clapperton respecting the course of this river [the Niger] has entangled the question more than before."[141]

Even before printing began on *Narrative of Travels*, Clapperton had returned to Africa with instructions from the Colonial Office to extend "the
legitimate commerce of Great Britain" and to put "an effectual check" on
the "infamous traffic [i.e., slavery] carried on in the Bight of Benin."[142] Clapperton's instructions from Barrow were more straightforward: "complete
the job he had left unfinished" and determine the Niger's termination.[143]
With three officers and his Cornish servant, Richard Lander, Clapperton
proceeded northward from Badagry on the Bight of Benin, eventually crossing the Niger at Bussa in January 1826, where Mungo Park had drowned
two decades previously. The expedition's attrition rate was high: Clapperton's companions died one by one, struck down by illness or exhaustion.
Throughout the journey, and despite its physical demands, Clapperton
maintained a "pocket journal and remark book," the latter serving as a log,
written at intervals during the day, on which the former would expand,
usually in the evening.[144] Clapperton's and Lander's journal keeping was
a source of suspicion. As Lander recorded, "The natives of the regions traversed by Captain Clapperton and myself ever regarded our writing apparatus with mingled sensations of alarm and jealousy; and fancied, when
they observed us using them, that we were making *fetishes* (charms) and
enchantments prejudicial to their lives and interest."[145]

In October 1826, Clapperton reached Sokoto where, as before, he was
detained by Sultan Bello. Ill with dysentery, Clapperton died there in April
1827. Richard Lander, the expedition's sole European survivor, made a perilous return to Britain with Clapperton's journals, a feat made more remarkable by the fact that he, Lander, "carried no authority, could not sign
bills, could draw no money, [and] could command no men."[146] Although
Clapperton and Lander had failed to determine the Niger's termination, the
information they collected regarding its course served to disprove Barrow's
assumption that it drained into Lake Chad. For Barrow, this presented a
problem. While he applauded Lander's endurance and tenacity, describing
him as "a very intelligent young man," Barrow's task of preparing Clapperton's journal for the press was marked by anger and frustration and recognition that his beliefs in the course and ending of the Niger were erroneous
and would be publically dismissed.[147] The resultant volume, *Journal of a
Second Expedition into the Interior of Africa* (app.: 109), typified the hybrid
nature of travel writing and of its authorship: the book was composed of
Clapperton's posthumous journal, heavily edited by Barrow; a short biography of Clapperton, supplied by his uncle; translations of various Arabic letters by Abraham Salamé (another of Murray's traveler-authors [app.: 58]);
and Lander's own journal.

The "breathtaking depth of animosity" that Barrow held for Clapperton was communicated with vociferous enthusiasm in Barrow's introduction to the volume.[148] Clapperton was, in Barrow's view, "evidently a man of no education; he nowhere disturbs the progress of the day's narrative by any recollection of his own, but contents himself with noticing objects as they appear before him, and conversations just as they were held; setting down both in his Journal without order, or any kind of arrangement. . . . There is no theory, no speculation, scarcely an opinion advanced."[149] Worse still, according to Barrow, Clapperton's journal was "written throughout in the most loose and careless manner; all orthography and grammar equally disregarded, and many of the proper names quite impossible to be made out" as a result of his handwriting. Barrow could not deny, however, that Clapperton's regular observations of latitude and longitude were "a most valuable addition to the geography of Northern Africa." But in preparing the journal for the press and despite protestations to the contrary, Barrow admitted that it was on his authority that "much . . . has been left out" and that he had imposed a structure on the narrative, breaking it into chapters— "always a relief to the reader."[150] Barrow thus shaped and defined Clapperton's posthumous reputation and qualified the relationship between authorship and authority in doing so.

If Barrow's public criticism of Clapperton was withering, privately it was positively cutting over Lander's stylistic and literary ability. Lander had approached Murray II with the intention of publishing his own account of the expedition replete with "a thousand amusing incidents" that his original account, having been "drawn up in haste," had omitted and, thus, to reclaim a degree of control of his authorial position.[151] Murray gave Barrow the manuscript to evaluate. Barrow confided to Murray that "it appears to me that Mr. Lander's 'Wanderings in Africa' which I have attentively read, would scarcely justify the experiment of substantive publication."[152] Lander was, in Barrow's view, guilty of "deficiencies of style" and "sins of egotism," of offering scientific insights that were "deplorably meagre" and submitting biographical information "utterly unimportant and uninteresting to any reader."[153] With the editorial wrangling over Clapperton's text fresh in his mind, Barrow concluded that Lander's text (despite having been improved stylistically by Lander's younger brother, John) would "require a great deal of toning before it could prudently be submitted to the public eye."[154] His censure of Lander's manuscript notwithstanding, Barrow thought that Lander had done "so much, so well and so sagaciously," despite the "disadvantage incidental to his station."[155] Having listened to Barrow, Murray declined to publish Lander's text. It was picked up the following year, 1830, by another London firm, Colburn and Bentley, and issued as a two-volume octavo to generally positive reviews.

Although the expedition's official quarto narrative was met with general approbation, the *Oriental Herald* commented on the dissonance between Barrow's preface and Clapperton's text: "We have received more pleasure from its [the journal's] perusal than the apologetical language of the editor in the preface led us to expect."[156] For one reviewer, Barrow's introduction read more poorly than did Clapperton's supposedly inferior journal: "This . . . is not very good English," the *Eclectic Review* remarked of Barrow's "maze of words."[157] There was a general feeling among the reviewing press that, while the physical and geographical achievements of Clapperton and Lander were praiseworthy, the Niger question had still not fully been resolved. The *British Critic*, for example, wished a "better fate to future adventurers" in their endeavors to "draw aside the veil which has from the earliest ages concealed the greater part of Africa from the rest of the world."[158]

And so the British scientific community turned once more to the Niger question. Together with his brother, John, Richard Lander was engaged by the Colonial Office late in 1829 to "proceed to Fundah, and trace the river from thence to Benin"— once and for all, it was hoped, to resolve the question of the River Niger's termination.[159] The party reached Bussa in the summer of 1830 and traveled the Niger first upstream and then down before, eventually, reaching the Gulf of Guinea, the Niger's terminus. On return to Britain in 1831, the brothers were feted: Richard Lander was awarded the Royal Geographical Society's first gold medal. Murray (having on Barrow's advice rejected Lander's earlier travel account) paid what the *Quarterly* called the "liberal sum of one thousand guineas" for copyright of the expedition's three-volume account.[160] Murray also took the decision— "with equal liberality"— to publish the Landers' account in three volumes in his Family Library, a popular and relatively inexpensive series that, "from the smallness of the price" (15s.), would ensure the book's wide circulation (app.: 124).[161]

Given Barrow's earlier hostility toward Richard Lander, Murray gave the task of editing the Landers' account, and producing a map based on it, to the Admiralty hydrographer, Lieutenant Alexander Becher.[162] Becher's task was complicated by the fact that as a result of an accident during their travels, Richard Lander admitted that he had lost "the whole of my journal, with the exception of a note-book with remarks," and John had lost his "memorandum, note and sketch-books, with a small part of his journal and other books."[163] This was frustrating, to say the least, because the Landers had been diligent in observing inscriptive discipline during their journey: "As soon as the day's journey is over, we mostly reach the halting place before twelve at noon, we take a little refreshment, and repose for an hour or two on our beds, or lay along a large mat, for the natives have nothing at all answering a chair or stool. Both of us keep a journal of our route and this occupies us a short time."[164] Becher's editorial role was, then, one of "blending" the surviving and reconstructed journals "into one."[165] Becher was unsuccessful in this: he was an "infuriating editor" whose "search for clarity" was simply the "cause of [textual] confusion."[166] Leaving aside his stylistic obfuscation, it is evident that Becher shared Barrow's pejorative assessment of the scientific and observational capabilities of the Lander brothers:

> A word or two may be said respecting the map which has been constructed from the journals. The accomplished surveyor will look in vain . . . for the instruments of his calling; and the man of science . . . need only be told, that a common compass was all they [the brothers] possessed to benefit geography, beyond the observation of their senses. . . . Too much faith must not therefore be reposed in the various serpentine courses of the river on the map, as it is neither warranted by the resources nor the ability of the travellers. The map, in its most favourable point of view, can be considered only as a sketch of the river, authenticated by personal observation, which will serve to assist future travellers, from whose superior attainments something nearer approaching to geographical precision may be expected.[167]

The implication of this statement was clear: for lowly Cornishmen, the Landers had done a satisfactory job, but their conclusions were far from sufficiently scientific or rigorous. It would take a man of science properly to authenticate their findings.

In a letter to Murray II, John Lander confided his indignation at Becher's treatment of him: "I cannot help regretting exceedingly that because accident has thrown me into a humble sphere of life, my veracity is questioned & my promises treated with indifference & contempt. I am very much afraid Mr Beecher [sic] has not behaved to me with the candour & Sincerity of a Gentleman."[168] So, whether the Landers' liked it or not—and they clearly did not—their twin, field-based accounts were put to order, later and elsewhere, by a third "author" (Becher) who, in concert with others (Barrow), supplemented the brothers' narrative with additional map work and, in so helping give their printed narrative an authoritative "voice," transformed

word and author alike. This example starkly highlights the difficulty of attributing authorship singly or simply in relation to narratives of travel. The
Landers' expedition that Murray published was clearly, albeit reluctantly,
a collaborative effort, and it, in turn, drew its authority and interest from
previous unsuccessful exploration, an enduring geographical problem, and
an expectant reading public.

Sarah Lushington ("Mrs Charles Lushington," as she conventionally if def-
erentially described herself on the title page), introducing her account of
her travels as she and her husband returned to Europe from colonial service
in Calcutta, made a point of proclaiming her modest role in its production:
"The Author is deeply sensible how much the defects of her Book will demand indulgence, as it has not been revised by any Literary person, but was
at once delivered by herself into the hands of the publisher; indeed, little
alteration has been made in the original journal, beyond adapting its contents to a narrative form, and omitting details that might prove tedious, and
descriptions which had been better executed by established authorities."[169]
Lushington's words carry more weight than she would have supposed in
light of the evidence presented in this chapter. Although Lushington did not
receive the guiding hands of others, many authors only became authors after having their travel notes revised. But as was commonplace, Lushington
did make changes to transform notes taken en route into a narrative form
and, in doing so, omitted matters that seemed to her "tedious." Just what
was taken out and on what grounds it was thought tedious we may never
know. But her admission then serves now as a note of caution about the
presumed completeness of works of travel, about their faithfulness to the
events themselves, and for the need to be attentive to the agents involved
in making a traveler an author. Robert Young, presumably with the first
John Murray's approval, used anonymity and a gap of some years to position himself as a writer of consequence. George Robertson let his name go
forward but only after establishing the credibility of his account by virtue
of declaring the instrumental means to its achievement. In contrast, George
Carter's 1791 account was seen by contemporaries as an example of book-
making: his words as author cum editor-compiler served to diminish the
book in readers' eyes. Scrutiny of *The Narrative of Robert Adams* (1816) and
James Riley's 1817 work has revealed yet further routes to authorship and
authorial credibility: Adams's by virtue of testimony repeated under interrogation, Riley by association — even as both men were, in their social position, far from gentlemen. George Francis Lyon struck the pose of modest

author well, in his words if not in his self-representation (see fig. 10), but at least he saw his books travel into print. Hugh Clapperton's words, much amended, assembled in association with those of several others, outlived him: his was authorship postmortem. He, at least, did not live to witness the emendation of his words in the ways endured by the Landers: they were authors almost despite themselves.

Making the Printed Work: Paratextual Material, Visual Images, and Book Production

The authority and appeal of works of travel rested not only on the qualities of the prose, the standing of the author, and his or her truth claims but also on matters of presentation. This, importantly, included maps and illustrations, which were seen as arbiters of truth by virtue of their authors being there and having drawn "on the spot" and as visual accompaniments to, and extensions of, the written text, even when the images were worked up by others after the explorer-authors' return. In this chapter, we examine some of the methods deployed by the house of Murray to turn an author's manuscript and drawings into a printed published work. Literary theorist Gérard Genette's evaluation of the importance of the paratext—prefaces, dedications, and other seemingly "peripheral" items—as part of the ways in which a printed work is both presented to and understood by its readers is predicated on the assumption that these are not merely supplements to the text but are intimately connected to its meaning and its effects. The text, for Genette, "is rarely presented in an unadorned state, unreinforced and unaccompanied by a certain number of verbal or other productions, such as an author's name, a title, a preface, illustrations." [1] Accordingly, such items have a crucial place in considering how works of travel and exploration were assembled and presented to readers. Given our central focus on questions of authorship and authority, we explore the paratextual matters of titles and their form, frontispiece illustrations, presentation of the author's name, and epigraphs. We then examine the place of illustrative material, notably maps, in producing books of travel and exploration before, in the final section of the chapter, discussing the Murrays' publication of the works of Austen Henry Layard as an example of the negotiations among author, publisher, and audience that lay behind the firm's publication of such works.

THE PARATEXT AND MURRAY
PRINT PRODUCTION

Of the thirteen texts of non-European travel with which John Murray I was associated between 1773 and his death in 1793, only six were his sole responsibility as publisher (app.: 2, 3, 7, 8, 11, 13). None was particularly outstanding for its production values. Most of Murray I's contributions to exploratory literature were presented in plain dress. An exception, perhaps, was the jointly undertaken production of Sydney Parkinson's *Journal of a Voyage* (1773), printed for the editor (Parkinson's brother) and sold by Murray and five other London booksellers (app.: 1, and see plate 6). With almost thirty illustrations, including a handsome frontispiece portrait of Parkinson, its printing would have been expensive, although there is no indication that Murray took much of a financial stake in its production, choosing rather to share in its sale and distribution, his preferred way of operating with respect to travel works. Murray bought a publishing share in George Carter's *Narrative of the Loss of the* Grosvenor *East Indiaman* (1791; app.: 12), which was neatly printed at the Minerva Press and copublished with that press's proprietor, William Lane. This had a dramatic frontispiece depicting the shipwreck and, throughout, other well-executed "copper plates descriptive of the catastrophe, engraved from Mr. Carter's designs" (as the title page explained). Murray was listed as the principal publisher, but it is probable that his role in its preparation was limited and that it was the printer who was most responsible, financially and stylistically, for the book's layout and embellishments: although, as Zachs has documented, Murray sold nearly five times as many copies as did Lane of this book, the very limited profit against sales was shared.[2] Murray's coproduction of Thomas Beddoes's *Alexander's Expedition* (1792; app.: 14) is noteworthy for the use of stylized woodcut map scrolls to frame the title on the title page. Murray employed this device to help place the narrative's setting with respect to the Indus River for "those who have not seen Mr. Rennel's [*sic*] maps"; several of the other illustrations in the book were taken from others' accounts or simply made up by the book's "uneducated and uninstructed artist" who worked not in London but in a remote village using as his style guide "Mr. Bewick's masterly engravings on wood."[3]

The house of Murray began to improve its presentation of travel work under the direction of Murray II, initially by extending its expenditure on engravings. From the 1810s on, there were notable experiments in fine printing and illustration, particularly in part-issue. In 1814, for example, Murray II published five volumes of William Alexander's *Picturesque Representations of the Dress and Manners* of various nations (app.: 28), each containing numerous colored engravings, a work that claimed to be "superior in point of elegance" to similar publications that the firm had hitherto undertaken.[4] In the volume on Turkish dress it was claimed that "the merits of this work depend on the accuracy and beauty of the drawings, and the truth of the colouring."[5] A further example was Edward Joshua Cooper's *Views in Egypt and Nubia*, published in parts by Murray between 1824 and 1827 with forty-two lithograph plates and seven folding plates, including a lithograph map (app.: 81). Murray II's use of the latest developments in printing sought to give additional authority to the typography as well as to illustration. One such initiative was John Francis Davis's *Chinese Moral Maxims* (1823), one of the first British titles to use Chinese type in innovative ways, although this was printed by Peter Perring Thoms at the East India Company's press in Macao "for the want of types in England" (fig. 13).[6]

Murray II and III could be conservative in their approaches to book production. Given the high cost of illustration and engraving, the firm was sometimes reluctant to comply with the wishes of authors ambitious for an expensively illustrated book. When George Catlin approached John Murray III to publish his two-volume *Letters and Notes on the Manners, Customs, and Condition of the North American Indians* (1841), for example, Murray declined and rejected the work on discovering that the portfolio was to include three hundred steel plate illustrations that "would require a very great outlay to artists to produce them": Murray advised Catlin to publish it himself by subscription.[7] When the proposal for Robert Fortune's *Three Years' Wanderings in the Northern Provinces of China* (1847) was submitted (app.: 185), Murray informed the author that, with an eye to balance profit and expense, he proposed "giving several of the Illustrations but not all."[8] The images that were included were calculated to appeal most to the prevailing interest in oriental exoticism among readers.

By the 1830s, there is evidence to suggest that the house of Murray's traditional production values had begun to lose their appeal, as readers called increasingly for cheaper books. In a growing and democratized market, commentators remarked on the firm's attitude toward lavish book production (see above, p. 121). Long after their competitors had shifted to the

XL.

Though a man may be utterly stupid, he is very perspicacious when reprehending the bad actions of others: though he may be very intelligent, he is dull enough while excusing his own faults: do you only correct yourselves on the same principle that you correct others; and excuse others on the same principle that you excuse yourselves.

Jin	人	A man,
suy	雖	though
chy	至	extremely
yu	愚	stupid,
tsě	責	reprehending, correcting
jin	人	other men,
tsě	則	then (he is)
ming:	明.	intelligent:
suy	雖	though
yew	有	he have
tsung	聰	} intelligence,
ming,	明	
shoo	恕	excusing
ky,	己	himself,
tsě	則	then (he is)

FIGURE 13 Typography and distant markets. In producing type of this sort, the house of Murray could advertise its capacity for innovation. Source: John Francis Davis, *Chinese Moral Maxims* (London: John Murray, 1823). By permission of the Trustees of the National Library of Scotland.

octavo format for travel works, Murray continued to produce highly priced books that some readers felt were gratuitous examples of "book-making." Janice Cavell has commented on the extent to which the Murray imprint came to be regarded in the mainstream press as anachronistic in its continued appeal to privileged readers, the expensive quarto format becoming a target for those satirists who saw it as "an inherently pretentious and elitist form of publication, reserved almost exclusively for wealthy bibliophiles."[9] Partly in response to the growing call for cheap literature, Murray II experimented with more modest production values in an attempt to appeal to a more popular readership. Where a book was aimed at the general public and sold at a relatively modest price, simplicity was called for in order to keep its production costs down. John Barrow's *Voyages of Discovery and Research within the Arctic Regions* (1846) was issued in a single volume with no illustrations apart from a frontispiece portrait of the author and a small map, an unusually modest presentation for a book associated with Barrow (app.: 178). A reader's digest of many of the volumes that had previously appeared on the subject, Barrow's volume sold for the comparatively low price of 15s., and was, according to its author, intended for the reader for whom the many individual accounts of Arctic discovery—usually, expensive quarto volumes—were beyond their means: "it did not escape me," Barrow wrote, "that something of this kind was also wanting, and might be acceptable, to supply the place of the official quarto volumes, whose costly size and decorations preclude them from the general and ordinary class of readers." While the "charts and prints by which they are illustrated" may have been "highly valuable to the man of science and taste, and well adapted for public libraries, or those generally found in the mansions of the wealthy . . . they are not exactly suited for general circulation."[10]

We show in the following chapter how, with the launch of cheap series such as the Family Library (1829) and the Colonial and Home Library (1843), John Murray II supplied reading matter for the nonspecialist general reader. With the boom in Australian emigration at midcentury, Murray III attempted to capitalize on the demand of an even humbler readership: potential emigrants hungry for information about prospects for Australian settlement. The firm's first foray into this niche was George Blakiston Wilkinson's *South Australia* (1848; app.: 192). Confessedly a book "in rough dress," it was intended, according to its author, for the benefit of his "poor countrymen."[11] The book's physical appearance, a modest and simple 10s. and 6d. post octavo volume, was in accordance with its literary style, which offered straightforward advice on topics ranging from mining, farming, and the cost of passage to the local flora and fauna. In the following year, Murray published, by the same author, *The Working Man's Handbook to South Australia* (1849), its title, format, and price leaving little doubt as to its intended audience (app.: 201). Wilkinson justified the book in its desire to supply "the Working Classes with all the practical information he possesses respecting the colony, in a cheap form."[12] In some instances, the house of Murray attempted to capture different audiences by recycling the same literary property in another form. As we shall show of Austen Henry Layard's works, titles were edited and re-presented in various versions to appeal to different pockets simultaneously. Some could be purchased in quarto and octavo; some with plates colored or plates plain; some bound or unbound. Léon de Laborde's *Voyage de l'Arabie Pétrée* (1830), for example, had already appeared in a lavish Paris folio edition with full-page illustrations by the artist Louis Linant de Bellefonds, when Murray II decided in 1831 to commission a translation, one of very few travel books the firm published in this way.[13] Keen though he was to capitalize on the public appetite for accounts of the recent finds at Petra, but mindful of the expense of including the original artwork, Murray's edition—published in 1836 as *Journey through Arabia Petræa* (app.: 133)—incorporated only a handful of Laborde's drawings—and those in reduced form (fig. 14).

According to the translator's preface, it differed from the French edition, which "from its unwieldy size and expensive form," is "inaccessible to general readers." Two additional chapters were added as the translator-editor had "endeavoured to exhibit, in one continuous narrative, the whole of the details which the translator had scattered through a preface, an introduction, an explanation of the plates, and a sort of itinerary."[14] In the end, it was claimed that "some incidental dissertations not intimately connected with the main object of the publication, I have omitted or abridged, with a view to render the work somewhat more attractive to the general reader than it probably would have been, had I confined my labours to a mere version of the original."[15] Laborde had originally consented to the English version, but he was not persuaded by Murray's down-market version, writing to the publisher that the volume required more illustrations if it was to convey an accurate view of the region.[16]

Even where the presswork was passably good, Murray's authors could be disappointed with the printed product. John Gardiner Kinnear's *Cairo, Petra, and Damascus* (1841; app.: 151), unusually for a book dealing with this geographical region, had no illustrations, maps, or charts, almost certainly because it was based on personal letters rather than on detailed exploration.

JOURNEY

THROUGH

ARABIA PETRÆA,

TO

MOUNT SINAI,

AND

THE EXCAVATED CITY OF PETRA,

THE EDOM OF THE PROPHECIES.

BY

M. LÉON DE LABORDE.

LONDON:

JOHN MURRAY, ALBEMARLE STREET.

MDCCCXXXVI.

FIGURE 14 Reducing the visual impact. In producing an English language version for the British market, Murray greatly reduced the number and size of the illustrations from Laborde's original account of his Arabian travels. Source: Léon Laborde, *Journey through Arabia Petræa* (London: John Murray, 1836). By permission of the Trustees of the National Library of Scotland.

Kinnear later regretted the absence of illustrative matter, mainly because he had been accompanied on his journey to Petra by the artist David Roberts, to whom Kinnear dedicated the book. For Kinnear, Roberts's visual depictions had the promise to be, in an emotional sense, more truthful than his own written account: "The interesting scenes through which we passed last year must become better known in this country from the admirable productions of your pencil . . . than any written description can make them. . . . Memory will fill up these meagre outlines, supply all the accessories of the picture, and impart to it the colouring of nature."[17] After its publication, Kinnear wrote despondently to Murray: "The result is certainly not very splendid."[18] (Murray's reply is not known.)

Title Pages: Declarations of Purpose

The title page was the bearer of the most fundamental information about the text and its status. The title often expanded into a description of the work's contents: the author's name, in its various forms and honorifics, an inventory of the features accompanying the text (illustrations, maps, and front and end matter), as well as bibliographical information regarding its place of publication, its date of issue, and the name of the publisher and/or printer with responsibility for its presence in the world. The title page functioned in three interrelated ways. It served, first, to give a unity to the work, offering a sense of completeness to the elements of the text and paratexts that followed, particularly through the use of the long title, which gathered together on a single page the different parts of the book and declared the purpose of the narrative. Second, the title page served an important hermeneutic function, providing both an entrée into the text and a means of orienting the reader as to how the text should be read, locating it in relation to already-established genres, whether scientific, autobiographical, or literary. Finally, the title page served a promotional function: as a short advertisement for its contents it brought to the potential consumer's attention features of the work that would give it additional appeal.

The house policy in this respect was and remained conservative, not least in the continued use of the long title, even in the face of alternative suggestions by the firm's authors. While other British publishers moved during the nineteenth century toward shorter-title designations, Murray continued the eighteenth-century tradition of presenting extended content summaries on their title pages. In 1853, Mansfield Parkyns—whose *Life in Abyssinia* was in preparation for the press—wrote to Murray III to say that, while he would prefer a shorter title, he was willing for commercial reasons to accede to the longer:

> I think that as a matter of taste the simpler the title of a work the better, though perhaps as a matter of sale, which is more important an addition would be better. I think we might say "from notes collected during nearly three years [*sic*] residence & travels in that country." . . . If I understand the meaning of personal narrative, I don't quite like it, as I disclaim any intention of exact journaling the events, explaining that I pretend only to give the reader as true an insight into "Abyssinian Life" as I can.[19]

As well as the promotional function alluded to by Parkyns, the use of the full title by the Murrays gave a sense of unity to sometimes highly complex works, many of which bore the traces of more than one author. Textual hybrids such as *Narrative of Events in Borneo and Celebes, Down to the Occupation of Labuan: From the Journals of James Brooke, Esq. Rajah of Saráwak, and Governor of Labuan. Together with a Narrative of the Operations of H.M.S. Iris. By Captain Rodney Mundy, R.N.* (1848; app.: 189) saw their title page become almost a preliminary table of contents. The miscellaneous titling of such works was not unusual on the Murray lists, especially before about 1812 (app.: 1, 2, 3, 6, 11, 12, 15, 19, 23), with some later exceptions (app.: 53, 81, e.g.). After all, exploratory expeditions could involve oceanic voyaging, overland journeying, and excursions and surveying on foot, and authorship could be a collective enterprise. From the 1820s, and particularly with narratives of exploration, Murray titles tend to be more essentially descriptive of the work undertaken and the place concerned, single themed if no less shorter. Because explorer-authors often included a wide range of different material between single covers—appendixes, maps, charts, tables, reports, illustrations—it was incumbent on the Murray firm to gather these separate elements together, and the long title was one economical way of achieving this.[20] Long titles reflected and represented the inescapably hybrid nature of the travel text.

A work's title could also signal its generic affiliation, identifying it as a scientific work, a personal memoir, or something in between, and thus served to position the work vis-à-vis its intended audience or to justify its stylistic or epistemic arrangement. Titles served to create a certain expectation among readers, while providing, in some, an almost apologetic function over the work's contents. Titles could give an indication not just of the genre but of the texture of the work as well: from the personal and the anecdotal ("A Personal Narrative," "Letters") through to the more general and descriptive ("A Journey," "A Voyage") to the more explicitly objective ("History of," "An Account") and so on. Some titles advertised their provisional, partial, or deliberately unpolished nature. The title of James Hamilton's *Wanderings in North Africa* (1856; app.: 227) gives an impression of the arbitrariness of his investigations, and Francis Bond Head's *Rough Notes Taken during Some Rapid Journeys across the Pampas and among the Andes* (1826; app.: 95) was hardly likely to capture the scientific imagination, positioning its appeal (as its author intended) as popular rather than prosaic and practical, although, as we have seen, its content and style were welcomed by critics and readers (see p. 37).

If we take a survey by general heading of the titles used by the house of Murray between the late eighteenth century and the later 1850s, the default position appears to have been "Narrative" (forty-three works had this as the single lead term in the title), followed closely by "Journal" (forty-one works) and "Travels" (thirty-five). Occasionally employed in combination, these terms indicate a univocal and unified discursive voice, though the latter terms, when pluralized, could indicate the works of several hands. "Journey," "Voyages," and even "An Account" provided alternatives, though less frequently. From the second quarter of the nineteenth century, the firm more commonly experimented with several new formulations, sometimes appealing to the highly personal ("Letters") and at others to the sensational, such as John Drummond-Hay's *Western Barbary: Its Wild Tribes and Savage Animals* (1844), issued in the populist Home and Colonial Library.

The correspondence between title and contents was scrutinized at different stages in production. Francis Bond Head, one of Murray's authors-cum-literary advisers, was asked to pass judgment in 1831 on the manuscript for Richard Vowell's three-volume *Campaigns and Cruises, in Venezuela and New Grenada, and in the Pacific Ocean; From 1817 to 1830: With the Narrative of a March from the River Oronoco to San Buenaventura on the Coast of Chocò; and Sketches of the West Coast of South America from the Gulf of California to the Archipelago of Chilöe. Also, Tales of Venezuela: Illustrative of Revolutionary Men, Manners, and Incidents* (which was eventually published by Longman in 1831). Although Bond Head could not recommend it for publication, he wrote to Murray II that he thought "the writer has performed all that his title page or advertisement promises—for he gives you 'campaigns,' a 'narrative of a march' and 'sketches of the western coast of South America.'"[21] Sometimes, where material was gathered together from different authors and sources, a title had to be devised for publication. In this instance, however, it was evident that the author had a clear idea of the title and even of the prefatory advertisement before the involvement of the publisher. As titles went into subsequent editions, the inclusion of additional or revised matter might enhance a book's appeal and convey to the reader a sense of the up-to-the-minute information contained within. In some instances, this provided the author and publisher with an opportunity to correct mistakes and clarify issues that had been overlooked in previous editions, but the commercial value of being seen to do so was also a primary objective for the market-wise Murrays. Additional illustrations, appendixes, charts, and maps were common ways of giving potential purchasers a sense of the added value that a new edition could bring and were often conspicuously advertised on the title page of the new work itself. Francis Galton had at first been indifferent to the title of his *Narrative of an Explorer in Tropical South Africa*, telling Murray III to "put what you like."[22] When Galton was called on to revise *The Art of Travel* (1855) for a second edition, readers were left in no doubt about the superiority of the updated version. In the preface to the first edition, Galton had invited contributions and corrections from fellow travelers, a technique he learned from his perusal of Murray's Handbooks. Incorporating much of the information gathered in this way, the new title page advertised itself, revised in content and modified in format, as a "Second Edition, Revised and Enlarged, with Many Additional Woodcuts."[23]

In these respects, the title page, so often overlooked by readers as merely conventional and secondary to the main part of the work itself, is crucial to understanding the text's function with regard to its ontology, its status, and, ultimately, its meaning. Genette argues that a title's first official appearance is its inscription on the publishing contract that was framed by the publisher.[24] True, decisions about the title page were generally regarded as falling under the jurisdiction of the publisher in connivance with the printer, not always with the author. Our evidence, in fact, has shown that the Murray firm was not above altering or proposing titles if the author was at all uncertain. In the end, what mattered was that the title—and other front matter—be arranged in accordance with the firm's instinct for authority, good taste, and capacity for sales.

Frontispieces: Picturing Authority

The practice of placing a frontispiece at the beginning of a work was a common convention by the eighteenth century, being found mostly in the works of "established writers, editions of classical authors, and, occasionally, high-profile biographies, histories, or travel narratives. For reasons both of custom and cost, the frontispiece portrait quickly evolved into a caste label."[25] The purpose of the frontispiece portrait was to convey a specific image of the author. On the one hand, and for commercial purposes, it served to extend his or her celebrity and so aimed at increasing the value of his or her literary property. On the other, it acted to legitimate the writer, constituting his or her voice as a guarantor of authentic discourse.

John Murray II and III treated this particular paratextual device in both ways. In most instances, Murray frontispieces were simply maps or illustrations that might otherwise have appeared later in the book. The house of Murray only rarely commissioned portraits to be copied by an engraver as an introduction to the work. Murray II, in particular, had firm opinions about how to present his authors to the world, several times provoking objections from writers about their own public images. On seeing a proof of his frontispiece portrait for *A Narrative of the Expedition to Algiers* (1819; app.: 58), the Egyptian poet, Abraham Salamé, wrote to Murray II in 1818 to note that he did not "approve of it at all, & I shall not have it in my Book— the only likeness in that engraving is the forehead, & the upper part of the Eyes; but from the under lash of the Eyes to the chin is not mine."[26] When the book appeared, it included a modified version of the portrait. In the original, Salamé is portrayed as a European in Arab garb; the revised version portrays him as an Egyptian in his native costume. Both portraits were executed to a high standard: having made the initial outlay, Murray was probably simply reluctant to waste an expensively engraved plate. Whether in consequence of this or for other reasons, the author wrote to Murray in anticipation of another edition to inform him that he was now "resolved to try another publisher."[27]

Where some authors sought verisimilitude in their printed likeness, others were less hesitant in having images worked up to make them or their subjects more appealing. After inspecting a preliminary sketch for his forthcoming *The Life of Bruce, the African Traveller* (1830; app.: 120), Francis Bond Head remarked on the shortcomings of the frontispiece portrait of Bruce: "Nature is to blame not the artist—the fact is that Bruce is in appearance a greasy Hippopotamus." The portrait, in Bond Head's opinion, was a little too true to life, and so might be improved with a little modification: "I think that the engraver might work his chin by broken shading a little more pointed, which would cut off about a pound and a half of his jowl" while "a little more fire in his eyes would improve him."[28] This is an unusually explicit example, but it exemplifies the extent to which writers, in connivance with their publishers, could fabricate a particular visual image—or have one fabricated for them—for the sake of favorable publicity. Other authors were more circumspect over exactly what sort of image accompanied and prefaced their words.

On publication of his *Travels into Bokhara* in 1834, Alexander Burnes became something of a celebrity, the popular press playing on his clandestine survey of the Indus and his travels in Afghanistan and parts of British India's northwest frontier in disguise.[29] In January 1834, Murray II wrote to Burnes, who had journeyed to Bukhara three years before as part of his diplomatic and geographical work, suggesting that a portrait of the author in native costume be inserted as a frontispiece to his *Travels*. Murray was no doubt prompted by the preface to Burnes's account, in which the author describes his adoption of local costume in order to avoid attracting undue attention. On his return to Britain, Burnes sat for a portrait in oils of himself in native costume (plate 13). To Murray's request, the author replied: "I am ready to comply with it on these terms that the likeness be given in costume with these words under it 'The Costume of Bokhara.' It will be known that it is a portrait & will save me from the appearance of vanity."[30]

Readers, Burnes felt, would be bound to recognize the portrait as illustrative of various episodes in the text that recounted the disguise that the author had used to ensure his invisibility in the field. The fact that Burnes sat for a portrait of himself in native costume is not wholly consistent with his views about "the appearance of vanity," but it is clear also that Burnes was nervous about overstepping the bounds of propriety by violating the polite codes that required the authorial voice to avoid self-aggrandizement in accordance with the conventional discourse of modesty. The portrait of Burnes in disguise was in stark contrast to a further contemporary portrait that showed him in modern European clothing, the Asian explorer as Regency gentleman (plate 14). Even so, Murray II ignored Burnes's wishes over his own selective self-representation.

When the book was advertised in the *Quarterly Review* in June 1834, it was announced that it would be accompanied by a portrait of the author, based on his portrait depiction, in oils, in native dress. Exasperated, Burnes wrote to Murray to say, "You would oblige me very much by altering this in the subsequent advertisement" by "leaving it out!"[31] When the book finally appeared, it was with a half-length portrait of the author bearing the legend "Sir Alexander Burnes, C.B. in the Costume of Bokhara" as the frontispiece portrait (fig. 15).

In December 1834, Burnes turned his mind to the preparation of a new edition of the book that, by then a critical success, had earned him honor

FIGURE 15 The author memorialized. This image was initially prepared for Murray by Edward Finden in about 1834 from an oil painting of Alexander Burnes (see plate 13). By the time it appeared as frontispiece to his 1842 book, Burnes had been dead a year, murdered in Kabul. Source: Alexander Burnes, *Cabool: Being a Personal Narrative of a Journey to, and Residence in That City, in the Years 1836, 7, and 8 with Numerous Illustrations* (London: John Murray, 1842). By permission of the Trustees of the National Library of Scotland.

and fame. Returning to the troubling question of the frontispiece, he informed Murray: "I would like however if you could make some alteration in *my visage* in the 'Costume of Bokhara' for it is said to be so rich and cunning that I shall be handed down to posterity as a real Tartar!! Suppose you strike it out altogether or get Mr. Mackie to touch it up a little, or suppose you substitute a lithograph of the miniature which I shewed to you & which my brother has."[32] The miniature in question was a portrait of Runjeet Singh, the raja of Lahore, which had recently been included in Henry Toby Prinsep's book on Singh.[33] On this occasion, Murray complied, and the Burnes portrait was removed from the second edition of 1835 despite the fact that Murray had paid the engraver Edward Finden £2 2s. to make "alterations" to Burnes's portrait in preparation.[34]

It is difficult to know if Murray II's explicit identification of the portrait with the author was a deliberate oversight in which the publisher ignored the author's wishes or was a determined act to promote his literary property. It is rather easier to understand that Burnes, if he wanted a picture of himself to be included in his book at all, should choose the one that validated his narrative accounts of disguise in native costume rather than a depiction as British gentleman (cf. plates 13 and 14 and fig. 15). Burnes's nervousness reveals much about the sense of propriety that Murray's travel writers could feel in the face of what might be construed as acts of shameless self-promotion. The replacement of an authorial portrait, artificially got up in native garb, with a more "authentic" alternative, also reveals something of the imperatives at work in the process of revision and correction, founded on the assumption that with each subsequent issue a greater proximity to authenticity is possible. Even so, Burnes was not impressed with the verisimilitude of his own portrait; he was careful to explain in his preface of 1835 that its substitute "is a good likeness, and, I believe, the only one ever taken with the front face."[35]

The more usual modus operandi was the oblique portrayal of the author in the field, unidentified but implied. Francis Galton's *The Narrative of an Explorer in Tropical South Africa* (1853) included two plates, both of which portray what the reader would take as the author, seated and surrounded with his equipment (see fig. 2). Although no mention is made in the captions of the pictorial subject, Galton's identity is clear from the narrative. What is more, the fetishization of equipment anticipates another text that Galton was preparing at the time for Murray and through which he was to become the leading authority of the day on the logistics of exploration: *The Art of Travel* (1855). Galton is not identified in the two 1853 images, but it is fair to assume that they were intended as anticipatory acts of self-fashioning.

The Author's Name: Declarations of Authorship and Authenticity

Reflecting on Philippe Lejeune's concept of the "autobiographical contract," Genette comments thus on the intimate relationship between the
author's name and the text that bears it: "The name of the author is not
a given that is external to and coexistent with this contract." The authorial name is instead "a constituent element of the contract"—it is, in other
words, an integral part of the paratext, a cue to the interpretation and value
of the work itself.[36] The Murray firm's choice of authorial designation can
tell us much about the way it wished to present the work and the writer to
the world. Rarely, if ever, was a Murray travel writer identified with his or
her name pure and simple. The use by Murray II of official and honorific
titles indicates that he published works that had official sanction (by the
Admiralty, the East India Company, the British government and so on),
and that he was, only secondarily, the publisher of the works of dilettante
writers. In prefatory statements, it was conventional to demur from any
appearance of professional authorship: the house of Murray rarely used
the common formula "by the author of," preferring instead to identify the
writers on his travel list with their official status, military, diplomatic, or
similar designations rather than as authors who might be construed as literary professionals.

Manuscript correspondence between the author and the publisher highlights just how delicate the matter of the author's name and its chosen formulation could be. As his work was being prepared for publication, James
Clark Ross was insistent that the title page to his 1847 book should have his
name in full—"the insertion of my name *in full* makes it more distinct from
that of any other Ross" (i.e., John Ross, of Arctic fame and disrepute). As
the book went through protracted processes of checking and proofing, Ross
insisted on having "a slight alteration made to the title—with the addition
of Hon. D. C. L. Oxford, after F. R. S. as the Oxonians might take offence"
at this being omitted.[37] In 1825, Frederick Beechey wrote to Murray II
requesting an arrangement for the title page of *Proceedings of the Expedition
to Explore the Northern Coast of Africa* (1828; app.: 102), which he had coau-
thored with his brother, Henry: "My brother & myself also wish you to say
By authority the narrative &c and will you be good enough to place FRS &
Member of Astronomical Society after my name—and let it succeed to my
brothers."[38] His motives behind the last request may have had to do with
the level of independence that the brothers had negotiated for their account
from the Admiralty for its contents, and it was presumably on this basis

that they had been granted permission for an unofficial though sanctioned publication. When the book appeared, it was clear, however, that Murray as publisher had made very few of the desired concessions: the phrase "by authority" was nowhere to be seen; the reference to the Astronomical Society was missing; even more irksome to the authors was the ordering of their names, Frederick's name appearing ahead of his brother's. Following the book's publication, Frederick returned, in writing to Murray, to the problem of authorial priority, having seen a number of periodical notices that had attributed the writing of the book to him alone: "I am not entitled [to this credit], my Brother, Mr Henry Beechey, having had *at least* an equal share of the labour of its compilation."[39] Murray II's desire to afford priority to "Captain F.W. Beechey, R.N., F.R.S." over "H.W. Beechey, Esq., F.S.A." is fairly clear. Years before, Frederick had achieved celebrity having accompanied Franklin and Parry on their polar expeditions. Henry was then a relatively unknown artist.

Ever keen to promote the association of his travel list with naval officialdom, it was unsurprising that Murray wanted to promote Frederick's role in the venture, not least because he undertook the African expedition on a government commission while his brother had, to the publisher's mind, merely accompanied Frederick on his own initiative. Had Murray honored the request, it would have been a departure from his usual protocol, which required that the individual "being charged with the expedition" was considered as the authorial "principal."[40] Murray's treatment of the Beecheys was no doubt also informed by Naval policy regarding the proprietorship of texts composed in the course of official duty, which required that all writing undertaken in the field be handed over to the presiding officer on return, although as we saw for the case of James Kingston Tuckey, unofficial narratives could sometimes slip unseen into the public domain and cause anxiety before being corralled for approval and publication.

On rare occasions, author's names would be omitted, particularly when the writer was a woman. In the late eighteenth century it was not unusual for the literary identity of polite female writers to be veiled in order to protect their reputations. In 1784, Murray I was jointly involved in the publication of Lady Mary Wortley Montagu's *Letters*—her name in the title being disguised by means of dashes (app.: 6). It is evidence of the conservatism of the time that, two decades after her death, it was still felt necessary to obscure the author's identity. Later publishers were less reticent: in 1803, the London bookseller Richard Phillips issued *The Works of the Right Honourable Lady Mary Wortley Montagu* in five volumes, without any attempt to dissemble. The house of Murray continued to disguise its female authors well into the nineteenth century: when Murray II published Julia Charlotte Maitland's *Letters from Madras* (1843), its authorship was simply, as the title page advised, "By a Lady." In a somewhat overscrupulous introduction, it was claimed that the letters had been "printed verbatim from the originals" but that it was necessary to omit all family details and "whenever European individuals are mentioned, fictitious names have been assigned to them, and other precautions taken to prevent the personal application of such passages."[41] It was not Murray's amour propre that was decisive: initially, Maitland's husband had requested that Murray include his wife's identity on the title page, but the author herself later wrote to request that her name be excluded from the title page since her mother objected to it and she was reluctant "to vex her."[42]

By the time Murray II turned to Maria Graham to compile a work "from the various journals and notes made by some of the officers and other gentlemen that accompanied Lord Byron [cousin of the poet] on his interesting voyage to the Sandwich Islands," Graham had already achieved a reputation as a celebrated author (see chap. 3, pages 69–72). Having gathered together the expeditionary journals for her *Voyage of the H.M.S.* Blonde (1826), Graham reworked them, supplementing them with additional information in order to present a univocal coherent narrative. But such was her contribution that she was accused by contemporaries of being more a "fabricator" than an "editor."[43] Appearing only as the anonymous editor throughout, Graham's name appears nowhere in the book let alone on the title page, the author line of which reads simply "Captain the Right Hon. Lord Byron, Commander" (app.: 91). Authorship was intimately associated with authority: the commanding officer was the individual responsible for the results, written or otherwise, of an expedition. As literary property, the Sandwich journals generated on the *Blonde* did not belong to the men who wrote them, nor did they belong to the woman who shaped them into a book. The proprietary "author" was, legally speaking, the Admiralty, the expedition's commander its symbolic representative.

At times the house of Murray was induced to treat the identities of even its male authors with delicacy. Where there was a conflict between authorship and their social standing, both author and publisher had to tread carefully. Godfrey Mundy, an Army captain who had served in India as aide-de-camp to Lord Combermere, assembled two volumes of his observations as *Pen and Pencil Sketches* (1832; app.: 125). Being the untutored author and soldier that he was as we saw in chapter 3, Mundy's book is full of unbuttoned gossip about army personnel and Indian potentates. Mundy was appalled to find that Murray had advertised the book before its publication, identifying him as the author: "The prospect of being advertised by name so long before publication, has, I confess, rather shattered me, particularly as I do not recollect having told Mr Murray that I had made up my mind to affix my name to the work."[44] Anonymity might have served a useful purpose: Mundy felt professionally awkward that his associates should find out that he was planning to make his experiences and frank opinions known to public scrutiny. In a further instance, Murray II did invoke anonymous authorship to avoid controversy. When *Sketches of Persia* first appeared in 1827, it was mysteriously attributed to "A Traveller in the East" (app.: 100). This anonymity perhaps had to do with its indiscreet tone: among other things, the text refers to "the rulers of the East" as "despots, and their subjects slaves; that the former are cruel and the latter degraded and miserable, and both totally ignorant."[45] Another reason for Murray's discretion was because, in the year of the book's publication, its author, John Malcolm, had become embroiled in political controversy. Having failed to secure the governorship of Madras, he was appointed governor of Bombay and, later that year, was at the center of an acrimonious constitutional crisis that placed him in disfavor with native and colonial governments alike despite his key role in having earlier advanced Britain's diplomatic interests in Persia and Central Asia.[46] Two decades later, after the dust had settled and the scandal was long forgotten, it was no longer necessary to disguise the book's authorship. When it was reissued in the Home and Colonial Library in 1845, twelve years after the author's death, the title page finally attributed it to "the late Major-General Sir John Malcolm, Author of 'History of Persia,' 'Life of Clive,' &c."

In general, the house of Murray routinely engaged in the gratuitous use of honorific titles that would lend credibility to the writer, authorizing him and, less commonly, her as a sacralized figure, with access to a privileged view of the various subjects being addressed. In this way, the firm was—or, more properly, it became—a master of ostentatious display, parading the capacity of their authors to speak knowledgeably through publicly acknowledged qualifications.

Epigraphs, Dedications, and Acknowledgments: Legitimacy in Brief

Genette identifies a dual function for the epigraph: acting as "a signal (intended as a *sign*) of culture, a password of intellectuality" and as a means of consecration, while the author (and, we should add, the publisher) is awaiting reviews "and other official recognitions, the epigraph is already, a bit, his consecration."[47] This combined function can be seen in many of the epigraphs included in Murray's title pages. The task of finding an apposite passage, where it was required, chiefly fell to Murray's literary advisers. As pendants to the title, some epigraphs were more allusively oblique than others. John Martin's edition of William Mariner's *An Account of the Natives of the Tonga Islands* (1817; app.: 39) includes an unusually long quotation under the title that functions both as advertisement and as a means of legitimating its content through its association with an already reputed authority: "The savages of America inspire less interest . . . since celebrated navigators have made known to us the inhabitants of the islands of the South Sea. . . . The state of half-civilization in which those islanders are found gives a peculiar charm to the description of their manners. . . . Such pictures, no doubt, have more attraction than those which pourtrey [*sic*] the solemn gravity of the inhabitant of the banks of the Missouri or the Maranon." Taken from the first volume of Alexander von Humboldt's *Personal Narrative*, which had been translated into English and published by Murray and others in 1814 (app.: 30), this epigraph sought to capitalize on Humboldt's growing reputation, while asserting the comparative importance of its subject matter.

Longer, more allusive epigraphs, too extensive to be inserted between the long title and the publication details, were often included on the verso facing other prefatory material. The verse accompanying Richard and John Lander's *Journal of an Expedition to Explore the Course and Termination of the Niger* (1832), for example, was taken from Richard Millikin's long narrative poem *The River-Side* (1807) and was unusually poignant.

> We pass o'er Africk's sultry clime,
> To where the Niger rolls his mighty stream
> With doubtful current, whether bent his course
> Or to the rising or the setting sun:
> Till one advent'rous man, thro' perils great
> And toil immense, hunger and thirst, and pain,
> The question solved, and saw him eastward flow
> Majestic through the woods.[48]

Although Millikin's poem had all but fallen from sight by 1832, the passage was carefully chosen to elevate the expedition's leader. Ordinarily the Murray firm and its authors would have avoided such gestures of self-aggrandizement, but, while the book was in production, Richard Lander had returned to Africa with a view to setting up a trading post on the Niger. Soon after arrival he was to die from gunshot wounds inflicted by locals while he was traveling upstream by canoe. Hidden in the epigraph is a veiled reference to the later fateful expedition: a few lines later Millikin had written: "Upon his banks, the savage Moor usurp / A haughty rule and bow the vassal necks / Of Monarchs to the ground and on them tread."[49]

A similarly highly nuanced epigraph appears at the beginning of Mansfield Parkyns's *Life in Abyssinia* (1853; app.: 218) and includes two quotations. The second is a generic text that might have been applied to any work, the famous lines from Walter Scott's *The Lay of the Last Minstrel* (1805): "This is my own, my native land! / Whose heart hath ne'er within him burn'd, / As home his wandering footsteps he hath turn'd / From wandering on a foreign strand?" The first, comprising two lines from Shakespeare's *Winter's Tale* (1623), is more suggestive: "Like an old tale still; which will have matter to rehearse, / though credit be asleep, and not an ear open." Parkyns's book was not merely an account of his travels: it was, more sensationally, a vindication of the observations of James Bruce's travels in the region, the Scot having earlier been defamed for what were thought to be fantastical claims about the people and their culture.[50]

The dedication, which was closely related to the function of the epigraph in its attempt to legitimate the newly published text through its association with fame and power, was given unusual prominence in many of Murray's travel books. In eighteenth-century Europe, it was not unusual for the dedicatee of a published work to pay to have his or her name in a prominent position at the beginning of the text. Noting the change that had taken place in the intervening years, Charles Dickens looked back in the 1850s to an age characterized by what he called "the shame of the purchased dedication," which he associated with "the dirty work of Grub Street [and] the dependent seat on sufferance at my Lord Duke's table." Traditionally, dedications to the great and the good had been acknowledgments of direct personal patronage as well as guarantors of the legitimacy of the book and its contents. By the nineteenth century, argued Dickens, the rise of a new reading public had "set literature free."[51] Yet even after this direct and economic function of the dedication had receded, "the indirect and symbolic social function of the dedication" remained significant.[52] Where the relationship between the modest author and the dedicatee was professional or official—which was often the case with Murray authors—then it could be an immediate means of consolidating the success of a publication within official circles, as well as of attaching to it associations of legitimate authority.

In considering the Murray firm's use of the form in this context, the dedication page served three important ulterior functions. The first was to create influence for the author and the text by association with influential individuals. The second was to cultivate and to demonstrate patronage, either real or symbolic. The third was to consolidate political, professional, and personal alliances with individuals of future benefit to the author, the publisher, and the text. In these combined senses, the Murrays' use of literary dedications fulfills Bruno Latour's observation that "a document becomes scientific when its claims stop being isolated and when the number of people engaged in publishing it are many and explicitly indicated in the text."[53] So, too, did the authorial acknowledgment, although this sometimes had a different rhetorical status. Usually inserted in the preface, or in a passing footnote, these are tributes to scholars and assistants who, it is claimed, helped shape the text or the mind of the writer. While the epigraph to John Martin's *An Account of the Natives of the Tonga Islands* (1817; app.: 39) paid tribute to Humboldt as a scientific authority, the book's dedication page was also calculated to establish the book's place in a sound British intellectual lineage: "To the Right Honourable Sir Joseph Banks, Bart. K. B. &c. &c."[54] As the ultimate symbolic sanction, this tribute endorsed the work not only under the imprimatur of Banks's scientific genius but also as a nationally recognized celebrity under the Order of the Bath, the highest honor conferred by the crown. James Wellsted's *Travels in Arabia* (1838; app.: 139) illustrates just such associations with authority as part of the book's social value. Among those identified as having a hand in the preparation of the book were John Arrowsmith, the mapmaker; the Reverend J. Reynolds, secretary to the Oriental Translation Committee, for translation of manuscripts; Captain R. Moresby for affording Wellsted access to charts of the Red Sea and the Gulf of Arabia; and Admiral Sir Charles Malcolm "to whose personal regard, as well as his enthusiastic zeal for the extension of geographical science, the Author was indebted." The work's bound-in map also bore a dedication, to the then recently retired founder and president of the Royal Geographical Society and influential patron of geographical expeditions: "To Sir John Barrow, Bt. &c &c through whose influence and exertions MODERN GEOGRAPHY is so pre-eminently indebted." [55]

In many respects, the social and political milieus of Murray's authors and the firm itself are one and the same. Because after about 1813 the firm was at the center of London's official and colonial networks, the publisher and its authors engaged in strategies of mutual legitimation. Murray's works of exploration found success in their appeal to common sociabilities, which cemented the status relationships on which the book, the author, and the social class of the readership depended. For Wellsted, the real coup was on the dedication page itself where he presented his work to Queen Victoria, only recently come to the throne. His book, he is telling us, has the royal imprimatur. Wellsted's use of acknowledgments and dedicatory statements is perhaps the most explicit example from Murray's list of 239 such works. But he was far from alone in so offering approbation by association through epilogue and dedication.

ILLUSTRATIONS AND MAPS:
ACCURACY AND COMPLETENESS

Very few of the Murray firm's explorer-authors traveled with map in hand. A map of the area in question or of the route of travel was for some, of course, among the principal objects of the exploration. Most maps appeared as the book was being readied for the press. Maps in that sense were part of the final printed product of exploration and authorship even if they were also a prompt to the initial undertaking. The explorer regarded the map as a warrant of authenticity and of accomplishment. The publisher saw it as an important supplemental element to the text of the narrative but, if costs were high or technical matters a problem, not always a vital and necessary feature. For these reasons, the assessment of publishers' archives with respect to the mapped representation of exploration and travel is particularly valuable. Where map historians tend to study the mimetic authority and content and meaning of maps, and book historians concentrate on text to the relative neglect of cartographic images as forms of text produced in particular relation to narrative, publishers could and did take a more strictly pragmatic view, mindful of costs, the market, and the value of books of travel that contained maps as opposed to those that did not.

Referring to the illustration revolution of the early nineteenth century, Michael Twyman has observed how the increased use of visual material from that period began to raise editorial and technical questions about the relationship between text and image. "At one extreme," observes Twyman, "pictures or plate books continued to be issued (often in large formats) with sets of pictures bound up in volume form, with or without explanatory text. . . . At the other extreme, books began to appear in which pictures and text were physically integrated . . . so that the picture and its related text fall together on the same page."[56] There are numerous examples throughout the first half of the nineteenth century of the Murray firm's use of illustrations at both of these extremes, and everywhere in between. A third way of presenting illustrations in a volume, and the one most favored by the house of Murray, was to tip in plates (occasionally as foldouts) or, where books were issued unbound, to provide plates along with the sheets and a page of instructions to the binder for the proper placing of the illustrative material corresponding with the printed text (fig. 16).

The adoption of new printing techniques from the 1820s onward made the production of illustrations cheaper and more versatile. The move from woodblock and copperplate to steel engraving and lithography provided material that was more durable, allowing for the continued use of the plates and maps from edition to edition and the easier integration of text and image on the same page, ensuring a greater correspondence between letter-

DIRECTIONS TO THE BINDER FOR INSERTING THE PLATES.

FIGURE 16 Planning the layout of the narrative. The publisher's guidance to the binder sought to place illustrations in relation to the text and, here, to position the charts at the end of the volumes rather than at the beginning (unlike King in 1827: see fig. 3). Source: William E. Parry, *Journal of a Second Voyage for the Discovery of a North-West Passage from the Atlantic to the Pacific* (London: John Murray, 1824). By permission of the Trustees of the National Library of Scotland.

press and illustration. Twyman has also observed how, as the nineteenth century progressed, there was an increased emphasis on the illustration of "representations of the visible or imagined world" that, in an age of science, increasingly required "the quality of appearing true or real."[57] Illustrations were often based on sketches taken in the field (by either the author or another) for later execution by artists and engravers. In some instances, the degree of artistic license involved as landscape was rendered more sublime, or as local inhabitants were presented so as to appeal to the spectacle of exoticism, could be considerable. As Bernard Smith has observed, the contribution of engravers at this stage in the process mediated "between perception and representation in the secondary acts of draughtsmanship."[58] For Robert David, "The requirement to market the final image, within the parameters of accepted canons of taste, was as apparent in visual representation as it was in written text." [59] The legacy of eighteenth-century theories of the picturesque was still strong in topographical imagery, in accordance with William Gilpin's assertion that images could be rendered more affective on the imagination when they were "properly disposed for the pencil."[60] Some authors were happy to comply with these constructed spectacles. Others objected on the grounds that they did not convey their own view in the field with suitable accuracy. In texts that were at pains to prove their verisimilitude, questions might well be raised about the impression that such reworked images might convey. The issue for the author and explorer was whether the work was done "on the spot," undertaken in the field and at first hand, if not necessarily by the author him- or herself. But truth from the field, so to say, might count for little if the engraver worked up the image under the publisher's directions and did so to embellish the account and address the market rather than let "unadorned" representation pass for reality.[61]

One copiously illustrated book issued under Murray III, for example, was Joseph Hooker's *Himalayan Journals* (1854; app.: 220), which included five lithographed landscapes and forty-five woodcut engravings. When the book was being readied for the press, Hooker wrote to Murray to complain that, of the plates that already been prepared, one was totally inadequate and so, as a result, "the whole scene seems thrown out of perspective," while another was "not well copied" from his original drawings.[62] In his preface, Hooker offered a detailed critique of contemporary illustration and its seeming truthfulness:

> The landscapes &c. have been prepared chiefly from my own drawings, and will, I hope, be found to be tolerably faithful representations of the scenes.

I have always endeavoured to overcome that tendency to exaggerate heights, and increase the angle of slopes, which is I believe the besetting sin, not of amateurs only, but of our most accomplished artists. As, however, I did not use instruments in projecting the outlines, I do not pretend to have wholly avoided this snare; nor, I regret to say, has the lithographer, in all cases, been content to abide by his copy. My drawings will be considered tame compared with most mountain landscapes, though the subjects comprise some of the grandest scenes in nature. Considering how conventional the treatment of such subjects is, and how unanimous artists seem to be as to the propriety of exaggerating those features which should predominate in the landscape, it may fairly be doubted whether the total effect of steepness and elevation, especially in a mountain view, can, on a small scale, be conveyed by a strict adherence to truth.[63]

What lay behind Hooker's anxiety was a sense that the reading public, saturated with sublime imagery, had become indifferent to faithful representation. That lithographers had become accustomed to exaggeration in the light of such aesthetic considerations presented an additional problem for the scientific artist whose chief objective was verisimilitude. There was also a strong sense that an impressive and colorful landscape could not adequately be rendered in black and white and in two dimensions. In order to resolve these problems, the principal desideratum for publisher and author alike was a combination of accuracy and aesthetic appeal. Noting how Hooker used his text and illustrations to construct a particular notion of "tropicality," David Arnold observes that Hooker's *Himalayan Journals* demonstrate "the extent to which a scientific writer in the early Victorian age was prepared to go, and the range of devices he was prepared to employ, in order to render his tropical travels and natural history narrative accessible to a wider public. Apart from maps and engravings, the printed word (and the multiplicity of images and allusions it could be used to conjure up) was consciously deployed to engage, entertain, and inform."[64] This is true of the first edition, but less so of the second edition: Murray III sold all fifteen hundred copies of the first run, but sales of the 1855 edition flagged, leaving Murray with a loss.[65]

David Livingstone's acclaimed *Missionary Travels and Researches in South Africa* (1857; app.: 235), which included forty-seven illustrations, including frontispiece and maps, was also the subject of illustrative manipulation. While it was in preparation, Livingstone expressed his objections to the pictorial treatment that had been given to some of the key episodes in his narrative (fig. 17). Livingstone was particularly agitated about the image of

FIGURE 17 The author amends the artist. In this proof of an engraving that appeared modified in his 1857 *Missionary Travels and Researches in South Africa*, David Livingstone instructs the engraver to "place more rings on the chiefs [*sic*] arm & legs" and to do likewise to other figures in order to get the picture correct as he recalled the scene. Source: National Library of Scotland, MS 42432). By permission of the Trustees of the National Library of Scotland.

his escape from the lion, complaining that it was "abominable" in its execution, and would lead readers familiar with the physiognomy of a lion to "die with laughing at it." [66] There is evidence throughout the marked-up proofs for the book of Livingstone's wrangling over the book's illustrations.[67]

The admissions of Hooker and Livingstone over the inaccuracy of their illustrations are unusually frank. By and large, it was incumbent on authors to maintain that the accuracy of their illustrations matched the veracity of their narratives. Archibald Edmonstone paid tribute in *A Journey to Two of the Oases of Upper Egypt* (1822; app.: 70) to the pencil of Robert Master, who had provided the illustrations, for which he could "most willingly vouch for their faithfulness and accuracy."[68] Dixon Denham's sketches, drawn on the spot during his travels through Africa in the 1820s, were worked up for publication by his friend Robert Ker Porter. Dixon confessed to having provided only "sketchy" drawings from his travels, but he was at pains to insist that the results were nevertheless "faithful," as were Porter's reworked illustrations. Directing his readers to the latter's *Travels in Georgia* (1821–22) as evidence, Dixon asserted that Porter's eye was "nearly as familiar as my own with the picturesque objects they display."[69] On occasion, illustrations had an impact beyond the text they were intended to supplement. Working on Christmas Day 1822, looking over the pictures that would accompany Franklin's 1823 *Narrative of a Journey* (app.: 74), John Barrow implored Murray II to run off the illustrations separately in addition to their being placed in the book: "I have seen Finden's Etchings for Franklin's book and they are superlatively beautiful, and being so, I think you would do well to desire that half a dozen of each plate, on India paper should be taken off . . . I should like to have a set myself."[70] A year later, Barrow was railing at the delay in publication of Parry's latest polar work—what eventually appeared as *Journal of a Second Voyage* (1824) the following year (app.: 84)—because the botanical illustrations being undertaken by Robert Brown were holding up the entire publication: "I pretend not to be competent to decide what length of time, what degree of attention and laborious research, what deep thought and superior intellect may be required to arrange and describe a little bunch of plants, but I cannot imagine that Capt. Parry, any more than myself, could have the least idea that two years and a half should pass away and nothing be produced."[71]

Perhaps naturally, Murray authors could be reticent about their maps' accuracy, given the circumstances of their in-the-field making. On their Niger journey, Richard and John Lander had, according to their editor, surveyed the accompanying map with only the aid of "a common compass," and even this they eventually lost. As Becher noted, the map "in its most favorable point of view, can be considered only as a sketch of the river authenticated by personal observation, which will serve to assist future travellers." [72] Unlike most of the pictorial illustrations, the house of Murray often derived its maps from already-existing publications, not least rendered-down images from atlases. The firm also bought copies of maps by quantity from reputable mapmakers, commonly from the Arrowsmiths, the London-based firm of cartographers, or from the engravers J. & C. Walker. On return from the field, some authors dealt directly with mapmakers, who would produce copies based on the latest geographical evidence. Such maps were produced in more than one format to operate in several commercial ways: to be sold individually, inserted in atlases, or sold to publishers for inclusion in travel accounts and reference works. Maps produced by the Arrowsmith firm for Murray travel texts were often reused in the former's own atlases: a prudent financial strategy that allowed the cost of a map to be spread between more than one publication. As Woodbine Parish observed in a letter over the map for his *Buenos Ayres* (1838; app.: 137), "Arrowsmith will prepare a new map which will eventually form one of his own Atlas & which he will therefore be able . . . to furnish at a more moderate expense than if intended solely for my Book."[73]

These processes disclose the close but complex relationships among mapmakers, engravers, and explorers in producing maps as part of the final product, but they do not reveal exactly how rough field sketches and maps were put to order in printed form. As the evidence of the map work of Richard Wilbraham reveals, the final engraved map, for which J. & C. Walker were paid the sum of £17 9s. for the print run of 750 copies, could lend an authority and a certainty to the narrative that was simply not apparent in the map's manuscript form (figs. 18 and 19).[74]

Some authors brokered the arrangements for the maps themselves. Francis Galton worked closely with the artist who was drawing up a map for inclusion in his *Narrative of an Explorer* (1853; app.: 214). In preparing his account of Ceylon in 1857 (app.: 232), George Barrow remarked to Murray III on the appeal of maps in geographical works: "I have heard so many people complain while reading a book, of there being no map, & I have felt this so strongly myself, that if you can possibly manage it, I think it would be most desireable to make some arrangement with Mr Arrowsmith." When he was writing to Murray, Barrow knew that Arrowsmith was often used by Murrays for its map work, had already completed a large map of Ceylon, and had plans for its reduction and inclusion in a small quarto atlas. As he understood it, the latter was not expected to appear for some time, "which are advantages to its publication in my little work." For his part, Barrow asked the publisher to acquire the map for insertion and thereafter to insert on the title page the phrase "With a Map."[75] Murray went one step further and included under the author's name "With a Map by John Arrowsmith," in clear reference to that cartographer on a title page. As Wellsted had cited John Barrow, so in his preface George Barrow was careful to inform his readers that he wished "to call special attention to the Map accompanying this work . . . the most complete and authentic map of the island which has yet appeared."[76]

For Alexander Burnes, the significance of his Indus and Afghan work was apparent in his text and in his public acclaim, not in the authenticity of his map material. On his return to Britain, Burnes had set about preparing a map for publication to illustrate his 1834 *Travels into Bokhara* showing

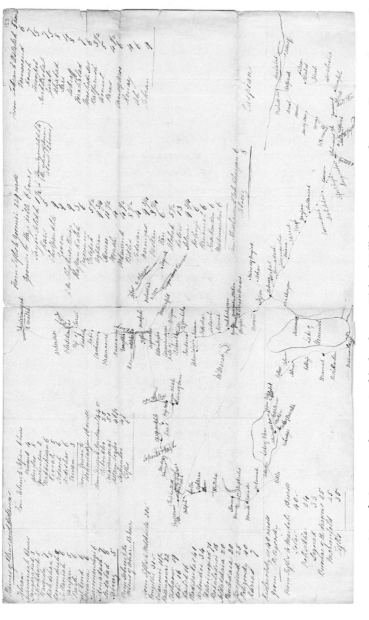

FIGURE 18 A traveler's manuscript map. In this sketch map, Richard Wilbraham appears to focus on distances (measured in hours of travel) rather than on topographical accuracy. Compare this "in-the-field" work with the engraved map that accompanied his published work (fig. 19). Note, too, Wilbraham's sketch depiction of "Mt Ararat" (*left of center here*). Source: National Library of Scotland, MS 42173, fols. 152v–153r.

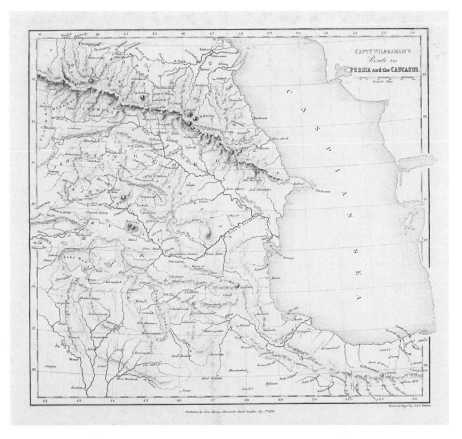

FIGURE 19   The presumed authority of maps. The engraved copy of Richard Wilbra-ham's route of his travels, by the leading London firm of J. and C. Walker, lends an au-thority to the published work that is not apparent from the author's own sketches—see figure 18. Source: *Travels in the Trans-Caucasian Provinces of Russia and along the Southern Shores of the Lakes of Van and Urumiah in the Autumn and Winter of 1837* (Lon-don: John Murray, 1839). By permission of the Trustees of the National Library of Scotland.

his travels and the course of the Indus River. But this map was much de-layed, possibly by Burnes's own attention to detail in transferring requisite information from field notebooks, possibly by John Arrowsmith to whom Murray II had contracted the task of engraving and producing the final map (fig. 20). As Burnes's text approached completion, it was passed, like so many others, to John Barrow for his scrutiny and judgment.

FIGURE 20 The incomplete map of travels. The illustration shows the final preparation of the map depicting Alexander Burnes's extensive travels along the Indus and in Sinde province. But the map was not ready in time to appear in his 1834 *Travels into Bokhara* and so was used elsewhere. Source: John Arrowsmith, *The London Atlas of Universal Geography* (London: Arrowsmith, 1834). By permission of the Royal Geographical Society (with the IBG).

Writing to Murray II, Barrow expressed his concern that there was no map: "And a book of Travels without a Map is very unsatisfactory and unintelligible." What had been sent, Barrow further observed, was Arrowsmith's map of Persia, "scratched and unfinished." "Is it a mistake that there is no map in the Book?" queried Barrow: "I hope it is" (fig. 21). Mistake or accident of timing, Burnes's 1834 book went to press without this map, the finished version being employed the same year by Arrowsmith in his *London Atlas of Universal Geography* (1834).[77]

Maps, particularly those that appeared as frontispieces, were meant to be read alongside the text as aids to interpretation. So, too, were the carefully placed images illustrative of episodes and scenes described within the accompanying narrative. The use, however, of other textual materials incorporating hydrographic readings, climatic tables, and other paratextual figures, for example, had a tendency to shift the register and interrupt the flow of realistic description if placed within the main text. Because many of Murray's volumes were calculated to appeal to scientific specialists as well as to general readers, the provision of technical data, as well as their placement in the publication, had to be considered carefully. One primary difficulty was the tendency of scientific data to interrupt the general narrative, which sought to present the journey as a matter-of-fact and truthful passage through various landscapes. That is why, in his *History of Java* (1817; app.: 41), Thomas Stamford Raffles informed his readers that the accompanying and very large foldout map was meant to "aid the reader in most of the geographical objects to which this volume will refer."[78] Raffles's other accompanying evidence, however, was consigned to a series of thirteen appendixes that together provided a summary of vocabulary, translations, government proclamations, weights and measures, and so on (fig. 22). In giving shape to the final product, the evidence was pushed to the appendixes, others' claims and discussions given in the footnotes, and the author's main narrative offered the interpretation.

In preparing his *Account of a Voyage of Discovery to the West Coast of Corea* (1818; app.: 44), Basil Hall similarly took care to position the supporting data for his findings, advertising as much on a title page that promised an "Appendix, Containing Charts, and Various Hydrographical and Scientific Notices" together with "A Vocabulary of the Loo-Choo Language." This tripartite arrangement of "Account," "Appendix," and "Vocabulary" was justified by its author in his prefatory claim that "All the Charts, Tables, and Nautical Notices have been placed in an Appendix, in order to avoid the interruption which such details are apt to occasion when inserted in a journal." [79] Scientific readers would, he felt, find it advantageous to be

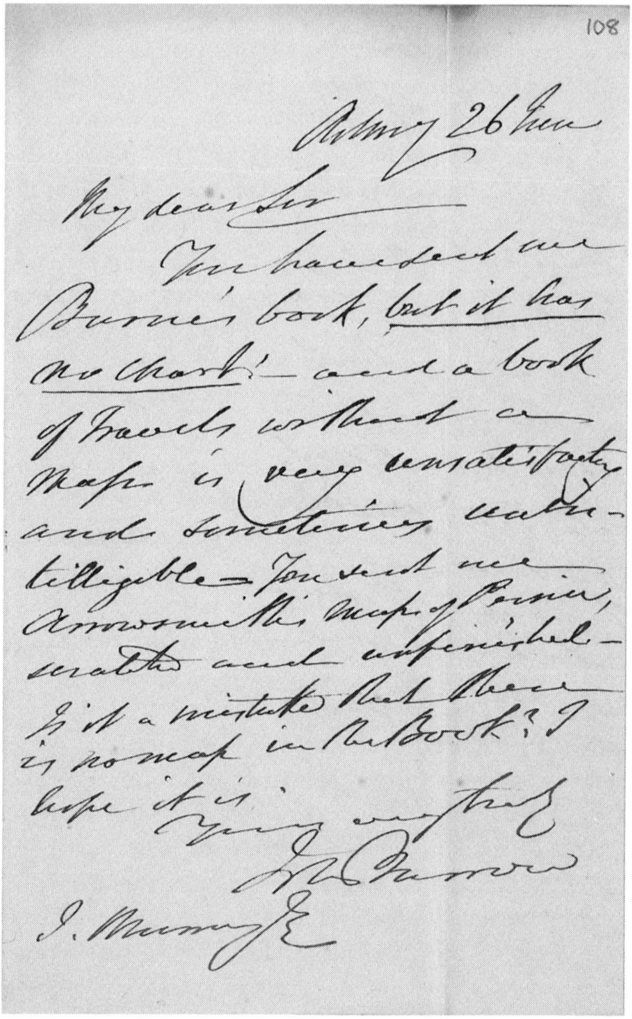

FIGURE 21 The incomplete book of travels. John Barrow here expresses his dismay at the lack of a map in Burnes's *Travels into Bokhara* (1834): see figure 20. Barrow laments: "You have sent me Burnes's book, *but it has no chart!*—and a book of Travels without a map is very unsatisfactory and sometimes unintelligible." Source: National Library of Scotland, MS 40057, fol. 108. By permission of the Trustees of the National Library of Scotland.

Hour.	Barom.	Thermom. Air.	Thermom. Sea.	Winds.	Lat.	Long.	
1				S by W			*Friday, July 26*, 1816.
2							
3				South			As the day broke, the breeze
4	29ⁱ.61ʰ	74°	72°	SSE			which had been light during the
5							night, freshened up, and the
6							weather, hitherto clear, became
7				South			suddenly quite foggy; this how-
8	29.61	74	72	SSW			ever lasted only half an hour,
9							and we enjoyed during the day
10	29.62						the same fine clear weather, with
11	29.62						the exception indeed of one
Noon	29.62	74	72	S by W	N 38°.07′	E 122°.0′	thunder squall, which lasted
1				SSW			only a few minutes, and passed
2	29.62			SW			over, going towards the SE.
3							N. B. This was the only in-
4	29.59	76	66	WNW	Yellow Sea.		stance of fog during the six
5							weeks that the ships were in the
6	29.60			SE by S			Yellow Sea.
7				SE			
8	29.60	73½	68				
9							
10							
11							
Mid.	29.60	75	74	SE by E			
1				South			*Saturday, July 27*, 1816.
2							
3							During the whole of this day
4				SSE			we had a fresh breeze from East
5							and ESE, with dark cloudy
6							weather. As we drew across
7							the Gulf of Petchelee we had
8	29.69	77	77½	SE			the wind much stronger.
9							
10	29.69			East			
11	29.70				N 38°.52′	E 117°.49′	
Noon.	29.70	76	76	ENE			
1							
2	29.68			E by S			
3					Yellow Sea.		We anchored at seven o'clock
4	29.61	76	77	E by N			in 3½ fathoms water.
5							
6	29.61			East			
7							
8	29.80	77	82	E by N			
9							In the night it blew hard
10	29.84			SE by E			from the East, and at sunrise
11							we had a violent thunder storm.
Mid.							

FIGURE 22 Presenting scientific evidence. Positioning the raw data and any associated explanatory remarks to the rear of volumes allowed the house of Murray to present the author's narrative unencumbered by its supporting evidence. Source: Basil Hall, *Account of a Voyage of Discovery to the West Coast of Corea, and the Great Loo-Choo Island* (London: John Murray, 1818). By permission of the Trustees of the National Library of Scotland.

able to consult such data without having to hunt through the narrative for it. Hall and Raffles, and many of their counterparts who produced works of polar and African exploration, were not only working to produce definitive narrative accounts structured in certain ways. They were also centrally placed in the emergence of that particular authorial category, the "scientific author," whose status at this time in producing books of exploratory and geographical science depended in part on separating out the narrative from its underlying evidence: the emergence of science, authorial credit, and intellectual property and the formatting of works of exploration were intimately connected.[80]

Gathered together in such ways under one title, travel works could thus be formally and structurally complex. The arrangement of paratextual elements, along with various framing devices (summary title page, extensive chapter contents, running headers, and so on) were what ultimately gave the text its intellectual depth, making it negotiable for, and navigable by, the reader. These elements were not incidental to the making of the book but were part and parcel of its meaning, as well as of its claim to authority and desirability as a product for ownership and readerly consumption. As the example of Austen Henry Layard's Near East texts shows, the use of paratextual material was an important element in producing travel narratives aimed at different sectors of the market.

ILLUMINATING THE PAST: AUSTEN HENRY LAYARD, BABYLON, AND THE NEAR EAST

The house of Murray's developing use of paratextual devices over time can be observed through its changing visualization of the Middle East, from its first forays into publications relating to the Holy Land to the production of the early titles of Austen Henry Layard, beginning with the phenomenally successful *Nineveh and Its Remains* in 1849, and its popular spin-offs. Unlike his successor, Murray II was not especially alert to the financial advantages of combining scholarship with spectacle: none of the titles was particularly well illustrated, the only such paratextual devices being frontispiece maps and a handful of small, in-text diagrams. One exception was Thomas Maurice's richly illustrated volume, *Observations on the Ruins of Babylon* (1816; app.: 36), although the firm neither arranged nor financed the book's illustrative content.[81] In 1819, Murray II published John Lewis Burckhardt's *Travels in Nubia*, an early and celebrated account of European explorations in Egypt (app.: 49). The following year, the firm published the first of several editions of Giovanni Belzoni's *Narrative of the Operations and Recent Discoveries* (1820; app.: 59) to coincide with the arrival of the colossal bust of Ramses II at the British Museum. The Murray firm's growing commitment to more sophisticated paratextual embellishments as part of its travel and exploration list was more evident under the guidance of John Murray III, with the translation of Laborde's *Journey's through Arabia Petræa* (1836; app.: 133) marking, as we have seen, the beginning of a new phase of publication.

Austen Henry Layard was by far the most celebrated and high earning of Murray's Middle Eastern authors. He had undertaken numerous expeditions in the 1840s, resulting in the acquisition, by the British Museum, of an impressive—not to say sensational—collection of archaeological remains. The unusually detailed archival evidence relating to the publication of Layard—in the form of surviving author-publisher correspondence, production ledgers, original manuscripts, and corrected proofs—sheds light on how one of the house of Murray's most celebrated travel writers was brought to market.[82] Initially, Layard had hoped to secure £4,000 from the British Museum to subsidize the cost of producing plates of his drawings. When these funds were not forthcoming, Murray III intervened and agreed to fund the publication of the images in a reduced form, on condition that the author provided an accompanying account of his discoveries in the form of the book that became *Nineveh and Its Remains* (1849). This was the first of several publishing ventures relating to Layard and his expeditions that the firm oversaw. Murray III presented Layard and his findings in various formats and modes of production calculated to appeal to different markets. In this regard, there are distinct parallels for John Murray III's work on the Middle and Near East with his father's earlier work in producing Arctic narratives in different formats for different sectors of the market.

Murray III's involvement with Layard was prompted by the publisher's realization that Layard's recent discoveries would have considerable commercial appeal; the publisher's initial investment was an indication of his commitment to the venture. But Murray III had to overcome his own initial misgivings, evident mainly in concerns about the cost of illustrations. At first, Murray had an altogether different sort of publication in mind: "I wish it were possible to induce you," he wrote to Layard, "to confine your thoughts at present to a work like that of Wilkinson [*Modern Egypt and Thebes* (1843; app.: 165)]—copiously illustrated with woodcuts—by the best artists from the bas relief—leaving the larger work to the British Museum at some future time."[83] From the title page to the illustrations, maps, and other figures, unusual attention was given to the book's paratextual features. Murray employed the artist George Scharf to make Layard's sketches ready for the market.[84] The visual impact of *Nineveh and Its Remains*, which sold in two octavo volumes for 36s., set it apart from most of the titles on Murray's list at the time, the sheer number of illustrations affording the work a decidedly artifactual quality. Shawn Malley has observed how not only were the illustrations "fundamental to the commercial success of the book" but they themselves also enacted the acquisitive nature of Layard's archaeological project.[85] The tipped-in frontispiece to the first volume, *Lowering the Great Winged Bull*, was one of several images that explicitly connected the book to the iconic artifacts that had made their way to the British Museum. Through its use of such images, *Nineveh and Its Remains* visually portrayed and symbolized a narrative of exploration, discovery, unearthing, and possession, with an array of subjects, characterized by titles such as the *Discovery of the Gigantic Head* to the *Procession of the Bull beneath the Mound of Nimroud*, the latter of which served as frontispiece to the second volume.

The title itself was a matter of considerable concern. Murray III had clear views as to how he wanted Layard to be presented. Both author and publisher were aware of the rambling nature of the narrative in manuscript, with its many diversions into anthropology and religious history. While he was inclined toward accepting Murray's suggested short title, Layard's preface suggests that the author was also anxious to preempt charges of false advertising: "With regard to my personal narrative, I may owe an apology to the reader for introducing subjects not included in the title of my work, for adding narratives of my visits to Tiyari and Yezidis, and a dissertation upon the Chaldæans of Kurdistan."[86] The long title on which author and publisher finally settled—*Nineveh and Its Remains, with an Account of a Visit to the Chaldæan Christians of Kurdistan, and the Yezidis, or Devil-Worshippers; and an Enquiry into the Manners and Arts of the Ancient Assyrians*—may have been cumbersome, but Murray got his short title in sensational short form and Layard's honor was served (app.: 196).[87] The illustrated title page advertised one of the book's most important selling points, namely, its appeal to a religious readership. With one well-chosen epigraph on the title page featuring an extract from Ezekiel 23:14-15—"She saw men pourtrayed [*sic*] upon the wall, the images of the Chaldees pourtrayed with vermillion, / Girded with girdles upon their loins, exceeding in dyed attire upon their heads, all of them princes to look to, after the manner of the Babylonians of Chaldes, the land of their nativity"—the book advertised its visual appeal, its on-the-spot authoritativeness, and its claim to provide conclusive evidence of biblical inerrancy.

The formula worked. In early 1849, Murray wrote to Layard, who was once more overseas, telling him of its success: by January 5, the publisher had disposed of a thousand copies, reporting later that month that the book had been received with "universal and unvarying approbation." By April, a second edition had almost sold out, and on May 3 the publisher informed the author that a further, third edition was also exhausted and proposed a new edition, offering Layard two-thirds of the profits, instead of the originally agreed on (and usual) half-profit arrangement.[88] The popularity and profitability of *Nineveh and Its Remains* (1849) is evidenced by its print runs and multiple editions. By the middle of 1849, 7,250 copies of four editions of the book had been printed, earning a combined profit for Layard and Murray of £3,600 13s. 10d. The success of this volume was capitalized on in 1851 through the publication, at 5s., of *A Popular Account of Discoveries at Nineveh* (app.: 196). By the end of the 1850s, 18,500 copies of that book had been printed, generating a combined profit of £1,026. Layard's popular account contrasted, in price and print run, with his more lavish folio counterpart, *The Monuments of Nineveh* (1849; app.: 197). Four hundred copies of the folio were printed by the end of 1852 — 250 on imperial-sized paper, and 150 on colombier-sized paper — generating a combined profit of £883 15s. 4d.[89] Layard's final volume, *Discoveries in the Ruins of Nineveh and Babylon* (1853) (app.: 216), enjoyed similar success: 13,500 copies were printed within a year, generating a combined profit (by the close of 1853) of £2,561 6s. 11d.[90]

Murray had initially written to Layard to decline the publication of a separate folio set of illustrations of *Nineveh*, reporting that the estimated cost of eleven guineas for each engraving was prohibitive. After the success of *Nineveh and Its Remains*, however, the publisher was persuaded to revisit the question of a set of deluxe plates illustrative of the *Remains*. In June 1848, Murray produced a prospectus for this work, promising a volume of one hundred plates, from drawings made on the spot. The cost to subscribers was eight guineas, with a price for members of the public of ten guineas, proofs to sell at ten guineas and 12, approximately £800 today. Subscribers included the Queen, Prince Albert, the king of Prussia, Earl Spencer, and other influential figures, as well as institutions such as the British Museum and the Royal Library of Berlin. This volume, *The Monuments of Nineveh*, and a second series published in 1853, was to become Murray's most distinguished production and one of the firm's most expensive (plate 15). The total cost for an initial three hundred copies of *The Monuments*, including half shares for the author and publisher of £198 14s. 2d. each, was a little over £2,435.

In an attempt to capture a different market, Murray issued another version of Layard's account in 1851—under the title *A Popular Account of Dis-*

coveries at Nineveh—as part of his Reading for the Rail series, priced at 5s. in a single post octavo volume and abridged by the author himself. While the title page promised "Numerous Woodcuts," this version was, in reality, pared down considerably. In the preface, Layard claimed that he had been induced to prepare it for publication "in a cheap and popular form," explaining that he had taken out much of the original second volume in order to concentrate on those biblical and historical aspects that he hoped might "be rendered more useful and complete" to the general reader.[91] This, too, was a formula that seemed to work well for the publisher. Reviews welcomed the availability of such a serious work to readers of limited means. "A five shilling volume will introduce them to rarer specimens of the ancient world," opined the *Eclectic Review*, "and teach them more of its artistic, social, and religious forms than could previously be gathered from many bulky tomes. We need not say that, 'the getting up' of the volume is creditable to the parties concerned. The name of the publisher is a sufficient guarantee of this."[92] Layard's returns on his *Popular Account* were healthy, if not spectacular. In July 1853, he wrote to a friend that his last work "has had great success," reporting that "nearly twelve thousand copies have already been sold, and three thousand more will be shortly printed."[93]

As he promoted his cheap abridgment, Murray was still advertising the two-volume *Nineveh and Its Remains*, now in its fifth edition and priced at 36s. But when he came to publish Layard's account of his second expedition as *Discoveries in the Ruins of Nineveh and Babylon* in the same year, it was announced that, "in consequence of the great interest felt in Mr. Layard's Discoveries, and the large demand for this work, not only in this country, but also in the United States and on the Continent, Mr. Murray has been induced to publish it at once, in this cheap form, COMPLETE, UNABRIDGED, & FULLY EMBELLISHED, in the hope of bringing it within the means of all classes of readers."[94] At the price of one guinea (21s.) for a single octavo volume, this was considerably cheaper than the first edition of *Nineveh and Its Remains*, yet the work still boasted three hundred maps, plates, and woodcuts. Murray's claim that its cheapness would appeal to "all classes of readers" may not have been quite as convincing as it had been for his 5s. Layard for the rail, but it does suggest the extent to which an old-fashioned publisher was, by then, experimenting with different formats and audiences.

In 1867 the firm issued abridged versions of both *Nineveh and Its Remains* and *Discoveries in the Ruins*. The preface of the first of these indicated that it had been "carefully revised by the Author" in order to take account of "the progress made in the interpretation of the cuneiform inscriptions," which had "enabled him to add to the text, and . . . led him to modify some of the

views which were expressed in his original work."[95] A comparison of the
1853 account of Layard's second expedition *Discoveries in the Ruins* with
its more popular 1867 version ("Abridged by the Author from his longer
work") indicates something of the extent to which, in the case of Layard at
least, Murray had come routinely to use redaction (the text block, too, was
much reduced) as a way of keeping costs down, while, at the same time,
introducing additional matter in order to give the cheaper version popular
contemporary appeal. Many of the 1853 chapters were conflated in the new
version; the sixteen plates were removed, with the exception of the map of
Assyria that now served as a frontispiece; the original 245 woodcuts were
reduced to 146. In order to give the book an up-to-the-minute feel, Murray
commissioned a new preface, referring explicitly there to the "further researches" that had taken place in the interim. This 1867 preface also made
it clear that this new version was intended to appeal to a less specialized
scholarly audience than its predecessor. The final chapter concerning the
"results of the discoveries" was heavily edited to take account of the additional work undertaken by others since the original account had appeared.
With this careful combination of abridgment and supplementation, Murray
was able to recycle Layard without giving the impression that his popular
readers were being shortchanged. They may not have had access to much of
the information that had been included in 1853, but they could rest assured
that they had the most recent and accurate information at their fingertips.

As we observe in our next chapter, it was customary for publishers to
"tranche down" their literary properties in such ways, initially selling in
large formats at relatively high prices and, over time, reissuing the same
properties in gradually smaller formats at increasingly lower prices. While
this was to an extent true in the case of Murray's treatment of Layard,
whose individual titles were repackaged in cheaper versions, the overall
picture is far more complicated than a simple tranche-down model might
suggest. At the height of his fame, Layard was being packaged by Murray in different formats and prices simultaneously, with the *Monuments* at
one end of the market and the Reading for the Rail volume at the other, in
order to appeal to a number of parallel audiences while "Nineveh mania"
was still in full flow—the Murray firm doing its best, of course, to maintain
the public's interest. Murray III's adept paratextual manipulations ensured
financial rewards for both himself and his author, lasting fame for Layard,
and distinction for himself as publisher.

Genette concludes his study of the paratext with the assertion that the principal function of the paratext is to ensure for the text "a destiny consistent

with the author's purpose."[96] Seeking to reinvest the writer with agency in the wake of structuralist theories of the death of the author, Genette overlooks the pressures that could be brought to bear by the publishing house on its literary property. Helpful as his analysis is, Genette's exclusive emphasis on printed evidence, along with his insistence that the authorial will was decisive in its paratextual presentation, demonstrates the necessity of considering the publishing context and, where necessary, the issues of the paratext on a case-by-case, book-by-book basis. A fuller story of the paratext is to be found in the correspondence between authors and publishing houses with long-standing lists and generations of trade practices behind them. The appeal of evidence was an important aspect of the truth claims of the Murray firm's travelers and explorers. To be able to show and tell, to give visual proof through illustrations, maps, charts, and graphs of claims elaborated on in prose, gave the text scholarly gravitas insofar as it offered the reader an intellectually tangible—if not always a strictly scientific— context for the narrative itself. While the house of Murray endeavored to respect its authors' wishes throughout this process, the final decision on the presentation of their literary property was more often than not the publisher's prerogative.

Over time, the paratexts that the Murray firm deployed in its travel list became the tried and tested means by which its texts were presented to the world. Murray's paratextual maneuvers were based on old-fashioned trade practices combined with a need to reinforce the authority and credibility of its writers in association with a keen eye for commercial success. The use of paratextual material was central to the significance and effect of those texts, framing them in meaningful ways and serving as mediators between authorial property and the consumers at which it was aimed. Most of all, these materials were crucial elements in the creation of authority, giving ultimate credibility to the works thus ushered into the world. This was especially so of maps and other topographic illustrations. But because the publisher had a powerful role in amending, placing, and omitting such artwork, we should suppose neither that the final inclusion of such visual material was straightforward nor that the images themselves straightforwardly depicted the worlds documented in the narrative. So often overlooked by geographers, historians, and literary scholars, paratextual materials were not mere elements of textual apparatus surrounding the discourses of truth and authenticity. They were, rather, integral to the truth claims of the works themselves and a vital means of distinguishing appropriate texts, amended where necessary, for appropriate markets.

CHAPTER SIX

Travel Writing in the Marketplace

We have shown in the preceding chapters how the house of Murray established authorial credibility for its writer-explorers and travelers and gave thought to the presentation of the paratextual elements of their books. This chapter examines something of the editorial and technical processes by which Murray's travel works were transformed from manuscript to print. The transformations in the technologies of print and its distribution and in copyright law that took place in the British and international book trades in the first quarter of the nineteenth century had considerable impact on the way in which publishers like the house of Murray negotiated with their authors, fellow tradesmen, and customers and on the way in which they brought their books to market. The advances made in printing technology in the early decades of the nineteenth century that affected the journey from manuscript to print brought financial advantages to the publisher, driving down the cost per unit of production and giving authors access to expanding markets. Allan C. Dooley has shown, however, that these same advances could come at a cost to authors, particularly to the control they felt over their texts after they entered what was more and more an industrial process, finding their works "partially controlled by printers, who had to operate within the limitations of their technologies, who strongly preferred to uphold established linguistic practices, and who attempted to make their work easier and more profitable by bending the author to their own needs."[1]

For the early nineteenth-century author and his or her readers, still operating imaginatively in the shadow of a belief that framed the author as the seat of expressive meaning, books were gifts to the world written under personal inspiration. This was a romantic myth that Karl Marx sought to explode when he claimed that the natural genius that had characterized Greek art and Shakespearean drama was no longer possible in an age of "self-acting mule spindles and railways and locomotives and electrical telegraphs." In the modern age, literature was no longer immune from the alienating effects of advanced industrialization: "What chance has Vulcan against Roberts and Co., Jupiter against the lightning-rod and Hermes against the Credit Mobilier? All mythology overcomes and dominates and shapes the forces of nature in the imagination and by the imagination; it therefore vanishes with the advent of real mastery over them. What becomes of Fama alongside Printing House Square? . . . Or the *Iliad* with the printing press, not to mention the printing machine?"[2]

For Marx, literature in an industrial economy had become a commodity for sale like any other, subject to the demands of the market and the economies of scale that turned a creative act into a material commodity. In such an economic context, authors were no longer solitary makers of meaning but could be regarded as little more than wage laborers. While Marx might have regarded the condition of the author as one that denied him or her access to the real means of production, others articulated the anxiety of authorship in terms that emphasized the lack of agency on the part of writers. It is against this context that we may understand the discursive strategies deployed by Murray and the firm's authors as they attempted to fashion themselves for readers who did not want to be reminded that their encounters with literary texts were part of an industrial process and mere commodity exchanges.

This chapter examines three themes that speak to these issues. We consider the importance of the house of Murray's literary advisers in regulating and shaping works of travel and the different strategies they employed to present the author and his or her work as a literary and marketable commodity. We discuss the processes of composition and stereotyping involved in the production of Murray's books, noting how, even at the proof stage, authors had considerable opportunity to rewrite elements of their narratives and so to present or re-present themselves in a particular light. Lastly, we explore the marketing strategies employed by the house of Murray from the later 1820s as the firm published works of travel in different formats, designed to meet the needs and interests of specific audiences.

In demonstrating in these ways the stylistic and physical construction of Murray's books of travel, we offer further insight into the collaborative nature of bookmaking and into credibility as a matter to do with the publisher's reputation, its standards in textual production, and the evaluation of audience demand. We can also see how conceptual models of book production may be amended by exposure to the complicated processes of mediation and production through which books were actually made and so positioned for the market. As scholars such as Leslie Howsam have shown for the British and Foreign Bible Society in the nineteenth century, and as Aileen Fyfe has demonstrated for the firm of W. & R. Chambers, changes in the technologies of production and the concomitant rise of cheaper forms of print were important, but they depended for their market reach on coordinating networks and processes of print production involving typefounders, compositors, paper manufacturers, and distributors.[3] As Robert Darnton's "communications circuit" reminds us, texts were not part of a simple gift economy from author to reader, nor were they authorized exclusively by the writer. As Darnton observes, the reader is not a simple end point in the production cycle of books but "influences the author both before and after the act of composition."[4] Here, the reader we have in view was in-house, employed by Murray to help shape the books of travel in various ways. This constitutive relationship between reader and author is conspicuously absent from Darnton's model, which is based on an earlier period of book production. The literary adviser, or publisher's reader, was a position that became increasingly important and more formalized in the nineteenth century as book capitalists came to employ trusted others to assess the commercial possibilities of manuscripts for the market, to determine whether there was demand for their content, and to judge whether they were to be relied on for their veracity.

A TRUSTED EYE: MURRAY'S IN-HOUSE READERS

From the moment a manuscript arrived at the house of Murray, it was subject to a regime of regulatory practices—passing through filters of technology, taste, and credit that would turn it into a Murray book. Like other quality publishers, Murray deployed a network of informed advisers to comment on a manuscript's suitability for publication and on the reliability of its content. Writing to Robert Fortune over the manuscript of his *Three Years' Wanderings* (1847; app.: 185), for example, Murray III warned the author that his text "greatly needs the revision of some literary Friend" who could supply "verbal correction before it goes to press."[5] The "literary friend" mentioned here was the standard euphemism for an anonymous and trusted adviser, the critical in-house reader, whose responsibility it was to work up or comment on a text before going into print. We have seen above how the Murray firm depended on a coterie of trusted literary stylists — the most prominent of whom included John Barrow (albeit that his prose amendments did not always find favor with the public or with the authors concerned), John Wilson Croker, William Gifford, Maria Graham, and Walter Hamilton.

In 1841 the house of Murray put the informal process of editorial redaction and commentary on a professional footing by hiring Henry Milton and, later, his son John as principal readers. Thereafter, other members of Milton's family were used for work of this kind, collectively commenting on "around fifteen hundred manuscripts and proofs" over a forty-year period.[6] As Angus Fraser has observed in his study of the Miltons and the Murrays, the details surrounding readers' activities are often patchy, their contributions treated by the publisher with discretion.[7] Nevertheless, some details survive, and there is an indication in the firm's copy day books of the rates paid for, as well as the number of hours worked on, specific titles. In 1856, for example, John Milton spent some 295 hours on William Napier's biography of his brother, Charles: *The Life and Opinions of General Sir Charles James Napier* (1857).[8] Henry Milton worked for twenty-four hours altogether in assessing for publication the manuscript of Herman Melville's *Narrative of a Four Months' Residence* (1846; app.: 180). When that work came to be edited, however, Milton was required to spend a further 162 hours on that task, receiving in payment £50 11s., a sum that "compared favourably . . . with what Melville was to receive for actually having written the book."[9] In this respect, the task of making a manuscript fit for publication could entail considerable time, effort, and financial expense, yet was an investment the Murray firm felt essential to the appropriate positioning of its literary output.

In response to this stage in the production process, some writers were more anxious than others to retain control of their work, and some were in a more powerful position to do so. George FitzClarence, for example, had to negotiate his reputation more vigilantly than most. The eldest, though illegitimate, son of the future king, William IV, FitzClarence had distinguished himself in the Peninsular War but had been stripped of his sword and dismissed from his regiment in 1813 for conspiracy against his commanding officer.[10] Sent to India in disgrace, FitzClarence worked hard to reestablish his reputation and spent the next few years reinventing himself as an oriental scholar, beginning with the publication of his *Journal of a Route across India* (app.: 51) in 1819 (fig. 23).

FIGURE 23 Enlightening discovery. Here the European traveler—perhaps George FitzClarence himself—encounters the enormity of the chambers beneath the Great Pyramid in Egypt. Source: George FitzClarence, *Journal of a Route across India, through Egypt, to England, in the Latter End of the Year 1817, and the Beginning of 1818* (London: John Murray, 1819). By permission of the Trustees of the National Library of Scotland.

Ever mindful of others' perception of his social standing, and bruised by his earlier experiences, FitzClarence was careful to assert the credibility of his Indian and Egyptian achievements. Anxious to win favor, FitzClarence took care to dedicate the book to the future king as "a humble token" of his "gratitude and attachment."[11] Something of the delicacy of his position is evident in a letter written to Murray II late in 1818 during the book's production: "Whatever alterations you may desire to make should be therefore, commenced directly as it must be understood, whatever they may be, they must all depend on my final instruction & approval."[12] Judging by the published preface to his account, in which he acknowledged that the work had "incurred the imputation of prolixity," it is likely that the publisher's requested redactions were not altogether implemented. In his defense, FitzClarence maintained that the contents were entirely true, had not been worked up, and adhered entirely to "fidelity."[13] Here, too, the author's word ought not to be regarded as the final one in this respect: Murray's ledger books show that a "Mr. Jordan," presumably with the knowledge of John Murray, had received payment of £75 for unspecified services in the book's editing and preparation prior to production.[14]

One of the strongest negative responses to editorial intervention came from Walter Hamilton, one of Murray's most trusted literary advisers, on the grounds that Murray's reader had an inadequate understanding of the facts that had informed his *East India Gazetteer* (1815; app.: 32). What was most distressing for Hamilton was the fact that what he saw as errors introduced by Murray's reader were not caught in time before publication. Confident of his own knowledge of the region, over and above that of his anonymous critic's, Hamilton objected to his publisher that

> the Gentleman who has made the corrections, is sometimes misled by errors in the maps which he has consulted, for instance, instead of "Ranjeshy" to read "Bettooriah"—Now there is no such district as Bettooriah—There is a large zemindarry or landed estate of that name (where I have been) comprehended in the district of Ranjeshy, one of the permanent subdivisions of Bengal—A great many of my facts, as you know, are derived from original manuscripts, which no pay constructor ever saw, & from personal observation on the spot. Much of what the annotator considers erroneous, is not so, & I could point out errors of magnitude that have escaped his research.[15]

Asserting his authority on the basis of personal experience and information acquired "on the spot," Hamilton sought to persuade Murray II that the emendations of the latter's expert were profoundly flawed. Hamilton was thus compelled to assemble a "very complete list of errata" that was used later as the basis for the revised version of the gazetteer, published five years later as *A Geographical, Statistical, and Historical Description of Hindostan* (1820).[16] Bettoriah was thus successfully transformed from being "a district in the Province of Bengal"—as it was described in the 1815 version of Hamilton's book—to being "a subdivision of the zemindarry of Rajeshahy, in the province of Bengal," in its 1820 rendering.[17]

On certain occasions, the stylistic changes to which a manuscript had been subjected by an (over)attentive and perhaps ignorant in-house reader or editor were objected to on the grounds that the overall tone of the authorial voice was lost. While his *Missionary Travels* (1857) was being prepared for the press (app.: 235), David Livingstone had assured Murray III that he would not be "cantankerous or difficult to deal with."[18] Once the process was under way, however, Livingstone's attitude toward his anonymous editor's interventions became decidedly frosty, as we have seen over his illustrative material (see figs. 11 and 17): he accused his editor—whom he suspected to be the *Quarterly Review*'s editor, Whitwell Elwin, though in fact it was John Milton—of the "emasculation" of the manuscript:

The liberties he has taken are most unwarrantable and I cannot really undertake to father them. I am willing to submit my style uncouth though it be to Mr Elwin or any friend you like and if it is not clearer more forcible and more popular than anything this man can give I shall then confess I am wrong. My letters written to the Geographical & Missionary societies are popular. Why must you pay for diluting what I say with namby pambyism. Excuse me, but you must give this man leave to quit. I really cannot afford to appear as he would make me.[19]

In this, and other instances, the house of Murray bowed to the wishes of an author whom it felt likely to prove a valuable literary property.[20]

On rare occasions, the desire of the publisher for revisions prior to publication caused a terminal break in relations. Negotiations with John Richardson broke down when he balked at the redactions requested by Murray III to the manuscript of his *Arctic Searching Expedition* (1851).[21] Richardson had previously enjoyed a good working relationship with the Murray firm. Having accompanied Parry to the Arctic, Richardson had contributed an extensive appendix to the *Journal of a Second Voyage for the Discovery of a North-West Passage* (1824; app.: 84). Between 1829 and 1831, the Murray firm had published the first two volumes of Richardson's *Fauna Boreali-Americana*. In 1848–49, Richardson traveled with John Rae in the search for John Franklin's lost expedition, and on his return submitted an account of the expedition to Murray for consideration. At this stage Murray requested substantial changes to the length and emphasis of the book. Thanking the publisher for his "friendly criticisms" and reflecting on Murray's suggestion that in its current state the manuscript would not be a worthwhile speculation, Richardson adopted a defensive tone: "As to abridging the work of a narrative of the latter part of the journey including our account of the natives, as you suggest, that would not meet with my wishes at all." Nor, he added, would the "very small remuneration that would accrue to me on the present such terms of publication . . . repay me for the trouble." Ultimately, Richardson told Murray that their "negotiation now ceases and I must trouble you to return the drawings at your earliest convenience."[22] Although Richardson informed Murray that he intended to try his luck with an American publisher, the book appeared the following year under the imprint of Murray's London competitor, Longman.[23]

In other instances, authors were eager to receive precisely the editorial guidance Richardson decried. Joseph Gurney, an evangelical Quaker who had published an antislavery volume, *A Winter in the West Indies*, with Murray in 1840 (app.: 147), wrote to the firm the following year over prospects for a new edition of the work, asking whether one of the firm's literary friends might "have the kindness to draw his pencil through such passages, whether in prose or verse, as he thinks it would be better to omit—& at the same time makes his marginal remarks with the utmost freedom, it would greatly assist me in determining the question of publication . . . & probably might lead to a considerable improvement of the volume."[24] Likewise, Mary Margaret Busk, when she received word from Murray that her manuscript titled *Manners and Customs of the Japanese* (1841; app.: 148) had been initially rejected, wrote to explain that it was still a work in progress—inviting the publisher to explain how she could "adapt it better to your views." She had been reading other works for Murray as a reviewer: "I have found the habit of reviewing is a bad training for original composition."[25] For the most part, like Busk and Gurney, Murray's writers sought a via media, acceding to requests for revision while seeking to retain the integrity of their authorial visions and displaying commitments to tell their stories in ways they felt most credible. If the requested revision ran contrary to the wishes of the author, such things could be negotiated even at the risk, as the example of Richardson shows, that the author would go elsewhere or, as the case of David Livingstone illustrates, that the author was published despite the reader's word.

NEGOTIATING BAD COPY AND THE PRINTER'S DEVIL

Notwithstanding the complex negotiations demanded by editorial intervention, once the manuscript had undergone revision in accordance with the publisher's (and to varying degrees, the author's) requirements, it would be sent to the printer for typesetting, or composition, where the text would go through various orthographic changes, in accordance with the convention of normalizing punctuation and spelling. In earlier periods it was customary for the publisher to be his own printer, and sometimes his own bookseller, insofar as bookmaking was a small-scale cottage industry that could take place under one roof. By the late eighteenth century, however, the book trade had become increasingly specialized, with publishers often acting as capital investors who served as go-betweens for authors, printers, and booksellers. In this capacity, Murray used a number of trade printers in the nineteenth century well known for the quality of their work (see app.), the most notable being the firm of William Clowes. In 1823 Clowes had become one of the first printers in London to acquire a steam-driven power press and, by the 1840s, had cemented its position as the largest printing works in the world. The Murray firm employed Clowes as their printers for a little over one-third of all of its non-European travel books, ninety-two texts (39 percent) being printed by them. In total, the house of Murray employed thirty-five different printing firms in production of its non-European travel and exploration narratives.[26]

An account written (and published anonymously) by Francis Bond Head for the *Quarterly Review*, one of the many publications for which Clowes also had responsibility as printer, provides insight into the working practices of the printing house. Head described the compositor's daily "heartache caused by 'bad copy'"—either the consequence of authors' indecipherable handwriting or stylistic impenetrability—which required continual interpretation, adjustment, and emendation on their part (a task which earned them 6d. per hour). Compositors were required, Bond Head observed, to attend continually to "a series of minute posthumous additions and subtractions."[27] Often called on to implement major corrections, too, compositors thus acted at times almost as coauthors who were required to be "competent to correct, not only the press, but the author. It is requisite not only that they should possess a microscopic eye, capable of detecting the minutest errors, but be also enlightened judges of the purity of their own language. The general style of the author cannot, of course, be interfered with; but tiresome repetitions, incorrect assertions, intoxicated hyperbole, faults in grammar, and above all, in punctuation, it is his especial duty to point out."[28] The role of printer-compositor in the process of making manuscripts ready for the press was thus more influential than is sometimes supposed and was intimately associated with the production of prestige works, which conveyed the quality and authority desired by both publisher and author.

Once the type was set, first proofs were printed to be corrected by an in-house reader. Often a first revise was run off, from which author's proofs were printed, to be forwarded to the writer for correction. If authors lived in proximity to the London printing firms used by Murray, proofs would typically be conveyed by one of the printing firm's apprentices—a so-called printer's devil—who carried his cargo "secured in a leathern bag, strapped round his waist."[29] The logistics of this stage in the process could vary greatly from printer to printer, and author to author. Sometimes the author had sight of one set of proofs, sometimes more: author's galley, author's revise, author's galley revise, author's page proof, author's revise.[30] Where a new edition of an already-existing title was required, the author was usually asked to mark up a copy of the previous edition so that it could be reset in accordance with his or her wishes. Opportunity thus existed to supplement the text with new manuscript copy and many Murray travel authors took this up even as their book neared publication.

In looking over his *Wanderings in North Africa* (1856; app.: 227) with a view to revisions, James Hamilton identified "a few pages containing gross misprints." These he proposed to make good and extend in the new edition with "a very interesting chapter containing an account of the barbarous, and still unavenged murder of a Sheikh who protected me."[31] From assessment of the manuscript evidence, it is clear that the number and significance of the changes that could be entered by the author at this stage varied considerably, depending on the nature of the errors to be corrected, the extent of the new matter to be added, his or her prestige, and the nature of the author's working relationship with the house of Murray. Given the financial implications to the firm (and, in some instances, to the author) of large-scale alterations, there was often resistance, understandably, on the part of Murray to wholesale proof revisions. Substantive changes could knock out the page or have consequences for many pages, necessitating expensive recomposition.

For most authors, reading and revising proofs was not an overly complicated phase of the publication process, particularly so from the 1840s when the new postal system made for efficient delivery and return of proofs—the printer's devil being superseded by the Royal Mail.[32] But the logistics could be complicated, especially where the explorer-author was overseas. Itinerant writers would sometimes have little time to correct proofs before they were off again on another long expedition. In such instances, authors could entrust the final corrections to the Murray firm or one of its agents or leave the job to a trusted friend or relative. Barron Field, author of *Geographical Memoirs on New South Wales* (1825; app.: 86), for example, wrote from Australia to inform Murray II that he had instructed his brother, Horace, "to offer his services in revising [the book] for the press"; placing trust in the production of fair copy in one who possessed "the eye of a scientific picker of weeds."[33] The distance from London to Sydney posed the problem of a mail journey of as many as three or four months, something that could potentially delay publication indefinitely; Barron's brother, a solicitor resident at Lincoln's Inn Fields, home to the city's leading legal practices, was close enough to call in at Albemarle Street, the Murray firm's London home, should it prove necessary.

In addition to preparing copy for final printing, the commissioning of accompanying illustrations was a central element in the production process, and one in which—as we saw in chapter 5—authors held varying degrees of agency. While some writers brought back detailed illustrations, maps, charts, and diagrams from their travels, others submitted the most rudimentary sketches to be worked up by skilled illustrators and engravers at a later stage. Still others employed artists in the field to take down scenes for inclusion in the final published text. For Edward Robinson, whose *Biblical Researches in Palestine* (1841; app.: 153) had been published first in the United States, the inconvenience of sending plates and maps so far for Murray's London edition was overcome, at the suggestion of the publisher, by Robinson's arrangement for his Berlin cartographer to send his drawings directly to Murray, so that the latter could have them "carefully engraved."[34] Subtle changes could creep in while the text was being made press ready and plates were being engraved from approved illustrations. This was especially so when substitutes for worn plates were being engraved for subsequent imprints or editions. Stylistic changes, aesthetic modifications, and, in some instances, visual content could undergo subtle as well as significant changes unanticipated by the author or illustrator at an earlier stage in a title's production, variables that made for instability over time. In the same way that Livingstone had been forthright about his preferences regarding proposed editorial changes to his text, so he had, as we have seen, unusually specific requirements about the illustrations to accompany his *Missionary Travels* (1857).

In 1836, Edward Strutt Abdy wrote to the Murray firm about just such matters as they concerned the substandard presswork of his recently published *Journal of a Residence and Tour in the United States of North America* (1835; app.: 128), which had been produced by the printing firm of George Woodfall. One of Abdy's friends had informed the author of several embarrassing typographical errors in the book: "Howard the philanthropist has been changed into Homard & as is substituted of us," Adby informed Murray. Worse still, the author had, in his book, "passed a few jokes on the Americans about grammar & spelling" and now found himself "laid . . . open to a similar charge" on account of the printer's compositional errors. Adby insisted that the mistakes were not his own—although, since he lived in London, we may assume that the author had sight of the final proof—claiming that the printer "has preferred his own mode of orthography" to that adopted by the author.[35] Again, it is clear how, in the process of composition, the correction of "errors" was an established activity and one that could subtract from, just as easily as it could add to, a text's credibility.

Tensions also arose between authors and printers and publishers over the slowness of the work's emergence into print, not just over the quality of its engraving, composing, and presswork. Phillip Parker King, for example, sent Murray II the last section of the manuscript of his *Narrative of a Survey of the Intertropical and Western Coasts of Australia* (1827; app.: 99) in February 1825.[36] More than a year later, in April 1826, he wrote to Murray to say that he would soon be setting sail for South America and would be gratified to see the book published before his departure.[37] A month later King wrote expressing even greater frustration: "I am quite disappointed at not seeing my book out before I sail. . . . Every body asks me why & I am tired of replying to it—I hope you will produce it immediately—for I am sure it answers no good purpose of keeping it back. People are tired of asking for it & I am heartily tired of hearing it mentioned. We are only waiting for a wind to leave Deptford."[38] The book was not published until 1827, by which time its author was again exploring in the Pacific: King did not return to Britain until 1830, a full five years after the submission of his manuscript.

The journey from manuscript to print was rarely smooth and seldom swift—a fact attested to by the torturous publication of one narrative of polar exploration. After a series of misfortunes in his attempt to traverse the Northwest Passage, William Parry returned to Britain in October 1825 with a view to seeing his account of the expedition in print. Throughout the following year, Parry petitioned the Admiralty for permission to undertake another expedition and so it was incumbent on him to give a public account of himself, not least in the face of rumors then circulating about his failure to complete the mission, after the beaching of one of the expedition's ships. Parry was clearly anxious about the situation and his letters to Murray II are full of complaints about the time that his *Journal of a Third Voyage* (1826; app.: 96) was taking in production. On February 9, 1826, he wrote to Murray in order to complain that Clowes, the printer, had not sent him a single proof sheet despite having had the text "*ten weeks* in hand."[39] Parry intimated that if the matter were not expedited, Murray should commission another printer. On March 7, Parry spoke in plainer terms: not only had the delays kept him in London longer than he had intended, but he was beginning to feel extreme "uneasiness," fearing that the delay would look to the Admiralty like "a dereliction of duty" on his part and to the public "as if I was ashamed to publish."[40] As time dragged on, and as the book made slow progress, Parry wrote again to Murray on August 14, in an attempt to engender action, citing reputational damage: "It has been hinted to me, in no very agreeable manner, that an idea exists abroad, and especially among those of my own profession, that my book is withheld because *I am ashamed to publish it*."[41] The journey into print, as King and Parry discovered, could take almost as long as the physical sojourn it sought to document. Moreover, as these examples show, the intervention of the printer, adding a further element to the compositional process, could further deauthorize a text that had already been subjected to intercessions that had little to do with the original writer's intentions.

Where a book was to be printed from stereotype plates (increasingly common during the nineteenth century), molds were taken from the final corrected text.[42] One advantage of the rise of stereotyping was a more "stable" text, as the same plates could be used from edition to edition. Whereas, previously, textual variants routinely occurred between and sometimes within individual editions as the text was corrected, modified, and updated, once the initial proof sheets were printed, the final corrected text was produced, and the plates were cast, the text was more "fixed" than it had ever before been. The existence of stereotype plates also made it easier for a different printer to be charged with producing a new or revised edition of an existing text. As the evidence in the appendix collectively illustrates, from the mid-1820s onward second and subsequent editions of a particular work were regularly printed by firms not involved in the original production. Henry Nelson Coleridge's *Six Months in the West Indies* (1826) was, for example, printed in its first edition by Charles Roworth; in its second (also in 1826) by Thomas Davidson; and in its third (in 1832) by William Clowes. While it was possible at these stages to make minor physical changes to stereotype plates, this was avoided if possible because it was a laborious task and the results were often unsatisfactory. Therefore, while stereotyping may have been an irresistible fiscal advantage for the publisher, for whom reprinting from existing plates was far cheaper than the commissioning of a newly composed edition, it had the inevitable effect of reducing the authorial control that had previously allowed writers to make significant changes to their texts as they were recomposed for a new impression or later edition.

MANAGING AUTHORS, MARKETING TEXTS

The final stage in bringing a Murray travel book to its public, from the publisher's perspective at least, was its marketing. While the text, with all of its accompanying apparatus, was being printed, bound, and made ready for the warehouse, its advertising was often a carefully orchestrated process. Copy had to be written for insertion in journals and newspapers and for notices of forthcoming work to be bound in with Murray's other titles, as the firm gathered together endorsements from reputable names in relevant fields, to be extracted in more detailed advertisements. The Murray firm kept a weather eye on reviews and for favorable comments that would lend its titles additional credibility. At the same time, advance copies were sent out for review, targeted strategically at individuals who might have good words to say about the book. Authors, too, were active in the process of soliciting positive comments on their work. Edward Robinson, author of *Biblical Researches in Palestine* (1841; app.: 153), had shared proofs of his book with the German geographer, Carl Ritter. As founder of the Berlin Geographical Society, professor of geography at the University of Berlin, and authority on the geography of the Holy Land, Ritter's opinion counted, and Robinson was aware of the potential intellectual (and financial) consequences of a positive notice. Happily for Robinson, Ritter's praise was so fulsome that he was compelled to reassure the author that it was not "*puff*, but . . . the result of . . . sincere & unbiased judgment."[43] Robinson's work, Ritter told him, would have lasting value for an understanding of the Holy Land and was a major advancement in the accuracy of the region's cartography. Extracts of Ritter's fulsome comments were incorporated in an advertisement placed by Murray in the firm's *Quarterly Literary Advertiser* (a pamphlet typically circulated alongside the *Quarterly Review*):

> I cannot often enough repeat what an uncommon amount of instruction I owe to the invaluable work you have left us here. It lays open, unquestionably, one of the richest discoveries, one of the most important scientific conquests, which has been made for a long time in the field of Geography and Biblical Archaeology. I can at present say this the more decidedly, because, having had opportunity to examine the printed sheets nearly to the end of the second volume, I can better judge of the connection of the whole than was before possible. Now, however, I perceive how one part sustains another, and what noble confirmation the truth of the Holy Scriptures receives from so many passages of your investigations; and that too in a manner altogether unexpected and often surprising, even in particulars seemingly the most trivial and unimportant. The accompanying maps, too, justify step by step the progress of the inquiry.[44]

While some publishers made it a policy not to review their own titles in their own periodicals, Murray was not above arranging for a prominent notice of the firm's own books in the *Quarterly Review*, despite that periodical's tradition of editorial independence. While its strategies might not compare with the worst excesses of eighteenth-century puffery, the house of Murray could nevertheless sail close to the wind in the methods that it employed to market its titles. Murray's authors understood this. This is clear from a letter sent in 1847 from James Clark Ross in which he thanked Murray III for a notice of his *A Voyage of Discovery and Research in the Southern and Antarctic Regions* (1847; app.: 187) in the *Quarterly*, saying that he was "indebted to the author of it to whom I should feel obliged by your conveying the expression of my sincere thanks."[45]

The influence that the house of Murray held over literary London, Edinburgh, and the provinces by the 1840s put it at an advantage when it came to having its books noticed by the reviewing press, with or without the knowledge of its authors. In some instances, authors sought themselves to initiate advertisements and reviews. Frederick Henniker advised Murray II to advertise his *Notes, during a Visit to Egypt* (1823; app.: 76) in "the following local papers, viz. Cambridge, Oxford, Bath, Ipswich, Colchester, Chelmsford and Ramsgate," the firm's name being "so well known in all of those places."[46] Similarly, in 1834, Alexander Burnes wrote to Murray that he had heard rumor that the *Edinburgh Review* intended to notice his *Travels into Bokhara* (1834; app.: 126) and requested that a copy be sent to the *Review*'s editor.[47] Only occasionally did authors stage manage the promotion of their works; the marketing of George Cumming's popular account of big game hunting, *Five Years of a Hunter's Life* (1850; app.: 202), was atypical in its extravagance. To coincide with the publication, Cumming organized an exhibition in London of hunting trophies, chiefly stuffed and mounted African animals, to which the public came in droves. Cumming delivered nightly lectures, with musical accompaniment, describing his adventures and the shooting of the creatures that, one was led to believe, now stood as mute testimony to the thrill of the hunt, an excitement captured in his book. Cumming's popular exhibition and lecture series ran for several years and significantly boosted the sales of his book: by the end of 1856, the work had amassed for author and publisher the substantial profit of £2,215 4s. 11d.[48]

For the most part, however, the house of Murray undertook the marketing process without consulting its authors and without resort to Cumming's style of popular self-promotion. This is attested to, of course, by the successful marketing in 1849 of Austen Henry Layard's *Nineveh and Its Remains* (chap. 5). Writing to Layard in February of that year, Murray III reported the success of the firm's marketing campaign: "If you were to step over to England at this moment, you would find yourself *famous*." Murray enclosed with his letter a number of reviews, including "one from *The Times* [which] was drawn up by a friend of mine."[49] The "friend" was, in fact, Sara Austen, the wife of Layard's uncle. Layard had taken Austen as his first name to please his uncle and had been close to his aunt from childhood, and on occasions had stayed with the family, while they had years before visited the Layards at their home on the Continent.[50] Knowing that Austen had an entrée to the *Times*, and having observed her close relations with her nephew, Murray showed her a damaging review of Layard that was then about to run in the newspaper, inciting her to write another that could be substituted for the offending item. Austen's anonymously published review was enthusiastic but it caused Layard difficulty because, in her enthusiasm, she had been too unbuttoned about the lack of government backing for Layard, which had led to severe difficulties in financing his archaeological work.[51] To his uncle, Layard wrote from Constantinople to say how embarrassing the review had been to him personally and expressed his fear that it might prove damaging in his relations with the embassy, to the extent that "I was ashamed to show it here."[52]

GENTLEMEN AMATEURS AND THE AUTHOR EFFECT

We have earlier shown the ways in which prefatory remarks that accompanied travel texts frequently included comments on the means by which the text had come into existence. While many of these remarks were connected to truth claims and the author's assertions of authenticity, there is another sense in which authorial disclaimers and references to the relative amateurism of the author's discourse were employed to disavow the real nature of production that surrounded the manufacture of what were, from their inception, commercial products. Despite all the behind-the-scenes interventions, arguing for the authenticity of accounts that had not been subject to an industrial process—characterized by the "working up" of illustrators, in-house editors, compositors, and advertisers—was a fairly routine practice in nineteenth-century authorial discourse. In many instances, even where the Murray firm's agents undoubtedly had a considerable hand, the conventions of the genre, in accordance with the protocols of a counter-industrial authorial discourse, required the disavowal of the very mechanisms that governed their presentation to the public.

This tendency toward what we might call the author effect is most evident in sometimes startlingly modest prefatory confessions. In the introduction to *Cairo, Petra, and Damascus* (1841; app.: 151), for example, John Kinnear confessed that "these are little more than a transcript of letters written to my own family during my absence." Although Kinnear went on to admit that he suppressed "those passages which were of a purely domestic character" as well as adding additional notes that he had taken on his journey, the overriding emphasis is on the unadorned, spontaneous, and uncommercial origins of the text.[53] William Hamilton, in preparing the manuscript of *Researches in Asia Minor* (1842; app.: 157), told Murray III that he had gone over the manuscript "very carefully and cut out as much as I could," recognizing that the book would "require considerable pruning."[54] Yet, when he came to write his preface, Hamilton claimed the opposite, telling his readers that "the form and style of my own Journal have been preserved as closely as possible."[55] Whatever the reality of the situation, and no matter how rigorous the constraints on authors and their texts, the fact that the display of authorial directness had become a stock convention in nineteenth-century preface writing requires us to treat such claims with a degree of skepticism. This, in part, is to reinforce our remarks above about the self-positioning of the modest author. It is also to make a point about authorial independence in the face of in-house intervention and industrialized production systems that might seem to denature the author.

The African journals of Richard and John Lander, which were, as we have seen, eventually combined and edited heavily by Alexander Becher as *Journal of an Expedition to Explore the Course and Termination of the Niger* (1832), bore a preface that claimed, "we have made no alterations, nor introduced a single sentence in the original manuscript of our travels."[56] Justifying a work that was confessedly "faulty in style" by claiming that, with all its stylistic shortcomings, it would retain its "accuracy and vividness of description," the text once again disguises the heavy extra-authorial hand behind the final version.[57] One of the most direct deployments of this technique, unusual for the directness with which it addresses the effect of the publishing process on the transformation of the manuscript, is evident in the preface to Sarah Gascoyne Lushington's *Narrative of a Journey from Calcutta to Europe* (1829): "The Author is deeply sensible how much the defects of her Book will demand indulgence, as it has not been revised by any literary person, but was at once delivered by herself into the hands of the publisher; indeed, little alteration has been made in the original journal, beyond adapting its contents to a narrative form, and omitting details that might prove tedious, and descriptions which had better been executed by established authorities."[58]

The wholesale reshaping of the text, identified by Lushington as adaptation to the "narrative form," as well as the implementation of redaction and excision, constitute, of course, more than a "little alteration." Nor do we know just how much influence others (literary or otherwise) might have had over the manuscript by the time it reached the hands of Murray. The author's husband, Charles Lushington, had been secretary to the governor of Bengal between 1823 and 1827, when his wife's diary had been composed. He was a published author himself, having written a history of British institutions in India in 1824 and may very well have advised her on the manuscript in the two years between its original completion and its final publication.[59] Another and more fundamental problem is that where the original manuscript is not extant, as is the case for Lushington, one cannot tell from such authorial statements how carefully the text was in fact "worked up" after the submission of the autograph in which these same claims are made. Statements affirming the authorial innocence of texts could thus mask the very mechanisms by which, in material terms, its discourse was actually framed.

One of the most excessive acts of dissembling by a Murray author is to be found in Frederick Henniker's *Notes, during à Visit to Egypt* (1823): "I have been persuaded to make a book:—but I have made it as short as possible, and to this accidents have contributed. Part of the following was written to a friend, to whom, verbum sat:—the amusements of drawing and shooting prevented me from the trouble of making long notes:—what I did write has but lately arrived in England: and part of my papers have been lost."[60] It seems remarkable today that an author would introduce an expensive work to readers by saying that he had too much of an appetite for leisure to offer them a work of more serious labor. More remarkable still is the confession that much of the original copy had become accidentally lost. The title of Henniker's volume alone bears witness to the fragmentary and incomplete state of the final text. Thus an overtly displayed lack of sophistication in writing belied the many acts of sophistication that the text underwent in its movement from writer to reader. Whatever the real relations of production, it was always incumbent on nineteenth-century authors to appear to speak honestly, directly, and, therefore, authentically to their readers.

Reflecting on the nature of nineteenth-century literary production, one contemporary commentator, Richard Horne, gave, in his aptly named *Exposition of the False Mediums and Barriers Excluding Men of Genius from the Public* (1833), the following piece of sardonic advice to publishers:

> The fame or reputation of a work, or a man's name, is what you purchase and speculate upon; the merit, whether real or assumed, is a question that belongs to the writer, critic, and public, and not to be meddled with by you. Your busi-

ness is solely to sell books. You are to cater for the public taste, and according as you see them "bite" so you are to provide as long as the craving lasts—but no longer. You are to look upon authors as the "raw material." You are to work them up by the machinery of your business, and apply them to such purposes as your peculiar line and connection require.[61]

Horne's polemic and its provocative title point to the continued purchase that the romantic myth of originary genius continued to have well into the nineteenth century. In their highly professional performances of amateur authorship, Murray's travel writers were operating within a modus operandi that required not only the disavowal of the real economics of literary production but also the need to present the work as an unmediated exchange between the writer and the reader, untrammeled by the complex, industrializing, and sometimes contradictory forces that gave shape to the final printed work as a commodity.

NEW MARKETS, NEW OPPORTUNITIES

Remarking on the way in which relatively expensive published works eventually came to be repackaged in cheaper editions, the process he refers to as "tranching down," William St. Clair observes how, in the late eighteenth and early nineteenth centuries, copyright holders, like other monopolists, preferred, initially at least, "to sell a smaller number of their goods at a higher price than they would in conditions of price competition."[62] The inordinately high prices charged by early nineteenth-century publishers for a first edition would seem to confirm St. Clair's argument. After the exhaustion of that first exclusive market, literary products moved "down the demand curve to a new position of less high prices and larger sales."[63] Such a pattern also corresponds with a move from "a larger to a smaller manufacturing format as a text was reprinted for a later edition."[64] Before the 1820s, this practice largely governed Murray's publishing policy.

Describing the response of the Edinburgh and London book trade to an increasingly literate populace in the 1820s and 1830s, Richard Altick remarks that publishers behaved "as if they stood on a peak in Darien, beholding for the first time a vast sea of common readers."[65] One of the effects of these changes, according to one contemporary commentator, was that large and expensive volumes were being replaced by "the small and low-priced volume which is accessible to all."[66] The appearance of series of cheap reprints at affordable prices grew conspicuously in this period, with ventures such as Chambers' Miscellany of Useful and Entertaining Tracts and Charles Knight's Library of Useful Knowledge among the new market leaders. Aimed at readers of more modest means with an appetite for improvement, their appeal was the seriousness of their tone, combined with a wide range of interesting and amusing content. In 1832, the *Penny Magazine* was launched by the Society for the Diffusion of Useful Knowledge. In the same year, the Edinburgh-based Chambers brothers inaugurated their successful *Chambers's Edinburgh Journal*. Both of these publications, which aimed at a less privileged readership, were to lead the way in the provision of miscellaneous articles on serious topics, from ancient architecture to the physical and natural sciences. Both publications identified and sought to satisfy an appetite among poorer readers for exotic accounts of travel and adventure, written from an informative point of view.

While tranching down continued throughout the first half of the nineteenth century, there were in this period several notable experiments in format, pricing policy, and marketing strategy. The house of Murray was among those that pioneered the systematic introduction of cheap series. The firm's first concerted attempt to exploit the newly emerging market for cheap editions began with the establishment by John Murray II of the Family Library in 1829. This was continued from the early 1840s by John Murray III with the Home and Colonial Library (initially the Colonial and Home Library) and reemerged with the Reading for the Rail series in the 1850s. Using these series, the firm attempted to work its travel backlist for ever-broader audiences, and its actions can, therefore, be seen to comply with St. Clair's model, based on the idea that titles in this period went through a life cycle, beginning with the expensive first edition, continuing through the moderately priced reprint, and concluding with a cheap popular edition, the size of print runs in inverse proportion to their price.

Many of the cheaper editions that emerged following the institution of the 1842 Copyright Act were bad reprints of out-of-copyright popular titles. Still others were drawn from backlists that had outworn their more expensive versions. The house of Murray, keen to maintain the firm's prestige in a democratizing market, did not always follow this standard model, sometimes to its financial detriment. We can further contextualize Murray's actions by consideration of Pierre Bourdieu's analysis of the marketing of symbolic goods.[67] Bourdieu describes two fields of cultural production. The first concerns cultural products in the field of restricted production, which he defines as prestige works produced for a privileged audience—those well-placed individuals whose influence in the field of cultural relations can cause an author's or a work's reputation to circulate among the cognoscenti. The logic of this field, according to Bourdieu, requires on the part of its producers and consumers a disavowal of the real economic relations involved in the production and circulation of the work, in which it finds legitimation principally in terms of its symbolic capital. The second field of cultural production is that defined by large-scale manufacture, epitomized by the market for cheap goods whose prime motive is the unashamed creation of economic capital. Applying Bourdieu's observations to the 1820s and 1830s, Patrick Brantlinger observes how the market for print came increasingly to be characterized by two distinct constituencies, reflecting the "social-class hierarchy" of readers themselves. It was a period in which expensive titles went upscale, against which emerged "a 'cheap literature' industry, catering mainly to the burgeoning working-class readership in the major urban centers, giving new cause for alarm to upper-class observers."[68]

Given the perceived class distinction between polite culture and mere trade in early nineteenth-century Britain, a world in which gentlemen could be distinguished for their amateurism and in which professionalism could be looked down on by an educated elite (notwithstanding the fact that the elite also bought cheap books), the house of Murray negotiated the reputation of its own class and market position with delicacy, often concealing hard-nosed business motives behind the more affable face of gentleman publishers. The firm's policy and the delicate ways in which it presented itself to the public may also be understood in the context of moral criticisms of cheap literature.[69] It has also to be understood in the context of the growth of popular publishing in a period in which some in the book trade, such as Thomas Tegg and James Lackington, came to be regarded as the bottom-feeders, dealing in cheap and badly made books, as well as remaindered stock at reduced prices.[70] Publishers specializing in reprints of out-of-copyright works often laid themselves open to charges of being second-rate and opportunistic, a charge from which the Murray firm was at pains to distance itself after about 1810. Murray I generated much of his income from the reprint trade, of course; a lucrative market that had opened up in the wake of the copyright reforms of 1774. Murray II, in accordance with his social status and that of his firm, invested in expensively produced original works of exploration and travel over which he often had a monopoly interest. We can see the firm's increasing desire to specialize in travel writing at the expense of fiction as an important part of its drive for legitimacy in the first quarter of the nineteenth century. The publishing of exploratory literature fed on, as well as consolidated and extended, Mur-

ray's polite credentials through his associations with influential figures and
public institutions: learned societies, military hierarchies, the Admiralty,
and the Colonial Office, among others. Such a list also had the advantage
of extending the reputation of an imprint coming to be associated with the
British imperial project and advancement in the sciences and in the realm
of international diplomacy as was so obviously the case of many of Murray's
travel authors. In much the same way that Murray strove to instill credibility in the works of travel that the firm published, those texts were central
to the shaping of the firm's reputation and status. In the face that it presented to the world, the house of Murray took every opportunity to show
itself to be not a money-grubbing business but a philanthropic, publically
minded endeavor. This fact concealed, of course, the hard-headed decisions
the firm was required daily to make with regard to public demand, market size, and the potential profitability of its wares. For a publisher whose
list consisted to a large degree of relatively high-brow travel narratives, the
negotiation of the newly burgeoning market for cheap literature proved a
difficult proposition.

The Family Library

It was in response to these challenges and opportunities that Murray II
launched the Family Library in 1829. This series appeared in 5s. monthly
parts and incorporated a total of fifty-three volumes by 1834. The series represented a major departure for Murray and shows the firm reaching out to a
growing reading public for whom it had not often catered in the past, while
still attempting to retain its reputation as a publisher of high-end prestige
titles. Something of Murray's attempt to negotiate concerns about the de-
grading effects of literature through the cheap reprint trade can be detected
from the way in which the Family Library was marketed. With a view to
inaugurating the series with a respectable title that he knew would find favor with a wide reading public, Murray approached John Gibson Lockhart
in 1828 to produce a one-volume digest of Walter Scott's nine-volume *The
Life of Napoleon Buonaparte* (1827).[71] Nervous about the possible cheapening
of Scott's magisterial study, Lockhart wrote to Scott (his father-in-law) to
sound him out. On October 30, 1828, Scott replied enthusiastically:

> Your scruples about doing an epitome of the Life of Bony, for the Family Li-
> brary that is to be, are a great deal over delicate. My book in nine thick volumes
> can never fill the place which our friend Murray wants you to fill, and which, if

you don't, some one else will right soon. . . . By all means do what the Emperor [Murray] asks. He is what the Emperor Nap. was not, much a gentleman, and, knowing our footing in all things, would not have proposed any thing that ought to have excited scruples on your side.[72]

Scott's response is revealing not least for the way in which it singles out Murray II's credentials as a "gentleman" whose literary taste made him a reliable route to a mass reading public. Whereas competitors like the Society for the Diffusion of Useful Knowledge were associated with the Liberal political cause, Murray could be counted on to frame his popular series in terms that were unthreateningly conservative.

In April 1829, Murray placed a three-page advertisement in the *Quarterly Literary Advertiser* to mark the launch of the Family Library—a series "handsomely printed in a pocket size, but with very legible type."[73] The advance press notices that Murray chose to reproduce in the advertisement are revealing for the way in which they illustrate his desire to promote the series, locating it at the center of the public debate about cheap literature and morality and presenting it as a product that would nevertheless appeal across class boundaries. Several press sources were deployed to refer to the superior production quality of the volumes. One effusive expression of commendation from the *London Literary Gazette* was typical: "We are very sure that if the *Family Library* go on as it begins, it will soon do more to put down the trade of literary trashery than arguments or reflections we could introduce here: and we therefore . . . [offer] our most hearty commendation of a design . . . which merits the highest encomium we can bestow on it."[74] In June, the Library's launch was commented on in a satirical tavern dialogue—part of a regular series, the *Noctes Ambrosianae*—published in *Blackwood's Edinburgh Magazine*. The dialogue has the Ettrick Shepherd—the non de plume by which James Hogg was occasionally known—commenting sardonically on Murray's decision to try his hand in the sphere of popular literature: "The Family Library, puttin' oot at Murray's, is hooever ae Tory speculation that lucks weel. I think they'll hae the heels of the Leeberals there."[75] The *Monthly Review* in a more prosaic commentary the following year took pains to commend the series for its quality in spite of its retail price: "No one, we should think, can take up these beautiful volumes and compare the insignificant price at which they are sold, with the rare union of intellectual and mechanical excellence which they display, without acknowledging that they originate in great good sense joined to a spirit of great commercial liberality."[76]

Murray seems to have been keen to recruit reviews that distanced the series from lowbrow cheap literary content. Not only would these volumes "form a valuable adjunct to the library of the rich" (in the opinion of the *Birmingham Journal*), they would also appeal to the "humble 'book-shelf' of the poor man" and "*must* become a favourite of all classes and benefit to society in general."[77] A similar view had been peddled in Murray's *Quarterly Review*, in which it was claimed that the series would furnish evidence that "the very highest talent no longer disdains to labour for those who can buy cheap books only," going on to confirm that such a gesture would "infuse and strengthen right principles and feelings . . . among those classes."[78] As this last sentiment indicates, Murray was at pains to sport his conservative literary credentials. The care that the Murray firm and its authors took over the presentation of their books, from quality printing, typography, and design to the style and integrity of the texts themselves, played a significant role in the reputation of the house and its list. In an environment in which cheap reprints and shoddy, mass production were becoming increasingly conspicuous, a Murray travel book, even at a reduced price, had to present itself as a quality product.

John Feather has observed that "Murray was not slow in adopting the technological changes that had taken place in book production since the turn of the century . . . but unlike other popular series the contents of the Family Library were distinguished for its titles."[79] Our evidence would suggest that while the firm attempted throughout the early nineteenth century to extend its markets into the realm of popular editions at lower prices, it was often at pains to promote itself as the producer of "original" and "new" titles, disguising hard fiscal policy as philanthropy, patriotism, or, even, religious piety. Previous cheap series were focused mainly on the reprinting of drama, fiction, and poetry, but it was not until the Family Library and its ilk that contemporary nonfiction quality prose was made available to the common reader for the first time.[80] In building their cheap list partly on "new" titles, Murray often paid over the odds for copyright, not merely reissuing in cramped formats the same old works but taking pains that editions were tailor-made to the shorter formats in which they were issued.

The relative paucity of travel titles in the series would seem to indicate that, at this stage, or perhaps for this imprint in particular, travel writing was not regarded by the firm as a suitable subject with which to capture a democratized reading public. Only three of the titles chosen for Murray's Family Library could be construed as travel writing. Henry Nelson

Coleridge's *Six Months in the West Indies, in 1825*, as its title suggests, was not new to the list when it appeared in 1832 (app.: 92). Although the series aspired to offer "new" works rather than reprints to its customers, Coleridge's account had originally been published in 1826, with a second edition appearing that same year. Thereafter, sales slowed and the select audience for whom the book was originally intended had been exhausted. The third edition did not appear in the Family Library until 1832, inspired, perhaps, by the Jamaica slave rebellion that had taken place in the previous year.[81] John Murray turned here to an item in his back catalog whose currency he hoped might once again capture the public imagination and generate sales. The only other travel book to appear in the Family Library—with the exception of Francis Bond Head's *The Life of Bruce* (1830) and John Barrow's *Bounty* volume (app.: 120 and 122, respectively)—was Richard and John Lander's *Journal of an Expedition* in 1832 (app.: 124), whose authorship and making we have addressed in several contexts. With their inclusion in the Family Library, Murray was attempting to publish cheap editions while the facts to which they referred still had appeal and topicality (the Landers) or which spoke to episodes either of drama (the *Bounty*) or personal significance (Bruce) in the undertaking of exploration.

While Coleridge's *Six Months in the West Indies* can be seen as a classic case of tranching down, Murray paid several of his Family Library authors generous sums for copyright, a factor that, as John Feather notes, "contributed substantially to his expenses. In the end, the Family Library failed, partly because the books were perhaps a little too serious for the popular market for which they were intended, and partly because of the comparatively high costs incurred by Murray's standards."[82] A third reason for the demise of the series, suggests Aileen Fyfe, is that the retail price of the volumes, while lower than much of the new literature of the day, was "only 'cheap' in comparison to the high price of original literature" and that the cost of five or six shillings "was unaffordable for the working classes and many of the lower middle classes."[83] In December 1834, Murray wrote to Lockhart to inform him that he was considering remaindering ten thousand copies of the series and subsequently sold off much of the remaining stock to Thomas Tegg for the bargain price of 1s. a copy. Before clearing them from his warehouse, Murray had 2,580 volumes of the Landers' *Journal* on hand and 2,023 volumes of Coleridge's *Six Months*.[84] While the sale brought him a slight profit overall, Murray III must have felt that the time involved in the experiment, and the potential damage to his reputation that it caused, was not worth the effort.

Colonial and Home Library

Recognition of the desirability of cheaper books for an international audience led Murray III in 1843 to launch the Colonial and Home Library (retitled in 1844 the Home and Colonial Library). Books were selected for the library on the basis of their "acknowledged merit, [and] the ability of their authors," texts which, in the firm's view, would be received with enthusiasm in the "Cottage of the Peasant and the Log-hut of the Colonist."[85]

It is interesting in this regard that Murray included the second edition of Charles Darwin's *Journal of Researches* (1845) within his Home and Colonial imprint. The first edition had been published by Henry Colburn in London in 1839. Colburn retained the right to reprint the work but not any abridged variant of it, a fact that was an opportunity for Murray. The suggestion that it might appear in Murray's Home and Colonial series seems to have come from Darwin. Murray, for his part, was anxious that the length of the work, and thus the price needed to offset costs, would preclude its publication by him: "The Price of the Library is so low & so large a Circulation is required to cover the outlay, that any sum which I could offer for the Copyright would I fear appear to you very small." Murray nevertheless pressed his case, informing Darwin that "I could not devote more than one volume or two numbers of the Library to your journal, and a curtailment of about [a] fifth part of your Volume would be required to reduce it to one of mine of 370 pages." Ending his letter to Darwin, Murray pointed to more general issues about circulation to which this instance, the Home and Colonial imprint in general, and the strategy of the firm more broadly all spoke: "The Journal would not fail I am sure to be popular; it is at present locked up as it were, from the world.—& a more extensive circulation would add greatly to its Author's reputation."[86] Within weeks, illness notwithstanding, Darwin was at work on the abridged version, writing to his mentor John Henslow how "I find a good deal to alter in the scientific part." [87] Darwin sent parts of the remaining scientific sections to Joseph Hooker to check for gross mistakes. As others have shown, Darwin's reshaping of his *Beagle* voyage *Journals* was in fact a thoroughgoing process, one that involved recasting elements of the chronology, omitting some personal matter, and placing himself in relation (subordinate) to that earlier global traveler whom Darwin much admired, Alexander von Humboldt.[88] It is noteworthy, too, that Herman Melville's *Typee* (1846) was also published within this series, after Melville agreed to add chapters of anthropological description and otherwise amend his initial text.[89] In this sense, Darwin and Melville both had to be shortened and cheapened to bring their works to market in affordable forms.

It is clear that the increasing internationalization of the book market brought with it, on the part of publishers such as Murray with a keen eye to expansion, a desire to extend operations into the global marketplace. In the past, the firm had experienced considerable difficulty with foreign publishers, both on the Continent and in North America, who were quick to cash in on the reputations of Murray authors, flooding overseas markets with cheap editions that undercut Murray's relatively high prices. In launching the Colonial and Home Library, Murray was prompted by recent changes to copyright law governing the protection of British books throughout the empire. In its conception, the 1842 Copyright Act served to make illegal the distribution of unauthorized works in the dominions, offering protection for publishers at home against their overseas competitors, a factor Murray identified in his September 1843 prospectus for the series: "It [the Library] is called for in consequence of the Acts which have recently passed the British Parliament for the protection of the rights of British authors and publishers, by the rigid and entire exclusion of foreign pirated editions. These Acts, if properly enforced, will, for the first time, direct into the right channel the demand of the Colonies for English Literature."[90]

Seeking to capitalize on the financial opportunities for global expansion that he felt the new legislation would bring, Murray may well have been concerned about the way in which the enterprise might reflect on his reputation as a serious publisher of quality books. Like his father, Murray III was keen to distance the firm, in its promotion of affordable reading, from those liberal advocates who were coming increasingly to identify "useful knowledge" with a widening of the franchise as well as from those who saw popular literacy as an instrument for more radical social change. Throughout its life, the series was to be associated with the imperial cause (fig. 24).

Citing a letter sent by Murray to Francis Bond Head, Angus Fraser also detects in the publisher's motives for the series "a forthright antirepublican slant": "besides damaging the copy rights of British Authors by the piracies of their works," Bond Head told Murray, "[the Americans] are sapping the principles & loyalty of the subjects of the Queen by the democratic tendency of the native American publications."[91] Bond Head, then editor of Murray's *Quarterly Review*, reflected in its pages on the new possibilities offered by mass production in terms that aimed to rescue the industrial press for a conservative imperialist agenda: "It is impossible for the mind to contemplate also, for a single moment, the *moral* force of the British Press,

CHEAP LITERATURE FOR ALL CLASSES.

THE

HOME AND COLONIAL LIBRARY.

DEDICATED, BY PERMISSION, TO THE SECRETARY OF THE COLONIES, AND THE PRESIDENT OF
THE BOARD OF TRADE.

MR. MURRAY'S
HOME AND COLONIAL LIBRARY,

Designed to furnish all classes of Readers in Great Britain and her Colonies with the
highest Literature of the day, consists partly of Original Works, partly of New Editions of
popular publications, at the lowest possible price. Mr. Murray has commenced the publication of a series of attractive and useful works, by approved authors, at a rate which
places them within the reach of the means not only of the Colonists, but also of a large
portion of the less wealthy classes at home, who will thus benefit by the widening of the
market for our literature : and the 'Colonial Library' will consequently be so conducted
that it may claim to be considered as a 'Library for the Empire.'

Thirteen Numbers of the 'Home and Colonial Library' have already appeared, and have
received the approbation of Critics and readers of all classes, in all parts of the British
dominions.

The recommendations of this series are ;—on the *score of smallness of cost ;* each Number
contains more than two ordinary 8vo. volumes, at one fifth or sixth of the usual price of
such works. On the *score of novelty, interest, and merit,* it includes works by Southey,
Heber, Borrow, Irby and Mangles, Drinkwater, &c.; *in variety,* it comprises Geography,
Voyages and Travels, History, Biography; Manners and Customs; rendering it equally
acceptable to the Cottage and Boudoir; while the utmost care is exercised in the selection
of works, so that they shall contain nothing offensive to morals or taste.

The popularity of the subjects, and the moderation of the price of the 'Home and
Colonial Library' *(an annual outlay of only twenty-five shillings),* recommend it to the following classes of persons :—

To the Clergy—*as fitted for Parochial and Lending Libraries.*

To Masters of Families and Manufacturers—as suited for the Libraries
of Factories, Workshops, and Servants' Halls.

To Managers of Book Societies, Book Clubs, &c.

To School Inspectors, Schoolmasters, and Tutors,—as suitable gifts
for the young as prizes, or adapted for School Libraries.

Travellers on a Journey will find in these portable and cheap volumes something to read on the road, adapted to fill a corner in a portmanteau or carpet-bag.

To Passengers on Board a Ship, here are ample materials in a narrow
compass for whiling away the monotonous hours of a sea voyage.

To Officers in the Army and Navy, and to all Economists
in *space* or *pocket,* who, having limited chambers, and small book-shelves, desire to lay up
for themselves a *concentrated Library,* at a moderate expenditure.

To all who have Friends in Distant Countries—as an acceptable
present to send out to them.

The 'Home and Colonial Library' will yield to the *Settler* on the *Plains*
of Australia, and in the Backwoods of America, and to the *occupant* of the remotest *can-
tonments* of our Indian dominions, the resources of recreation and instruction, at a moderate
price, together with many new books within a short period of their appearance in England.

The Student and Lover of Literature at Home, who has hitherto been
content with the loan of a book from a book-club, or has been compelled to wait for its
tardy perusal from the shelves of a circulating library, or perhaps has satisfied his curiosity

FIGURE 24 Publication for Britain's empire. Murray's Home and Colonial Library
sought to provide affordable works in order to provide a "Library for the Empire."
Source: *Quarterly Literary Advertiser* (October 1844, p. 16). By permission of the Trustees of the National Library of Scotland.

without reflecting, and without acknowledging that, under Providence, it is the only engine that can save the glorious institutions of the British empire from the impending ruin that inevitably awaits them, unless the merchants, the yeomanry and the British people, aroused by the loud warning of the said press, shall constitutionally disarm the hands of the destroyers."[92]

Murray's intention to address a colonial audience specifically was subject nevertheless to criticism in the press. The *Examiner* was particularly scathing: "Whether the Empire will be content to restrict its 'Library' to Mr Murray's copyrights, is quite another matter. As a bookseller's speculation—the pretence to anything else being altogether out of place—we must frankly say that the *Colonial Library* seems to us anything but a scheme discreetly devised, or well begun. It is a blow against high-priced books at *home*, which will tell where least desired."[93] It was as a result of such criticism that the series underwent a subtle rebranding in 1844, becoming the Home and Colonial Library. An advertisement placed in October that year gives some indication of Murray's ambition for the series. It was promoted as something "Designed to furnish all classes of Readers in Great Britain and her Colonies with the highest Literature of the day." The series would not simply be old books repackaged but would consist "partly of Original Works, partly of New Editions of popular publications, at the lowest possible price." Murray's stated aim was thus to offer a series of "attractive and useful works, by approved authors, at a rate which places them within the reach of the means not only of the Colonists, but also of a large portion of the less wealthy classes at home, who will thus benefit by the widening of the market for our literature."[94] The advertisement went on to list those audiences who would benefit most from the enterprise (see fig. 24).

What is most evident from the potential subscribers identified here is the range of reading categories to which Murray felt the series might appeal. As the prospectus and advertisement together suggest, there were three major constituencies that the Home and Colonial Library was calculated to cultivate: what we might call "cultural monitors" (factory masters, school inspectors, and librarians); itinerant readers (travelers, emigrants, and military personnel); and domestic readers with limited means. Intended as a series that would find audiences across not only different social classes but also geographical boundaries, the Murray firm considered travel a theme particularly suited to readers who were themselves on the move. It was only logical, after all, that tales of far-flung adventure would be associated in the publisher's mind with mobile audiences throughout the empire. The imperial project in which many of these texts were temperamentally embedded would, it was felt, make them relevant to readers who were themselves living in and between Britain's overseas dominions.

Unlike the Family Library, the conspicuous presence of travel writing in the Colonial and Home Library was evident from the beginning. The title with which the series was launched in September 1843 in two parts was George Borrow's *The Bible in Spain; or, The Journeys, Adventures, and Imprisonments of an Englishman, in an Attempt to Circulate the Scriptures in the Peninsula*. Of the next twenty titles to be issued, all but six were works of travel and exploration. By Priya Joshi's reckoning, the generic breakdown of Home and Colonial titles was overwhelmingly in favor of travel titles: travel writing (45 percent); history (20 percent); memoirs and biographies (18 percent); poetry and miscellaneous (9 percent); and fiction (8 percent).[95] Yet these figures do not tell the whole story, especially if we take the category of travel in its broadest sense. Many of the titles assigned to history, as well as memoirs and biographies in Joshi's analysis — for example, Louisa Anna Meredith's *Notes and Sketches of New South Wales* (1844) — might be construed as travel writing, albeit with a biographical or historical bias.[96] Taking this into account, a more liberal estimate would give us thirty out of forty-nine titles dedicated to travel and exploration — somewhere in the region of 60 percent.

The works included in the Home and Colonial Library were, from the first, a combination of newly published titles, reprints of works that had already achieved a degree of success, and, in a few instances, translations of foreign-language works. Among the established travel books included were Reginald Heber's *Narrative of a Journey* (1843-44; first published by Murray in 1828 [app.: 104]), Matthew Gregory Lewis's *Journal of a Residence among the Negroes in the West Indies* (1844; first published by Murray in 1834 as *Journal of a West India Proprietor* [app.: 127]), John Malcolm's *Sketches of Persia* (1845; first published by Murray in 1827 [app.: 100]), and, as we have seen, Darwin's *Journal of Researches*. New travel works selected for the series included John Drummond Hay's *Western Barbary* (1844), Louisa Anne Meredith's *Notes and Sketches of New South Wales* (1844), Herman Melville's *Narrative of a Four Months' Residence among the Natives of a Valley of the Marquesas Islands* (1846) and *Omoo* (1847), and Henry Haygarth's *Recollections of Bush Life in Australia* (1848; app.: 169, 172, 180, 186, and 191, respectively).

Ultimately, the series did not prove a financial success. Murray published the last title in 1849, a mere five years after its launch, although the firm was still advertising it up to 1861, at which point he was reducing the cost of the volumes to 2s. 6d. and so disguising as philanthropic gesture what was little more than damage limitation. Richard Altick has noted that "the unctuous trade-journal advertisement heralding this public spirited gesture is typical of those that sought to disguise publishers' efforts to liquidate an unlucky speculation as a contribution to the grand cause of cheap literature."[97] Not all of the titles were unremunerative. Although the final tally recorded in the Murray financial ledgers does not fully account for some of the overheads incurred over the years, Heber's *Narrative of a Journey* (app.: 104) earned the firm £642 16s. 6d in half profits by 1873. In many cases, Murray over time made a modest income from the titles: Melville's *Omoo* yielded him £205 17s. 9d. by 1877, and Meredith's *Notes and Sketches* £169 13s. 9d. by 1883. Other titles made an outright loss—Drummond Hay's *Western Barbary* and Lewis's *Journal of a Residence* still showed deficits in 1883.[98]

There were several reasons why the series did not fulfill its potential. Lack of political will to enforce the 1842 Copyright Act meant that Murray's works did not enjoy the protection, even in the British dominions, that he had anticipated. There was also, in the colonies, a feeling in some quarters that, in remarketing works whose appeal at home had already been exhausted, Murray was engaging in the familiar practice of "dumping" books. Such, at least, was the opinion of *Simmonds's Colonial Magazine*, which complained that "the selection of works announced for publication does not conform to the new and current works of the day, but only those which have been extensively circulated at home, until no further sale can be obtained for them."[99] Despite the inclusion of such distinguished men of science and of letters, and the fact that the scheme was an innovative and rapid response to the 1842 Copyright Act, the Home and Colonial Library, which contained an entire class of texts advertised under the title "Voyages, Travels, and Adventures," failed to realize its potential: the firm's backlist "lacked appeal for a popular international audience."[100] By the later 1840s and into the 1850s, however, the standing of the house of Murray was high enough to withstand the relative failure of its Home and Colonial venture.

Reading for the Rail

In 1851, undeterred by the relative failure of the firm's other cheap series, Murray III launched a third: Reading for the Rail. Once again, anxieties about the deleterious effect of cheap reading matter on the part of the quality press came to the fore in the advertisement placed to mark the series' appearance. Murray was careful to co-opt testimonials from the leading reviews in an attempt to show that he had compromised neither his production values nor his attention to the superiority of its contents. One advertisement for the series gave prominence to a quote from the *Athenaeum*, in which it was claimed that this was "a series of cheap and healthy publications, to supplant the deleterious mixtures sold too frequently from the want of more wholesome food." The *Economist* was also invoked, reinforcing the difference between Murray's quality productions and the more corrosive production of his less discriminating competitors: "We hail Murray's 'Reading for the Rail,' with much pleasure, as one of the many efforts now making to supply the public with books at once cheap and good. This is the only legitimate means by which literature that is cheap and worthless, or positively mischievous, can be fairly and efficiently put down." Other testimonials were recruited in order to assure readers that Murray had not compromised production values in the interests of cheapness. A review from the *Atlas*, in high patriotic mode, was cited in order to distance the firm from those "Cheap novels" that condensed "some five volumes of French nonsense into one small duodecimo, must necessarily be in small print and on bad paper. This bothers most eyes and wears many out. The same objection applies to many even of the really useful books sold at railway stations. Mr. Murray, whose subjects do not require the same compression, gives a good readable type, quite large enough for ordinary eyes, even in express trains. For railway reading this is really the chief consideration of all."[101] Despite the brave face that Murray III sought to put on, one of the aspects that comes through in the press reviews he chose to reproduce in advertisements for Reading for the Rail was that the launch of the new series was regarded as another exercise in damage limitation—an attempt, in other words, to rescue his reputation as a publisher of cheap series for a mass reading public.

The expensive lessons of the past had been learned. It was clear that a publisher could not pay expensive copyright fees for serious literature and still expect to make a handsome profit on a cheap product. Almost all of the titles in the railway series were, then, works that had been previously published, some of them many years before, whose expensive editions were losing their appeal. What is more, the character of the books chosen for inclusion was quite different from the titles that comprised the Home and Colonial Library. There was a tendency, for instance, toward more domestic works; the series included treatises on music and dress, the keeping of bees, the flower garden, and the art of dining and began with a collection of essays by Charles Apperley from the *Quarterly Review*, which had originally been issued in parts in 1837 as *The Chase, the Turf, and the Road*. The relative absence of travel writing in the Reading for the Rail series — the exceptions for non-European travel and exploration are Laurence Oliphant's *Journey to Katmandu* and Layard's *A Popular Account of Discoveries at Nineveh* (1851; app.: 211, 196 respectively), and the series also included Francis Bond Head's *The Emigrant* (1852) and John George Holloway's *Month in Norway* (1853) — would seem to indicate that the firm now considered such works unsuitable for a relatively lightweight list, for quick and ready consumption.

There is again a sense that Murray III was working the firm's backlist a little too hard. Damning it with faint praise, the *Literary Gazette* was not altogether persuaded by the addition of *The Emigrant*, since "the work originally appeared at a time when Canada occupied more public attention than now." The book had initially been well received and had gone into five editions by 1847, when it was still selling for 12s. Remarking on the 2s. 6d. volume in the railway edition, the *Gazette* regretted that "much of the importance of these questions has passed away."[102] With the exception of Layard's *A Popular Account of Discoveries at Nineveh*, which was highly remunerative for its author and its publisher, the Reading for the Rail experiment, like its predecessors, was not an overwhelming success. As one commentator has observed, the somewhat ad hoc arrangement of Murray's Reading for the Rail, with volumes "issued irregularly from 1851 and varying in size, price, and binding," gave it an improvisatory flavor that was not well suited to the cultivation of customer loyalty.[103] The performance of Murray's railway series was far from impressive in comparison with that of its near contemporary, Routledge's Railway Library, whose inclusion of popular fiction guaranteed it wide appeal. In many respects Reading for the Rail was less ambitious than its two predecessors. Unlike the two earlier series, it was not reliant on investment in new literary property but, instead, chose to work the firm's backlist. Unlike the Home and Colonial Library, ambitions for Reading for the Rail were domestic rather than global. For all that, the series remained an unspectacular but steady earner.

Late in 1859, John Murray III was busy laying plans for the marketing of the book whose date of publication marks the end of our period of study of the firm's imprint of non-European travel and exploration narratives. The Murray firm's decision to print twelve thousand copies of McClintock's *Voyage of the 'Fox'* (app.: 238), an exploration narrative based on the search for Franklin's lost Arctic expedition, revealed a significantly larger print run than the other texts within the firm's travel imprint. On November 22, 1859, more than seven thousand copies of the book were sold to well-lubricated booksellers at Murray's annual trade dinner at the Albion Tavern in London's Aldersgate. Within a decade of the book's publication, it had generated for Murray and McClintock a total profit of £3,700 13s. 6d. (two-thirds of which went to McClintock). Murray had chosen his moment, the topic, and his author well. Sir John Barrow's mantle as promoter of British polar endeavor had by then passed to Clements Markham of the Royal Geographical Society. In addressing the Dublin meeting of the British Association for the Advancement of Science in 1857, Markham had stressed the importance of polar work to geography, to science, and to Britain's sense of itself as a maritime and exploring nation and lent his voice to that public call for a search for Franklin of which McClintock's *Voyage* was one expression. Similar sentiments were expressed at the meeting of the British Association for the Advancement of Science in Leeds in the following year, at Aberdeen in 1859, and in Oxford in 1860, in which last setting poetry written for the by-then long dead Franklin and his crew echoed the grief of the wider British public.[104] Murray III's McClintock volume sold because it spoke to a nation still in shock.

The changing profile of John Murray's popular series in particular reveals much about the kinds of audience that the firm, over time, identified with its imprint and, in particular, how, at different moments, it felt it could position its travel writing in relation to a mass market. Under Murray I, the house did not have firm control over the marketing of books of travel, for the reason that it published relatively few and most of those in association with other publishers or printers. Under Murray II, there was an evident focus on quarto and octavo volumes in the first two decades of the nineteenth century (see table 1, p. 25) but a reluctance to experiment with cheaper format popular travel at that time, a fact that suggests that the publisher regarded it as a genre reserved for a relatively specialized and privileged readership. With the Home and Colonial Library, however, and the changes of format of the 1830s and later, Murray III implemented specific publishing and marketing strategies in order to reach out to a broader class of audience with his travel list, while also extending his influence through an appeal to popular imperialism. With one or two exceptions, the relegation of travel books in Reading for the Rail suggests a conscious decision as to the unsuitability of the genre, and a vast part of the company's list, for the popular reader. In the process, an important lesson had been learned. It was not merely cheap prices that sold books to the common reader—their contents had to have popular appeal as well. Murray's list, it was now clear, was far too serious for the new reading audiences and reading cultures that were coming to the fore in the 1830s and 1840s.[105]

CHAPTER SEVEN

Assembling Words and Worlds

We have said enough, perhaps, to lead our readers to reflect on the influence of geographical exploration on the life of our century. After meditating the whole sweep of change, some will fix their admiration and gratitude on the advancement of science in many departments; others on the arts which are reciprocally the cause and effect of commerce; all being under a common obligation to the same benefactors—the travellers.

Harriet Martineau, "Travel during the Last Half Century"

Harriet Martineau's reflections on the "shrinking" of the planet as a result of travel and exploration, with which we began the book, and her further observations about the importance of exploration to advances in science and the arts with which we preface this final chapter, spoke then, as they speak now, to major issues. The facts of travel did indeed change the ways the world thought of itself from the later eighteenth century and during the first half of the nineteenth century. Exploration did advance science and the arts. The commerce in ideas was greatly increased as a result of travel and exploration. The facts of exploration and travel did not have these effects, however, because everyone traveled but because, in a variety of ways and by different routes, the process and results of exploration and travel were made to work in print. Getting words and pictures into print involved processes of authorship and acts of editorial artifice. Explaining to his readers the practices of composition and synthesis that rendered complete his *Picturesque Views on the River Niger* (1840; app.: 146), the naval officer and explorer William Allen professed to the exercise of artistic license: "In the accompanying plate, I have grouped together all the principal characters of whom I had individual sketches. They are, I believe, likenesses; though I must confess that the foreheads have expanded to a more noble contour under the pencil."[1] Allen's illustration, depicting a group of Nigerian dignitaries, was, by his own admission, unreal—a collage of separate elements brought together, unified, and exaggerated—but it was, nevertheless, representative of similar scenes to which he *had* been witness (fig. 25). Artifice was, for Allen, a route to securing verisimilitude. His production of this illustration stands as an emblem for the making of books of travel between the late eighteenth century and the mid-nineteenth: these were acts of assemblage, of craft, and of truth making.

Above all else, Allen's sketch points to the important interaction among witnessing, creativity, and correspondence in efforts made by the authors of travel texts to demonstrate their own credibility and that of their written representations.

FIGURE 25 The assembled image. William Allen's depiction of "The Palaver" drew together individual field sketches from the early 1830s to form, later, an arranged collective portrait. Source: William Allen, *Picturesque Views on the River Niger, Sketched during Lander's Last Visit in 1832–33* (London: John Murray, 1840). By permission of the Trustees of the National Library of Scotland.

PLATE 1 Portrait of
John Murray I (1745–93),
by David Allan, ca. 1777.
By permission of the
Murray Collection.

PLATE 2 Portrait of
John Murray II (1778–
1843), by Henry William
Pickersgill, early 1830s.
By permission of the
Murray Collection.

PLATE 3 Portrait of John Murray III (1808–92), by Sir George Reid, 1881. By permission of the Murray Collection.

PLATE 4 Portrait of Sir John Barrow (1764–1848), by John Jackson, 1825. In his role as second secretary to the Admiralty, Barrow was in an influential position, promoting voyages of exploration and overseeing and amending the resultant narrative accounts, often working directly with John Murray II and his literary advisers in doing so. By permission of the Murray Collection.

PLATE 5 The sociable world of print culture associated with the house of Murray is captured in this painting of an imagined gathering of authors in Murray's drawing room at 50 Albemarle Street, London, in a painting by L. Werner ca. 1850. From left to right, the figures are Isaac D'Israeli, John Murray II (*seated at the desk*), Sir John Barrow, George Canning, William Gifford, Sir Walter Scott, and Lord Byron. By permission of the Murray Collection.

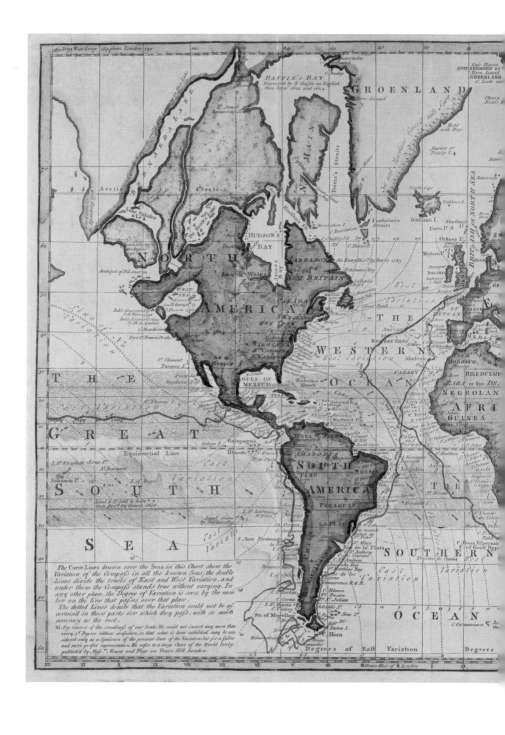

180 Deg West Long. 160 from London 140   120   100   80   60   40   20

GROENLAND

RAFFIN'S BAY
Discovered by W. Baffin an English
Man Anno 1622 and 1624

Fair Haven
SPITZBERGEN or
Born Sound
GREENLAND
C. Look out

Cherry
Bear's I.

Arctic   Circle

Davis's Straits

Hudson St.

North Cape

ICELAND I.
Tarro I²

Shetland
Orkney I²

NORTH

HUDSON'S
BAY

LABRADOR

NEW BRITAIN

An Entry Disc. by Davis 1585

IRELAND

WEST
SEA

CANADA

AMERICA

NEW
ENGLAND

Albion

New
Mexico

St. Clement
Tuturee I.

GULF OF
MEXICO

Bahama
Islands

WESTERN

OCEAN

CANARY

Madera

MOROCCO

BILEDULG
ZARA or the DE.

THE

NEGROLAN

AFRI

GUINEA

TERRA FIRMA

Galapagos

Equinoctial Line

Islands

SOUTH

PERU

BRASIL

AMERICA

PARAGUAY

LA PLATA

SOUTH

SEA

Rio de la Plata

SOUTHERN

C. Bona Esperanza
or Good Hope

The Curve Lines drawn over the Seas in this Chart shew the
Variation of the Compass in all the known Seas; the double
Lines divide the tracts of East and West Variation, and
under them the Compass stands true without varying. In
any other place, the Degree of Variation is seen by the number on the line that passes over that place.
  The dotted Lines denote that the Variation could not be af-
certain'd in those parts over which they pass, with so much
accuracy as the rest.
  N. By reason of the smallness of our Scale, We could not insert any more than
every 5ᵗʰ Degree without confusion, so that what is here exhibited may be considered only as a Specimen of the present State of the Variation but for a fuller
and more perfect representation. We refer to a large Chart of the World lately
publish'd by Messʳˢ. Mount and Page on Tower Hill London.

Str. of Magellan

Horn

Degrees of East Variation   Degrees

OCEAN

XII   XI   X   IX   VIII   VII   VI   V   IV   III   Hours West of London   II   I   O

PLATE 6 The still uncertain state of geographical knowledge concerning the shape and the dimensions of the New World—Australia or "New Holland" as well as the northwest of the American continent—is clear from Thomas Kitchin's "An Accurate Chart of the World," originally published in The London Magazine (February 1758), but here colored and tipped in a copy of Sydney Parkinson's *A Journal of a Voyage to the South Seas* (London: Printed for Stansfield Parkinson, the Editor, 1773): see also fig. 3. Source: National Library of Scotland, FB.M.214(4). This copy, originally owned by Thomas Percy (1729–1811), Bishop of Dromore, is a unique foliation of letters and other material relating to the content and the production of Parkinson's narrative. By permission of the Trustees of the National Library of Scotland.

PLATE 7 The dependence of explorers on local knowledge was often part of a broader culture of gift and exchange. Here, the reciprocity of Arctic discovery is evident in the goods being inspected and in the gestures between Inuit and British naval officers. Source: John Ross, *A Voyage of Discovery, Made under the Orders of the Admiralty* (London: John Murray, 1819). By permission of the Trustees of the National Library of Scotland.

PLATE 8 China and the Chinese were largely unknown to Western explorers and travelers. Illustrated books of travel helped make this part of the world more familiar to European audiences. Source: Henry Ellis, *Journal of the Proceedings of the Late Embassy to China* (London: John Murray, 1817). By permission of the Trustees of the National Library of Scotland.

PLATE 9 Portrait of Maria Graham (Maria, Lady Callcott), by Sir Thomas Lawrence. Oil on canvas, ca. 1819. By permission of the National Portrait Gallery (NPG 954).

PLATE 10 The coloring of the Crimson Cliffs depicted by John Ross on his first Arctic voyage "presented an appearance both novel and interesting" and penetrated the snow to a depth of several feet. It was later proved to be vegetable matter. Source: John Ross, *A Voyage of Discovery, Made under the Orders of the Admiralty* (London: John Murray, 1819). By permission of the Trustees of the National Library of Scotland.

PLATE 11 The hazards of Saharan travel are exemplified in this scene "Drawn from Nature" by George Francis Lyon, who in the 1820s had experience of the dangers of exploration in the Arctic as well as in North Africa (see also fig. 6). Source: George Francis Lyon, *A Narrative of Travels in Northern Africa* (London: John Murray, 1821). By permission of the Trustees of the National Library of Scotland.

PLATE 12 The *Tuaricks of Ghraat*. Explorers' depictions of people were often undertaken with a view to representing "typecasts," offering a sort of visual ethnographic catalog of the different cultures and peoples encountered. Source: George Francis Lyon, *A Narrative of Travels in Northern Africa* (London: John Murray, 1821). By permission of the Trustees of the National Library of Scotland.

PLATE 13 Contrasting images of the explorer-author: the case of Alexander Burnes (1805-41). This portrait in oils, by Daniel Maclise in 1834, was used as the basis to the engraving by Edward Finden that appeared as the frontispiece to the first edition of Burnes's *Travels into Bokhara* (1834)—see also figure 15 and plate 14. Source: Alexander Burnes, *Travels into Bokhara* (London: John Murray, 1834). By permission of the Murray Collection.

PLATE 14 Alexander Burnes as a Regency gentleman. Portrait of Alexander Burnes, ca. 1835, oil on canvas. By permission of the Royal Geographical Society (with IBG).

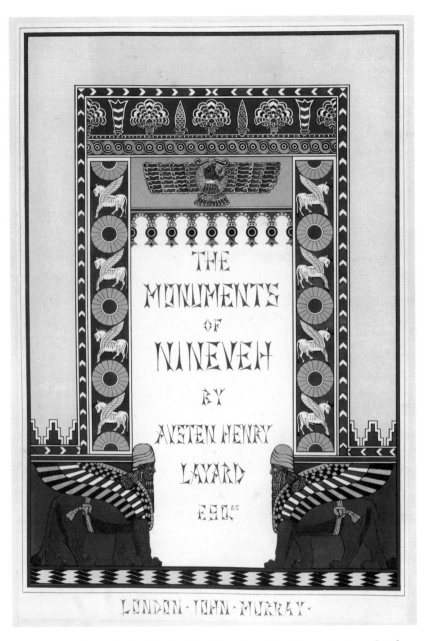

PLATE 15 Book production for the elite. The color plate title page to Layard's *The Monuments of Nineveh* echoed in its style, layout, and epigraphy the monuments discussed by Layard in his best-selling and popular text. Source: Austen Henry Layard, *The Monuments of Nineveh: From Drawings Made on the Spot* (London: John Murray, 1849). By permission of the Trustees of the National Library of Scotland.

This book has been concerned with precisely those processes—with what Richard Sher has called the "complicated, creative, and contingent" mechanisms by which in-the-field writing became authoritative print and by which travelers became authors.[2] We have examined the ways in which the world was put into words by the house of John Murray and that firm's authors; how texts made truth; and how decisions taken with respect to literary style and scholarly significance had profound implications for the making of books, their authors, and their audiences. Our focus has been less on routes in the geographical sense—tracing work on the ground—and much more in disclosing the routes that texts took from notebook to published book and, correspondingly, that travelers took in becoming authors—tracing the world in print.

The almost 240 non-European travel texts issued under Murray's imprint between Sydney Parkinson's *A Journal of a Voyage to the South Seas* in 1773 and Francis McClintock's *The Voyage of the 'Fox'* in 1859 represent—in their varying purpose, style, and geographical focus—the larger-scale intellectual, political, and literary contexts in which they took shape. The eighty-six years that separated Parkinson's book of Pacific travel from McClintock's account of Arctic voyaging witnessed the emergence of a mass reading public in Britain (and elsewhere in Europe and North America), the making of science as a global endeavor and its increasing codification around clearly defined disciplines, and the high watermark of Britain's imperial endeavors.[3] It would be unwise to claim straightforwardly causal relationships, yet, as Martineau and many others recognized, exploration and travel, and writing about exploration and travel, were central to, and were shaped by, these circumstances: just as travel and exploration provoked popular interest in the world, so they facilitated its commercial, political, and scientific exploitation. Alongside the legislative and technological changes that led to the expansion of the British publishing industry during this period, imperial ambitions (particularly in relation to Africa, the Arctic, and Central Asia) created the conditions for growth in travel literature. Perhaps more so than for any other period, this was a time in which the world turned *to* print as the world was turned *into* print.

We have been concerned to illuminate the relationships that connect travel to questions of science and imperialism, literature and print culture, and gender and the bodily performance of the explorer-author. We have seen that the nature of travel and its written products was shaped by the demands of correct scientific practice; by particular political or mercantile ambitions; by the expectations of critical and popular audiences; by the economics of the literary marketplace; and by assumptions made about class- and gender-appropriate behaviors. Writing about travel was never a straightforward and uncontested process; it was always and inescapably mediated, manipulated, and managed. In-the-field jottings became polished prose through an iterative process of moderation in just the same way that the sweat-stained and mud-bespattered traveler assumed the role of crisply authoritative witness and author. Our focus has been on related processes: the making of texts, the making of authors, and the making of truth as they came together in the making of books. We have sought to place greater emphasis than has been evident before now on the contingencies of travel writing: on travelers' inscriptive practices (and, for some, the demands of prior instructions on where to go and why); on authorship as a singular and a collective responsibility; on the function of textual editing, authorial mediation, and literary staging; on the publisher's role in making both texts and authors; and on the processes by which the truth of travel was claimed and tested. In closing, we reflect on these issues and offer some thoughts on the implications of taking seriously travel's print culture and its relationship with the making of science and truth, audience and empire.

BECOMING A TRAVELER-AUTHOR

In considering the growth of Britain's print industry during the first quarter of the nineteenth century, a commentator in the *Quarterly Review* in 1828 speculated as to whether "from a nation of shopkeepers, we shall [in time] become a nation of book-makers." Tongue in cheek as this enquiry was, it spoke nevertheless to that awareness among contemporaries of the rapidity with which the business of print had altered in the previous decades. As the *Quarterly Review* went on to note, "Some two generations ago, the market for intellect [i.e., books] was a monopoly, chiefly confined to the garrets of Grub-street; but now . . . the book-making trade, like all other trades, which political economy has set free, will spread unshackled far and wide."[4]

The house of Murray played a central role in these changes. In an era when London's printing presses abounded with books of travel and exploration, and accusations of book-making were regularly leveled by critics against publishers and authors whom they suspected of the sin of compilation, the terms "traveler" and "travel writer" appeared, superficially at least, to be synonymous. No journey, it seemed, was undertaken to the non-European world without subsequently being committed to print. Although the appetite for, and possibility of, authorship was greater in this period than it had ever before been, the intellectual and practical capacities demanded of the role of traveler, traveler-writer, and traveler-author were distinct and differently acquired: competence in one was no guarantee of competence in the other.

The making of texts and the making of travelers shared a number of common characteristics, not the least of which were replication and mimicry. Defining what was an appropriate literary style or correct in-the-field performance was often a question of travelers repeating—or rebelling against—what others before had done. Existing texts of travel provided Murray's authors with a template to follow, or to ignore, depending on the character they wished to inhabit and the characteristics they wished to convey to their audience. This decision was bound up with the expectations and demonstration of credibility and trustworthiness. Maria Graham, for example, turned *to* (and, ultimately, turned away *from*) the work of Alexander von Humboldt for guidance both in observation and inscription, witnessing and writing. For other travelers, inspiration came often from existing expertise and the social and professional context in which they operated. This was true, particularly, for the many half-pay naval officers who applied the disciplines of maritime survey and logbook writing to their land-based travel. In tracing their transcontinental journey from Lima in Peru to Pará in Brazil, for example, the British naval officers William Smyth and Frederick Lowe were assiduous in their attention to correct determination of location, altitude, and distance traversed—even when broken and malfunctioning instruments rendered those tasks impossible. Smyth and Lowe nevertheless clung to the promise of what they felt to be necessary: movement across space was to be punctuated at regular intervals by instrumental observations designed to fix their location.

While Smyth and Lowe felt the credibility of their endeavor to chart a navigable river route between the Andes and the Atlantic to depend on their survey work alone, they were keenly aware of the scientific interest that might attach to the flora and fauna encountered en route. The pair thus performed additional roles with which neither felt entirely comfortable: those of botanist and zoologist. As their *Narrative of a Journey from Lima to Para* (1836; app.: 134) at one point records, "Birds and insects are very numerous, more especially the latter. We commenced a collection of both, but . . . being novices in the art of curing them, we only preserved a few."[5] Despite lacking formal training in biology, Smyth and Lowe recognized the expectations that might attach to them in their role as travelers in South

America and sought, albeit in a limited way, to do what they felt their position demanded.

Maria Graham, and Smyth and Lowe, and many of the other explorers and travelers whose words and works we have studied here, had to negotiate their identity as travelers by attempting to resolve both what was expected of them and the ways in which they wished to present themselves. For a few individuals, existing texts of travel presented a corpus against which to position and test themselves—models to follow or strategies to avoid, procedures to replicate or ideas to refute. At least until the publication of practical guides to "correct" travel from the mid-nineteenth century—including Julian Jackson's *What to Observe* (1841), the Royal Geographical Society's *Hints to Travellers* (1854), and Francis Galton's *The Art of Travel* (1855; app.: 221)—the making of the category of traveler was an activity that depended on a series of often ad hoc judgments. Naval traditions of observation and inscription—codified in books such as John Bettesworth's *The Seaman's Sure Guide* (1783; app.: 5)—may not have influenced the conduct of many of Murray's travelers and their vision of what the intellectual and practical purpose of their journeys should be. But, even if they did, such books could not guarantee safety in the field or ensure that travel would provide easy means for writing about travel, and they could not at all influence what would happen to the explorer on his or her return and as he or she was turned, often gradually and in similarly ad hoc ways, into "the author."

The process was something akin to mapping. Very few Murray travelers journeyed with a map in hand to guide them. For several, production of a map of the region or place in question may have been an important end in view, but doing that work was often more faltering and uncertain than the final map suggested, if it was produced at all. As Archibald Edmonstone noted of his travels in Egypt in the early 1820s, "As we had not the means of taking observations, they [the maps] are merely laid down from calculations derived from comparing the camels' march with the points of the compass, and by conferring our own remarks with information obtained from the natives."[6] Making one's way as a traveler or explorer commonly involved such negotiation and compromise, effecting daily acts of correspondence among what one saw, was told, or could glean from instruments such as one's compass or camel.[7]

As the diversity of Murray's travelers makes clear, the category traveler was never homogenous and the responsibilities attached to it were never wholly agreed on. For some individuals, the pervading spirit of scientific and political curiosity in the first half of the nineteenth century rendered their role almost that of encyclopedist: whether one felt capable or not, being a traveler seemed to necessitate assuming the role of agricultural- ist, botanist, economist, geologist, meteorologist, political strategist, and zoologist in passing authoritative comment on the physical and cultural characteristics of the lands traversed. A moral duty—seeing all, reporting all—attended the privilege of geographical mobility. In such cases, how- ever, the undertaking of travel was influenced by the traveler's understand- ing of what constituted their duty: whether duty to country, science, or religion or duty to verisimilitude. For some, the exercise of duty was itself often an internally validating process: travelers who wished to become au- thors could make the claim that their desire to write and to publish came not from individual aspiration but from obligation owed in the service of Britain's political and mercantile interests or, more generally, to the geo- graphical and scientific understanding of the world. This was especially so of those explorers and Murray authors undertaking work in West Africa, the Arctic, and in Central Asia between about 1813 and about 1836 (app.: 26-135) who owed their travels and, sometimes, their accounts of it, to the patronage of John Barrow, John Croker, and the Admiralty. Duty was, thus, both obligation and license: something imposed on travelers but also often the means by which to validate and justify the act of writing travels. It is wrong to characterize Murray's travelers as scheming and manipulative, but it is clear equally that they were not passive and naive when it came to positioning themselves as travelers, even if they were also made such by the actions of others.

In the same way that what it took to be a traveler in the period un- der investigation was rarely fixed and explicit, what it took to become an author—and what was expected of authorship of a text of travels—was multiple and varied. The tendency of would-be travelers to read already- published narratives in practical preparation for their own journeys had a disciplining effect on styles of writing. Particular tropes of expression be- came associated with, and codified around, specific types of travel. Stylistic variation between accounts of government-backed high-latitude explora- tion was, for example, relatively limited (especially in the case of official expedition narratives) as explorer-authors were constrained by Admiralty- approved inscriptive practices, were guided in their stylistic composition by the published accounts that populated their shipboard libraries, and were moderated by the editorial hand of Barrow or Croker. Even in instances where travelers were untrammeled by officialdom, stylistic and structural similarities—as well as common rhetorical staging—were apparent across a range of Murray's other travel texts. This is evident, for example, in both the encyclopedic mode of observation and writing, where travelers commented with apparent authority on everything from religious practices to domestic economy and recorded instrumentally derived data with dedicated precision. And it is apparent in the avowedly antiencyclopedic mode, where authors such as Francis Bond Head, Godfrey Mundy, Sarah Lushington, and even Mansfield Parkyns eschewed claims to rigor and comprehensiveness in favor of writing that spoke of intimacy, immediacy, personal uncertainty, and, from that, emotional truth.

Authors' decisions with regard to style and genre were never made in isolation. The house of Murray and its reader-editors were instrumental in shaping travelers' written products and authorial personae. Replying to a Mr. Kinglake in November 1805 over a proposed book later rejected, John Murray I hinted that publishers knew better than their authors what was likely to find success: "A Bookseller is to an Author's mind what a Physician is to our own Body—we reason from Facts & Experience & know better than our Patients who have had occasion to attend to them."[8] The making of traveler-authors and of travel texts themselves was always profoundly collaborative, although it was not necessarily always mutual. The Murray firm's literary reputation and emergence as a geographical authority during the first half of the nineteenth century were, for certain authors, a sufficient basis to devolve responsibility for textual and authorial positioning to Murray and the firm's editors alone. George Lyon was more than happy to leave the correction of the proofs of his *Journal of a Residence and Tour in the Republic of Mexico* (1828; app.: 106) to "Mr Lacken, Mr C. Hokis, & Mr Taylor, my principal friends in the Real del Monte Company" so that he might depart Britain for his final, ill-fated journey.[9] Writing to John Murray III over the publication of his brother's *Political and Statistical Account of the British Settlements in the Straits of Malacca* (1839; app.: 142), F. S. Newbold expressed an anticipation common among prospective traveler-authors: "your Father's 'good name' will afford us the respectability, which my brother chiefly desires."[10] For Newbold, and many others, the house of Murray's judgment with regard to format, price, textual content, and stylistic rendering—brought to an acme under the direction of Murray II—were sufficiently sound to be taken as read. Even when authors had a clear opinion as to a desirable price and format, they were often willing to be guided by Murray's expertise. Francis Bond Head felt that a proposed octavo format for his South American book would, for example, be "too small, and would cramp me"; nevertheless, he assured Murray II, "I will be guided by you."[11]

The house of Murray was, in turn, influenced in its decisions on books of exploration by its close associations with the Admiralty and the Royal Geographical Society and with John Barrow in particular. The firm's strategic relationship with institutional and disciplinary authorities mattered to the ways in which its authoritativeness was demonstrated and managed and the impact this had on the texts published under the firm's imprint. That is not to say that this relationship was without its tensions over the publication of exploration accounts. John Murray II had occasion to write stiffly to John Croker at the Admiralty in 1821 over the publication that year of William Edward Parry's first Arctic voyage given what he, Murray, saw as the Admiralty's laxness in allowing a separate account to appear in print (perhaps he had in mind Barrow's experiences in 1818 with Tuckey's volume):

> I beg leave most respectfully to state for the information of the Lords Commissioners of the Admiralty that I have paid to Captain W. E. Parry the sum of One Thousand Guineas for the entire Copyright of His "Journal of the Late Voyage for the Discovery of a North West Passage" and also for the exclusive right of first Publication on that subject (which I was assured their Lordships had vested in me and would secure to Captain Parry.)—That I have further been at a considerable additional expense in procuring the necessary & suitable Charts, Maps & other engravings—and in printing the Work with a view in all respects to render its appearance not unworthy of their Lordships' "authority" with which the imprint has been officially honored.
>
> That I have used every means in my power to effect the speedy publication of this Work—I beg you Sir further to state to their Lordships that I now find wt. astonishment and mortification that all my expense and labour will terminate most probably in loss and disappointment in consequence of the unexpected and as I consider illegal publication by another account of the same Voyage by Mr Alexander Fisher who accompanied the expedition in quality of Surgeon on board H. M. S. Griper.—I therefore throw myself on the protection of their Lordships and solicit such remedy as the case may appear to require—and their Lordships may be disposed in their Wisdom to grant me.[12]

Croker's reply is not recorded, but the incident is illustrative of many of the themes we have explored: Murray II's recognition of the Admiralty's "authority" and his firm's need to publish works in keeping with its status, as, of course, the house did for many of the Arctic and African narratives in particular; anxieties over any loss of control of the intended publication, here through another publisher, more often by authorial delays in producing on time or the failure to write with an eye to the intended audience; and the costs, economic and reputational, that could attend such delays and mishaps.

In general, the relationship between the Murrays and the firm's traveler-authors was often guided by deference (whether or not such sentiments were genuinely held on the part of the author). While the majority of Murray's authors were deferential and expressed satisfaction with the volumes thus produced, the example of Richard and John Lander—and the slight they felt at being traduced by the editorial introduction to their own volume—makes clear that the firm's judgment was not, at least from an author's point of view, always to be met with satisfaction. This would have been unavoidable given the range of opinion that existed as to what counted as appropriate and desirable in a work of travel. Even so, the Murray firm became generally more skillful and nuanced in its adjudication of travel writing and travel writers: judgments that served as the basis for the development of its reputation as a geographical authority.[13]

Our attention to the epistemological and physical movement of in-the-field travels to printed text, and to the construction of books and their authors, has revealed the persistent importance of negotiation: not just between author and publisher, and publisher and public, but also at a more personal level as travelers made choices, in different circumstances, about what was demanded of them in terms of bodily performance, inscriptive practice, and observational regime. In almost all instances, what guided these decisions, for authors and publisher alike, was a concern for credibility—that what was done in the field, what was committed to paper, and what became print had the potential to render them trustworthy and their texts truthful. At base, truth in texts of travel was a question of correspondence: of demonstrating a correlative relationship between the world as experienced and the world as written about. Such a seemingly simple relationship concealed a complex series of practices and judgments designed to shepherd the process of correspondence through the varying demands of audience, genre, class, and gender. For that reason, one issue above all others has emerged as central to the making of Murray's authors and their texts: truth and claims to credibility.

TRUTH AND THE CREDIT ECONOMY

Perhaps the most significant consequence of this investigation into the production of travel writing between the late Enlightenment and the mid-nineteenth century is what it tells us about truth, trust, and credibility. Our attention to questions of credibility has shown the making of travel texts to be inescapably the work of a credit economy—credibility being a form of cultural capital acquired through shrewd and judicious practices, invested in particular claims of stylistic or observational appropriateness and turned either to profit or loss when subject to evaluation by editor, publisher, or audience. Like any economy, there was more than one way to earn credit through travel writing and thus to demonstrate truth and to earn trust. What counted as credible—and, therefore, what counted as true—was inherently variable, and its correct demonstration required careful judgment on the part of Murray and the firm's travelers as to what would be expected by different audiences of different kinds of travel writing.

Credibility could be earned not just differently but, often, in entirely antithetical ways. Depending on the audience to whom a text was addressed, a traveler might demonstrate his or her credibility equally well by claiming to be naive, unskilled, and ingenuous (Godfrey Mundy, or George Lyon, for example) or by advertising warrants of professional capability, exhibiting rigorous scientific practice, and asserting expert subject knowledge (William Martin Leake, William Hamilton, Francis Galton, to name only a few). That two such dissimilar routes to securing credibility could exist simultaneously points to the fact that what travel and writing about travel were assumed to necessitate varied according to the requirements and expectations of different reading publics. It varied also with respect to authorship. William St. Clair's influential discussion of reading cultures in Britain in this period is predicated on the connections among reading and mentalities and in tracing, backward as it were, the connective tissues between reading and book production: "To trace readership, we need to trace access. To trace access, we need to trace price. To trace price, we need to trace intellectual property, and to trace intellectual property, we need to trace the changing relationship between the book industry and the state."[14] Here, we have tried to trace the relationships linking published narratives and exploration and travel by tracing Murray's management of the authority of the firm's books and the firm's sense of its audience. To trace a book's authoritativeness, we need to trace its authorship. To trace its authorship, we need to trace its authoring, by one or more persons, and disclose the processes by which claims to authoritativeness in the text were secured. To trace authorization and to trace the author means we have to know what demands may have been required of them, by the state or others, as the act of exploration was undertaken or was being planned. Texts of travel were read, at turns, for their geographical and scientific revelations, their political and economic insights, and their exoticism and adventurousness. Each of these criteria demanded of the author (and of the publisher) different things in terms of content, style, and price. The role of Murray and the firm's editors and interlocutors was one of positioning and staging—of ensuring that the right text met with the right audience. It is clear from the opinions of the contemporary periodical press, however, that what was regarded as appropriate—in style and content—was never fixed and agreed on. Judgments of appropriateness varied not only through time but also between different periodicals and the audiences to whom they were addressed.

Certain travelers appeared supremely confident in their choice of style and content. Others agonized about their self-presentation and the substance of their writing. In some, this uncertainty or anxiety as to appropriate content reflected and engendered an encyclopedist mode of writing—the desire or perceived need to address as many topics as possible in an effort to ensure the satisfaction of the largest number of readers. It was as a reaction precisely to this highly procedural mode of observation and inscription that a desire emerged among certain authors and audiences for more immediate and subjective forms of writing. Quite apart from the mode of writing that particular audiences favored at specific times—"rough writing" that resulted from "rough travel" and produced books in "rough dress" such as Francis Bond Head's *Rough Notes* (1826; app.: 95) or Godfrey Mundy's *Pen and Pencil Sketches* (1832; app.: 125) was different from Mary Montagu's epistolary observations in her *Letters* (app.: 6) and from John Gardiner Kinnear's letter-based *Cairo, Petra, and Damascus* (1841; app.: 151)—what unified judgments of travel writing was the assessment and evaluation of their truthfulness. What mattered more than anything else in the process of gaining credit was assuring readers that the world as written about accurately corresponded with the world as it had been experienced. Verisimilitude was a question of testimony—of the social and epistemological capital of first-hand witnessing and of the ability of travelers to demonstrate appropriate discretion when their claims to truth relied on the reports of guides, indigenous informants, and other go-betweens. These issues were not distinct to authors of travel narratives or to their publishers in the period we have studied. In the legal world, Jeremy Bentham's *Treatise in Judicial Evidence* (1825) argued that testimony needed to be backed up by material proof—in things, not words alone—while in the emerging world of the systematic social sciences, authors such as Harriet Martineau in her *How to Observe* (1838) were stressing the need for rigor and method.[15] But precisely because travelers could not always bring back things (or, where they did, because such things did not accompany the written assessments of them) or because travelers did not always come back, written words had to suffice as surrogates of empirical encounter, statements of truth, and evidence of moral rectitude.

Ensuring correspondence between experience and inscription, witnessing and writing, was a matter, at turns, of observational procedure and stylistic flourish. Instrumentally derived observational data sat alongside affective and pulse-quickening prose as strategies employed by Murray's authors to communicate the quantitative and qualitative experience of being in the field. Verisimilitude could be communicated either through carefully composed illustrations and precise and detailed maps or, more subjectively, through writing that situated the traveler's emotional response to the fore. The imperative on travelers was to give the impression of communicating verbatim. The imperative on the publisher was to give the impression that the author was worth the effort made, by publisher and by the reader, even if it meant that maps and images and words were "got up" in the book's making. As a critic in the *Westminster Review* noted in 1865, "It is far the best for the traveller to tell us everything he sees or hears, just as he sees or hears it; and everything that happens, just as it happens." The onus would fall, then, on the reader to make a judgment as to the author's veracity. Irrespective of the observational regime and rhetorical strategies employed by authors, the ultimate judgment of their credibility lay in the hands of their readers. The *Review*'s critic went on to summarize the processes by which nineteenth-century readers might evaluate texts of travel. The factors identified reinforce the importance of observational and instructive strategies and underline how readers' expectations of texts might vary on account of their authors' personal circumstances and the geographical regions traversed. For the *Review*, "The estimate of a given traveller's tale will be made partly from a consideration of his own personal achievements, literary and locomotive—partly from regard had to the special situation or circumstances of the lands he has visited":

> A book of travels does not admit of the test of cross-examination. It will, therefore, be reliable and serviceable so far only as by its particularity of detail, by its varied and copious anticipation of every question and every want, and by a certain ingenious transfusion of the author's own personality into the story, it at once supersedes and forbids a judicial scrutiny. It depends on the traveller himself how he makes his observations, and what observations he makes; how he multiplies them, and how he records them. It is for his readers to classify those observations, to select from them, and turn them to account as best they may—each appropriating what is adapted to his own special needs and independent province of inquiry.[16]

Although their appropriation depended on satisfying particular expectations, texts of travel were, in an important sense, whatever their readers chose to make of them. Travelers (and their books) were thus "instruments in the hands of others"—geographical proxies whose observations and physical endeavors were generative of knowledge circulated, evaluated, and accumulated.[17] This is not to place questions of significance only in the hands and mind of the reader. As we have also shown, many travelers were no less "instruments in the hands of others" by virtue of the instructions they had received about how and where to travel and on what they were to observe and write.

Knowing what readers might want—and understanding what for them would constitute convincing testimony—was the responsibility of Murray and the firm's editors and advisers. Being able correctly to counsel prospective authors on content and style was crucial not only to the potential success of an individual text but also more generally for the credibility and authority of the firm. Just as Murray made authors, so authors made Murray—a mutual relationship that pivoted around the construction and demonstration of credibility. Although the house of Murray emerged as Britain's leading nineteenth-century publisher of travel texts, its ultimate status in that respect was hard-won and reflected the importance of its careful assessment of marketplace demand and critical expectation. Under the guidance of John Murray II and III, increasingly sophisticated efforts were made to evaluate audience demand, to market books appropriately based on price and format, and—through such choices—to manage the reputation of the firm's authors and, by so doing, the firm itself. John Murray III knew this as he wrote to Charles Darwin in the ways we have seen (chap. 6, p. 200) to secure the latter's *Journal of Researches* (app.: 174). Humphry Sandwith, traveler in Asia Minor and, later, author of works on landlordism and on religion, knew this to be the case when he approached Murray III in December 1859 over a further volume: "I think the success of the last (tho' depending almost altogether on the popularity of the subject) will help this off as far as to pay expenses at all events, but of this you are the best judge. The enclosed has been written in huts, caravanserais, steam boats, coffee rooms & railway stations, so that I fear it will not have much merit as a *literary* production, but I am not & never pretended to be a literary man."[18] Literary men and women or not, those persons who were Murray travel authors were so in no small measure because that firm managed them, and its literary productions in that genre, well.

TRAVELS IN PRINT AND BEYOND

Our concern in this book has been to treat texts of travel as more than simply repositories of geographical knowledge—to see them instead as contested objects of reputational and epistemic construction. While similar arguments could be made in relation to other genres of writing, travel texts are particularly rich in their potential to reveal the processes by which authorship, credibility, truth, and audience (among other important concepts) were defined, constructed, and contested. In our period of concern, at least, travel writing as a genre straddled the professional and the amateur, the disciplinary specialist and the popular author, the worlds of scientific instruction and of literary endorsement. For this reason, its analysis speaks to the nature of Britain's print culture between the late eighteenth and mid-nineteenth centuries and has important interdisciplinary implications for how the relationships between author and publisher, audience and marketplace, authorship and truth are to be understood in historical and geographical context.

In recognizing, with others, that travel writing is "best considered as a broad and ever-shifting genre, with a complex history which has yet to be properly studied," we have also demonstrated that it is a genre whose diverse geography, biography, and routes to material production all merit detailed attention.[19] The Murray travel imprint was, of course, not only about different parts of the world: its analysis has also shown how different parts of the world could influence how travel was undertaken, or what was written about the travel experience itself, and how authors and publisher put exploration accounts together in ways that were attentive to the literary marketplace for the work in question. Mary Baine Campbell has made the telling observation, in relation to travel writing and its theory, that "much of the theoretically informed writing on travel and travel writing has had to do with imperial periods of the later eighteenth, nineteenth, and early twentieth centuries, in which the geographical surveying of the globe as well as the anthropological investigation of its non-metropolitan 'or cityless' (*aporoi*) peoples produced so much knowledge in the service of so much desire for power and wealth."[20] We would not disagree, but our concern has been much more with the empirical production of works of travel, with their undertaking in the field, and with the mediating relationships between author and publisher, publisher and audience, and, thus, more with the content and narrative form of nearly 240 books of travel and exploration, than with the connections between this material and its imperial-cum-commercial context. Our concern has not been to "prove," one way or another, the applicability of others' notions of book history, although we hope that we have shown that geographers can work with literary scholars and historians of the book to map new possibilities for the study of works of travel and exploration as cultural artifacts.[21]

If, as others have noted, "two distinct journeys" must have occurred in order for a (travel) narrative to take its final form, "the first geographical, the second authorial," understanding the content and the material form of the "final" product requires that we should examine—as, indeed, we hope we have—the relationship between the two, not least in assessing the truth claims of the narrative.[22] This requires, and more so, perhaps, than has been done to date, that scholars should examine the motivation for travel, the impulse or instructions to explore, before the traveler set out: not just before he or she could be called "author" but also before even the epithet "traveler" or "explorer" could properly be used of them. Leah Price has proposed new ways of thinking about how books are received, including even their rejection, in Victorian Britain, noting rightly how making "book-historical claims on the basis of textual evidence is not the same as making cultural-historical claims on the basis of literary evidence."[23] Our focus has been rather with ways of thinking about book production, with authorship, authority, and authoritativeness, and in making book-geographical claims on the evidence of literary process as explorers became authors. For some of those men and women who became Murray authors, travel was an act and a process of moral refinement and cultural acquisition, not merely a matter of imperial authority or commercial gain. Their end in view was self-improvement. For others, quite the reverse was true: many of those who produced works of exploration did so because they were instructed to explore, to revise contemporary geographical knowledge, or to extend the reach of Europe in other ways. It is for that reason, of course, that scrutinizing the declared aims of the traveler or explorer once he or she was an author is also important in further studies of travel writing. The trope of reluctance, as also that of authoritativeness, was a commonplace: rather than read these at face value, we need to ask in whose interests were such statements made? What purpose did they serve?

Illuminating the relationships connecting the category of author, proclamations of authority, and the chosen means to textual authoritativeness is crucial in the study of travel writing. For numerous Murray authors-to-be, their authority in writing what they did, when they did, and in the ways they did (and, even, where they did) was vested in the facts of institutional association and, vitally, in direct personal encounter. But even for those who proclaimed immediacy, we must be alert not just to the textual strategies employed in the field but also, and vitally, to the retrospective embellishment of the text-to-be by the author, the publisher, or the publisher's agents. Exactly what truth is being advanced, on what basis, and by whom exactly? This is a theoretical matter. As Mary Baine Campbell has noted, "A text that generically proffers itself as 'true,' as a representation of unaltered reality, makes a perfect test case for analytical work that tries to posit or explain the fundamental fictionality of all representation."[24]

This issue of truth in texts and in authorial candor is also a practical, even a methodological, matter. As Ian MacLaren has pointed out, "Several matters round into focus if one dwells on this rather obvious habit of blurring the distinction between, on the one hand, exploring, and, on the other, inscribing, transcribing, editing, publishing, and identifying manuscripts as a whole rather than emphasizing the different stages of their composition." It is practical inasmuch as the detailed exegesis of texts of travels, as MacLaren further points out, may have to proceed on a case-by-case basis, book by book, author by author.[25] Again, we would not disagree. Ours has been a case-by-case, book-by-book, study. But we have also shown the value of looking at one publisher's imprint in order that individual texts may be placed in relation to one another given the common denominator of a shared publisher and of in-house practices. Key individuals—John Murray II, John Barrow, and sets of Murray readers such as Henry Milton, Francis Bond Head, and Walter Hamilton—certainly gave practical shape to authors' intentions in their judgments on, and alterations of, their manuscript accounts. Even so, it is hard to determine a "Murray model" of authorial redaction or editorial involvement as a matter of house policy. The process of authoring, of turning the Explorer into the Author, was involved and collaborative, but it was always attentive to the demands of given books and their authors-to-be and was never formulaic. Turning travels into print was often inescapably hard intellectual labor. Not for nothing did David Livingstone ruefully conclude, "I would rather cross the African continent again than undertake to write another book."[26]

Understanding the processes by which travelers' narratives took shape depends, too, on the nature of the archive that survives. Assessment of the prepublication incarnations of explorers' and travelers' printed narratives is only possible when and because these materials survive in one archive or another. But archives are not neutral spaces. Understood at a macroscale, they emerged from a context of metropolitan, national, and imperial record keeping and served the interests of state powers, a fact that makes it difficult to produce a historiography for the critical exegesis of travel writing that is not led by the existence of archival sources as remnants of imperial authority.[27] At a microscale, as scholarship in archival studies has demonstrated, archives and the processes of archiving reflect and produce particular structures of authority—in their classification schemes, in the authority of the archivist, and in the ways in which what survives is always partial, the residual traces of larger bodies of evidence.[28]

The greater import of these remarks is clear. Either from the confines of a disciplinary specialism or from the perspective of interdisciplinary enquiry, students of travel writing must no longer regard the printed issue of explorers' narratives and travelers' tales as the definitive version of the events they purport to describe without, at the very least, being alive to the possibility, and to the responsibility of showing, that an earlier version is extant, either in whole or in fragmentary traces. If a manuscript account or preprint version survives, it behooves us to show what changes were made, by whom, and, where we can, why. Because books of travel were powerful forms of "communicative action," as Harriet Martineau so well recognized, their study requires us, where we can, to expose the conditions of their making rather than to be content with the content alone.[29] Print travels as a powerful means of information exchange. But it does not do so in ways of its own choosing, somehow beyond human agency or divorced from determining material circumstances. As Adrian Johns has proclaimed, "The history of the book is consequential because it addresses the conditions in which knowledge has been made and utilized."[30] Precisely because all print "travels" to impart its message, we need to show more widely, as we have here demonstrated for works of geographical exploration and travel between the 1770s and the late 1850s, how the printed word was part of more complex circumstances of writing and publishing; what it meant to be called an explorer or an author; and what it took for the world to become words.

Appendix

BOOKS OF NON-EUROPEAN TRAVEL AND
EXPLORATION PUBLISHED BY JOHN MURRAY
BETWEEN 1773 AND 1859: BY DATE OF
FIRST IMPRINT, WITH NOTES ON
EDITION HISTORY BEFORE 1901

Note to users: Despite the literary significance of the house of Murray, the majority of its published output has not been subject to systematic description and identification. William Zachs, in *The First John Murray* (1998), assembled a comprehensive checklist of titles issued under the supervision of Murray I. No similar bibliography exists for the subsequent two hundred years of the firm's history. The following list enumerates John Murray's 239 books of non-European travel and exploration in the period between 1773 and 1859. As in all such bibliographical enterprises, particularly given the generic vagaries of travel texts, accuracy and comprehensiveness are imperatives but impossible to guarantee.

The information assembled here relating to authorship, titling, printing, price, format, and edition history of the Murray firm's texts of non-European travel will, we hope, serve as a working bibliographical resource for book historians, literary scholars, and geographers as well as being the database for our study here. Wherever possible, authors' names have been verified by reference to, and rendered as they appear in, the *Oxford Dictionary of National Biography*. In other instances, names and attributions have been verified by reference to the OCLC Virtual International Authority File.

Long titles and imprint details are presented as they appear on the title page of first editions, retaining, verbatim, spelling and punctuation. Edition history has been traced using Copac and WorldCat union catalogs, and bibliographical details have been verified by inspection either of printed copies at the National Library of Scotland and the British Library, electronic equivalents (accessed using JISC Historic Books or the HathiTrust Digital Library), or with the assistance of librarians at a number of institutions (as noted in our preface and acknowledgments). Only numbered or revised editions of books are listed; reprints are not enumerated here. Details of price and format have been verified by reference to advertisements in contemporary periodicals, to the Murray firm's lists of published and forthcoming texts, and to the firm's ledger books. Book formats (which refer not to the absolute size of a volume but to the number of times the original printed sheet was folded to make each section, or signature, of the book) are abbreviated thus: 2° (folio); 4° (quarto); 8° (octavo); 12° (duodecimo); and 18° (octodecimo). In general terms, the greater the number of folds, the smaller and cheaper the resulting book: folio and quarto books were prestige items, while octavo, duodecimo, and octodecimo were produced with a wider audience in mind. For books printed between 1773 and 1800, the English Short Title Catalogue (ESTC) citation number is provided. For books printed between 1801 and 1859, the Nineteenth-Century Short Title Catalogue (NSTC) number is provided. In relation to authors' biographical details, the following abbreviations have been used: b. (born); bap. (baptized); d. (died).

1. 1773. Parkinson, Sydney (d. 1771). Note: "Faithfully transcribed" and edited by Parkinson, Stanfield (d. 1776).

A journal of a voyage to the South Seas, in His Majesty's ship, the Endeavour. *Faithfully transcribed from the papers of Sydney Parkinson, draughtsman to Joseph Banks, Esq. on his late expedition, with Dr. Solander, round the world. Embellished with views and designs, delineated by the author, and engraved by capital artists.*

1 vol. 4° £1 5s. London: Printed for Stanfield Parkinson, the Editor: And sold by Messrs Richardson and Urquhart, at the Royal-Exchange; Evans, in Pater-noster Row; Hooper, on Ludgate-Hill; Murray, in Fleet-Street; Leacroft, at Charing-Cross; and Riley, in Curzon-Street, May-Fair. Printer: unrecorded. ESTC: T147793.

No revised editions.

2. 1779. Robertson, Robert (1742–1829).

A physical journal kept on board His Majesty's ship Rainbow, *during three voyages to the coast of Africa, and West Indies, in the years 1772, 1773, and 1774: to which is prefixed, a particular account of the remitting fever, which happened on board of His Majesty's sloop* Weasel, *on that coast, in 1769.*

1 vol. 4° 10s. London: Printed for J. Murray, No. 32, Fleet-Street. Printer: unrecorded. ESTC: T153715.

No revised editions.

3. 1782. [Macintosh, William (?–?)]. Note: published anonymously.

Travels in Europe, Asia, and Africa; describing characters, customs, manners, laws, and productions of nature and art: containing various remarks on the political and commercial interests of Great Britain: and delineating, in particular, a new system for the government and improvement of the British settlements in the East Indies: begun in the year 1777, and finished in 1781.

2 vols. 8° 12s. London: Printed for J. Murray, No. 32, Fleet-Street. Printer: unrecorded. ESTC: T97734.

No revised editions.

4. 1783. Benjamin of Tudela (1130–73). Note: "Faithfully translated from the original Hebrew" by Gerrans, B. (?–?).

Travels of Rabbi Benjamin, son of Jonah, of Tudela: through Europe, Asia, and Africa; from the ancient kingdom of Navarre, to the frontiers of China.

1 vol. 12° 5s. London: Printed for the translator; and sold by Messrs. Robson, New Bond-street; J. Murray, in Fleet-street; T. Davis, Holdborn; W. Law, Ave-maria-lane; and at No. 7 Canterbury-Square, Southwark. Printer: unrecorded. ESTC: T51601.

No revised editions.

5. 1783. Bettesworth, John (?–?).

The seaman's sure guide, or, navigator's pocket remembrancer: wherein are given such plain instructions in every useful branch of navigation, as will in a short time form the complete mariner.

1 vol. 12° 3s. London: Printed for the author, and S. Hooper, No. 212, High-Holborn, facing Southampton-Street, Bloomsbury-Square; J. Murray, Fleet-Street; and D. Steel, on Tower-Hill. Printer: unrecorded. ESTC: T114215.

No revised editions.

6. 1784. Montagu, Lady Mary Wortley [née Lady Mary Pierrepont] (bap. 1689, d. 1762). Note: authorship redacted on title page by means of dashes.

Letters of the Right Honourable Lady M—y W——y M——e: written during her travels in Europe, Asia and Africa, to persons of distinction, men of letters, &c. in different parts of Europe. Which contain among other curious relations, accounts of the policy and manners of the Turks; drawn from sources that have been inaccessible to other travellers. A new edition. To which are now first added, poems, by the same author.

2 vols. 8° 4s. London: Printed for T. Cadell, and T. Evans, in the Strand; J. Murray, in Fleet-Street; and R. Baldwin, in Pater-noster-Row. Printer: unrecorded. ESTC: T66781.

No revised editions.

7. 1787. [Young, Robert (?–?)]. Note: published anonymously.

An account of the loss of His Majesty's ship Deal Castle. *Commanded by Capt. James Hawkins, off the island of Porto Rico, during the hurricane in the West Indies, in the year 1780.*

1 vol. 4° 15s. London: Printed for J. Murray, No. 32, Fleet-Street. Printer: unrecorded. ESTC: T102124.

No revised editions.

8. 1788. [Robertson, George (d. 1791)]. Note: published anonymously.

A short account of a passage from China, late in the season, down the China Seas, through the southern Natuna Islands, along the west coast of Borneo, through the Straits of Billiton or (Clements Straits) to the Straits of Sunda.

1 vol. 4° £2 12s. 6d. London: Printed for J. Murray, No 32, Fleet-Street. Printer: unrecorded. ESTC: T181538.

No revised editions.

9. 1789. Equiano, Olaudah [Vassa, Gustavus] (ca. 1745–1797).

The interesting narrative of the life of Olaudah Equiano, or Gustavus Vassa, the African. Written by himself.

2 vols. 12° 7s. London: Printed for the Author, No. 10, Union-Street, Middlesex-Hospital. Sold also by Mr. Johnson, St. Paul's Church-Yard; Mr. Murray, Fleet-Street; Messrs. Robson and Clark, Bond-street; Mr. Davis, opposite Gray's-Inn, Holborn; Messrs. Shepperson and Reynolds, and

Mr. Jackson, Oxford-Street; Mr. Lackington, Chiswell-Street; Mr. Mathews, Strand; Mr. Murray, Prince's-Street, Soho; Mess. Taylor and Co. South Arch, Royal Exchange; Mr. Button, Newington-Causeway; Mr. Parsons, Paternoster-Row; and may be had of all the booksellers in Town and Country. Printer: unrecorded. ESTC: T140573.

No revised editions.

10. 1790. [Dallas, Robert Charles (1754–1824)]. Note: published anonymously.

A short journey in the West Indies, in which are interspersed, curious anecdotes and characters.

2 vols. 8° 5s. London: Printed for the author, and sold by J. Murray, Fleet-Street, and J. Forbes, Covent-Garden. Printer: unrecorded. ESTC: T110820.

No revised editions.

11. 1790. Le Couteur, John (1761–1835). Note: "Translated from the French."

Letters chiefly from India; containing an account of the military transactions on the coast of Malabar, during the late war: together with a short description of the religion, manners, and customs, of the inhabitants of Hindostan.

1 vol. 8° 6s. London: Printed for J. Murray, No 32, Fleet-Street. Printer: unrecorded. ESTC: T92283.

No revised editions.

12. 1791. Carter, George (bap. 1737, d. 1794).

A narrative of the loss of the Grosvenor *East Indiaman, which was unfortunately wrecked upon the coast of Caffraria, somewhere between the 27th and 32nd degrees of southern latitude, on the 4th of August, 1782, compiled from an examination of John Hynes, one of the unfortunate survivors.*

1 vol. 8° 3s. London: Printed for J. Murray, Fleet-Street, and William Lane, Leadenhall-Street. Printer: Minerva Press. ESTC: T136158.

No revised editions.

13. 1791. Pagès, Pierre Marie François de (1740–92). Note: "Translated from the French."

Travels round the world, in the years 1767, 1768, 1769, 1770, 1771.

3 vols. 8° 8s. (vols. 1 and 2); 5s. (vol. 3). Published 1791–92. Vol. 3 dated 1792. London: Printed for J. Murray, No 32, Fleet Street. Printer: unrecorded. ESTC: T93539.

1793: second edition, "corrected and enlarged" (printer: as first edition).

14. 1792. [Beddoes, Thomas (1760–1808)]. Note: published anonymously.

Alexander's expedition down the Hydaspes & the Indus to the Indian Ocean.

1 vol. 4° 7s. 9d. London: Sold by J. Murray, No. 32, Fleet-Street; and James Phillips, George-Yard, Lombard-Street. Printer: unrecorded (known to be J. Edmunds, Madeley, Shropshire). ESTC: T142758.

No revised editions.

15. 1795. Taylor, John (d. 1808).

Considerations on the practicability and advantages of a more speedy communication between Great Britain and her possessions in India: with the outline of a plan for the more ready conveyance of intelligence over-land by the way of Suez; and an appendix, containing instructions for travellers to India, by different routes, in Europe as well as Asia.

1 vol. 4° 4s. London: printed for J. Murray and S. Highley, (successors to the late Mr. Murray) No. 32, Fleet Street. Printer: unrecorded. ESTC: T146368.

No revised editions.

16. 1799. Taylor, John (d. 1808).

Travels from England to India, in the year 1789, by way of the Tyrol, Venice, Scandaroon, Aleppo, and over the Great Desert to Bussora; with instructions for travellers; and an account of the experience of travelling, &c. &c.

2 vols. 8° 15s. London: Printed for J. Carpenter, Old Bond-Street; and Murray and Highly [*sic*], Fleet-Street. Printer: S. Low, Berwick Street, Soho. ESTC: T146710.

No revised editions.

17. 1800. Taylor, John (d. 1808).

Letters on India, political, commercial, and military, relative to subjects important to the British interests in the East. Addressed to a proprietor of East-India stock.

1 vol. 4° £1 1s. London: Printed for Messrs. Carpenter and Co. Old Bond-Street; Egerton, Whitehall; Murray and Highley, Fleet-Street; Wallis, Paternoster-Row; Vernor and Hood, Poultry; and Black and Parry, Leadenhall-Street. Printer: S. Hamilton, Falcon-Court, Fleet-Street. ESTC: T153471.

No revised editions.

18. 1802. Pallas, Peter Simon (1741–1811). Note: "Translated from the German," by [Blagdon, Francis William, 1778–1819].

Travels through the southern provinces of the Russian empire, in the years 1793 and 1794.

2 vols. 4° £7 7s. (£9 9s. on "fine Royal paper, with early Impressions of the Plates"). London: printed for T. N. Longman and O. Rees, Paternoster-Row; T. Cadell Jun. and W. Davies, Strand; and J. Murray and S. Highley, Fleet-Street. Printer: A. Strahan, Printers-Street. NSTC: P164.

No revised editions.

19. 1805. Parkinson, Richard (bap. 1747, d. 1815).

A tour in America, in 1798, 1799, and 1800. Exhibiting sketches of society and manners, and a particular account of the American system of agriculture, with its recent improvements.

2 vols. 8° 15s. London: Printed for J. Harding, St James's-Street; and J. Murray, Fleet-Street. Printer: T. Davison, Whitefriars. NSTC: B3444.

No revised editions.

20. 1806. Bell, John (1691–1780).

Travels from St. Petersburgh in Russia, to various parts of Asia.

"A new edition, in one volume." 1 vol. 8° (price unrecorded). Edinburgh: Printed for William Creech, and sold by John Murray, 32, Fleet Street, London. Printer: Alex. Smellie. NSTC: B1379.

No revised editions.

21. 1807. Savage, John (1770–1838).

Some account of New Zealand; particularly the Bay of Islands, and surrounding country; with a description of the religion and government, language, arts, manufactures, manners, and customs of the natives, &c. &c.

1 vol. 8° 5s. 6d. London: Printed for J. Murray, Fleet-Street; and A. Constable and Co. Edinburgh. Printer: W. Wilson, at the Union Printing Office, St John's Square. NSTC: S549.

No revised editions.

22. 1811. Kerr, Robert (1757–1813).

A general history and collection of voyages and travels, arranged in systematic order: forming a complete history of the origin and progress of navigation, discovery, and commerce, by sea and land, from the earliest ages to the present time.

18 vols. 8° 6s. Published 1811–24. Murray's name appears only on vols. 1–16. Vols. 1–3 dated 1811; vols. 4–7 dated 1812; vols. 8–9 dated 1813; vols. 10–12 dated 1814; vols. 13–16 dated 1815; and vols. 17–18 dated 1824. Edinburgh: Printed for William Blackwood, South Bridge Street; J. Murray, Fleet Street, R. Baldwin, Paternoster Row, London; and J. Cuming [*sic*], Dublin. Printer: George Ramsay and Company. NSTC: H1848.

No revised editions.

23. 1812. Alcedo, Antonio de (1735–1812). Note: translated by Thompson, George Alexander (?–?).

The geographical and historical dictionary of America and the West Indies. Containing an entire translation of the Spanish work of Colonel Don Antonio de Alcedo, Captain of the Royal Spanish Guards, and member of the Royal Academy of History: with large additions and compilations from modern voyages and travels, and from original authentic information.

5 vols. 4° £10 10s. Published 1812–15. Vols. 1–3 dated 1812; vol. 4 dated 1814; and vol. 5 dated 1815. London: Printed for James Carpenter, Old Bond-Street; Longman, Hurst, Rees, Orme, and Brown, Paternoster-Row; White, Cochrane, and Co. and Murray, Fleet-Street, London; Parker, Oxford; and Deighton, Cambridge. Printer: Harding and Wright, St John's Square. NSTC: A771.

No revised editions.

24. 1812. Malcolm, Sir John (1769–1833).

Sketch of the Sikhs; a singular nation, who inhabit the provinces of the Penjab, situated between the rivers Jumna and Indus.

1 vol. 8° 8s. 6d. London: Printed for John Murray, Albemarle Street. Printer: James Moyes, Greville Street, Hatton Garden. NSTC: M803.

No revised editions.

25. 1813. Broughton, Thomas Duer (1778–1835).

Letters written in a Mahratta camp during the year 1809, descriptive of the character, manners, domestic habits, and religious ceremonies, of the Mahrattas.

1 vol. 4° £2 8s. London: Printed for John Murray, 50, Albemarle Street. Printer: T. Davison, Lombard-street, Whitefriars. NSTC: B4709.

1813: issued under the variant title *The costume, character, manners, domestic habits, and religious ceremonies of the Mahrattas* (printer: as first edition).

26. 1813. Kinneir, Sir John Macdonald (1782–1830).

A geographical memoir of the Persian empire.

1 vol. 4° £2 2s. (£3 13s. 6d. with map). London: Printed for John Murray, Albemarle-Street. Printer: Cox and Baylis, 75, Gt Queen-Street, Lincoln's-Inn-Fields. NSTC: K649.

No revised editions.

27. 1813. Kruzenshtern, Ivan Fedorovich [also known as Adam Johann] von (1770–1846). Note: "Translated from the original German" by Hoppner, (Richard) Belgrave (1786–1872).

Voyage round the world, in the years 1803, 1804, 1805, & 1806, by order of His Imperial Majesty Alexander the First, on board the ships Nadesha *and* Neva, *under the command of Captain A. J. von Krusenstern, of the Imperial Navy.*

2 vols. 4° £3 3s. London: Printed for John Murray, bookseller to the Admiralty and the Board of Longitude, 50, Albemarle Street. Printer: C. Roworth, Bell-yard, Temple-bar. NSTC: K915.

No revised editions.

28. 1814. Alexander, William (1767–1816).

Picturesque representations of the dress and manners of the Chinese.

5 vols. Royal 8° £15 15s. London: Printed for John Murray, Albemarle-Street. Printer: W. Bulmer and Co. Cleveland-Row. NSTC: A864. Issued as part of a series, whose other numbers included *Picturesque representations of the dress and manners of the Austrians* (NSTC: unrecorded); *Picturesque representations of the dress and manners of the English* (NSTC: E927); *Picturesque representations of the dress and manners of the Russians* (NSTC: R2104); and *Picturesque representations of the dress and manners of the Turks* (NSTC: T1847).

No revised editions.

29. 1814. Humboldt, Alexander von (1769–1859). Note: "Translated into English" by Williams, Helen Maria (1759–1827).

Researches, concerning the institutions & monuments of the ancient inhabitants of America, with descriptions & views of some of the most striking scenes in the Cordilleras!

2 vols. 8° £1 6s. 6d. London: Published by Longman, Hurst, Rees, Orme & Brown, J. Murray & H. Colburn. Printer: W. Pople, Chancery Lane. NSTC: H3024.

No revised editions.

30. 1814. Humboldt, Alexander von (1769–1859), and Bonpland, Aimé (1773–1858). Note: "Translated into English" by Williams, Helen Maria (1759–1827).

Personal narrative of travels to the equinoctial regions of the new continent, during the years 1799–1804.

7 vols. Published 1814–29. Murray's name appears only on vols. 1 and 3 (there being no vol. 2) 8° £1 6s. Vol. 3 dated 1818. London: Printed for Longman, Hurst, Rees, Orme, and Brown, Paternoster Row; J. Murray, Albemarle Street; and H. Colburn, Conduit Street. Printer: W. Pople, 67, Chancery Lane. NSTC: H3026.

No revised editions.

31. 1815. Elphinstone, Mountstuart (1779–1859).

An account of the kingdom of Caubul, and its dependencies in Persia, Tartary, and India; comprising a view of the Afghaun nation, and a history of the Dooraunee monarchy.

1 vol. 4° £3 13s. 6d. London: Printed for Longman, Hurst, Rees, Orme, and Brown, Paternoster-Row; and J. Murray, Albemarle-Street. Printer: A. Strahan, New-Street-Square. NSTC: E739.

1819: second edition, "with an entirely new map" (printer: as first edition).

32. 1815. Hamilton, Walter (bap. 1774, d. 1828).

The East India gazetteer; containing particular descriptions of the empires, kingdoms, principalities, provinces, cities, towns, districts, fortresses, harbours, rivers, lakes, &c. of Hindostan, and the adjacent countries, India beyond the Ganges, and the Eastern Archipelago; together with sketches of the manners, customs, institutions, agriculture, commerce, manufactures, revenues, population, castes, religion, history, &c. of their various inhabitants.

1 vol. 8° 25s. London: Printed for John Murray, Albemarle Street. Printer: Dove, St John's Square, Clerkenwell. NSTC: H360.

No revised editions.

33. 1815. Park, Mungo (1771–1806).

The journal of a mission to the interior of Africa, in the year 1805. By Mungo Park. Together with other documents, official and private, relating to the same mission. To which is prefixed an account of the life of Mr. Park.

"The second edition, revised and corrected, with additions." 1 vol. 4° £1 11s. 6d. London: Printed for John Murray, Albemarle-Street. Printer: W. Bulmer and Co. Cleveland-Row, St. James's. NSTC: P354.

No revised editions.

34. 1816. [Adams, Robert (?–?)]. Note: published anonymously; edited by Cock, Simon (?–?).

The narrative of Robert Adams, a sailor, who was wrecked on the western coast of Africa, in the year 1810, was detained three years in slavery by the Arabs of the Great Desert, and resided several months in the city of Tombuctoo.

1 vol. 4° £1 5s. London: Printed for John Murray, Albemarle-Street. Printer: William Bulmer and Co. Cleveland-Row. NSTC: 2A3203.

No revised editions.

35. 1816. Legh, Thomas (1793–1857).

Narrative of a journey in Egypt and the country beyond the cataracts.

1 vol. 8° 12s. London: Printed for John Murray, Albemarle-Street. Printer: C. Roworth, Bell-yard, Temple-bar. NSTC: 2L10210.

1817: second edition (printer: as first edition).

36. 1816. Maurice, Thomas (1754–1824).

Observations on the ruins of Babylon, as recently visited and described by Claudius James Rich, Esq. resident for the East India Company at Bagdad; with illustrative engravings.

1 vol. 4° £1 5s. London: Printed for the author, and sold by John Murray, Albemarle Street. Printer: Maurice, Fenchurch-street. NSTC: 2M20921.

1816: issued under the variant title *Observations connected with astronomy and ancient history, sacred and profane, on the ruins of Babylon, as recently visited and described by Claudius James Rich, Esq. resident for the East India Company at Bagdad; with illustrative engravings* (printer: as first edition); 1818: appendix issued (see below, item 46).

37. 1816. Park, Mungo (1771–1806).

Travels in the interior districts of Africa: performed in the years 1795, 1796, and 1797. With an account of a subsequent mission to that country in 1805. By Mungo Park, surgeon. To which is added an account of the life of Mr. Park.

"A new edition." 2 vols. 8° 24s. London: Printed for John Murray, Albemarle-Street. Printer: William Bulmer and Co. Cleveland-Row. NSTC: 2P3536.

No revised editions.

38. 1817. Ellis, Sir Henry (1788–1855).

Journal of the proceedings of the late embassy to China; comprising a correct narrative of the public transactions of the embassy, of the voyage to and from China, and of the journey from the mouth of the Pei-Ho to the return to Canton. Interspersed with observations upon the face of the country, the polity, moral character, and manners of the Chinese nation.

1 vol. 4° £2 2s. London: Printed for John Murray, Albemarle-Street. Printer: T. Davison, Lombard-street, Whitefriars. NSTC: 2E7913.

1818: second edition (2 vols., printer: as first edition).

39. 1817. Martin, John (1789–1869). Note: "Compiled and arranged from the extensive communications" of Mariner, William Charles (1791–1853).

An account of the natives of the Tonga Islands, in the South Pacific Ocean. With an original grammar and vocabulary of their language. Compiled and arranged from the extensive communications of Mr. William Mariner, several years resident of those islands.

2 vols. 8° 24s. London: Printed for the author, and sold by John Murray, Albemarle-Street. Printer: T. Davison, Lombard-street, Whitefriars. NSTC: 2M16717.

1818: second edition, "with additions" (printer: as first edition).

40. 1817. McLeod, John (ca. 1777–1820).

Narrative of a voyage in His Majesty's late ship Alceste, *to the Yellow Sea, along the coast of Corea, and through its numerous hitherto undiscovered islands, to the island of Lewchew; with an account of her shipwreck in the Straits of Gaspar.*

1 vol. 8° 12s. London: John Murray, Albemarle Street. Printer: W. Clowes, Northumberland-court, Strand. NSTC: 2M7123.

1818: second edition (printer: as first edition); 1819: third edition (printer: as first edition).

41. 1817. Raffles, Sir (Thomas) Stamford Bingley (1781–1826).

The history of Java.

2 vols. 4° £6 6s. London: Printed for Black, Parbury, and Allen, booksellers to the Hon. East-India Company, Leadenhall Street; and John Murray, Albemarle Street. Printer: Cox and Baylis, Great Queen Street, Lincoln's Inn Fields. NSTC: 2R814.

1830: second edition (printer: Gilbert and Rivington, St. John's Square).

42. 1817. Riley, James (1777–1840).

Loss of the American brig Commerce, *wrecked on the western coast of Africa, in the month of August, 1815. With an account of Tombuctoo, and of the hitherto undiscovered great city of Wassanah.*

1 vol. 4° £1 16s. London: John Murray, Albemarle Street. Printer: C. Roworth, Bell-yard, Temple-bar. NSTC: 2R11183.

No revised editions.

43. 1818. Barrow, Sir John, first baronet (1764–1848).

A chronological history of voyages into the Arctic regions; undertaken chiefly for the purpose of discovering a north-east, north-west, or polar passage between the Atlantic and the Pacific: from the earliest periods of Scandinavian navigation, to the departure of the recent expeditions, under the orders of Captains Ross and Buchan.

1 vol. 8° 12s. London: John Murray, Albemarle Street. Printer: C. Roworth, Bell Yard, Temple Bar. NSTC: 2B9857.

No revised editions.

44. 1818. Hall, Basil (1788–1844). Note: contains "a vocabulary of the Loo-Choo language" by Clifford, Herbert John (?–?).

Account of a voyage of discovery to the west coast of Corea, and the great Loo-Choo Island; with an appendix, containing charts, and various hydrographical and scientific notices.

1 vol. 4° £2 2s. London: John Murray, Albemarle Street. Printer: T. Davison, Lombard-street, Whitefriars. NSTC: 2H2617.

1820: "A new edition, with plates" issued under the variant title *Voyage to Corea, and the Island of Loo-Choo* (printer: as first edition).

45. 1818. Kinneir, Sir John Macdonald (1782–1830).

Journey through Asia Minor, Armenia, and Koordistan, in the years 1813 and 1814; with remarks on the marches of Alexander, and retreat of the Ten Thousand.

1 vol. 8° 18s. London: John Murray, Albemarle-Street. Printer: C. Roworth, Bell-yard, Temple-bar. NSTC: 2K6495.

No revised editions.

46. 1818. Maurice, Thomas (1754–1824).

Observations on the remains of ancient Egyptian grandeur and superstition, as connected with those of Assyria: forming an appendix to Observations on the ruins of Babylon.

1 vol. 4° £2 10s. London: Printed for the author, and sold by John Murray, Albemarle Street. Printer: Maurice, Fenchurch Street. NSTC: 2M20922.

No revised editions.

47. 1818. Tuckey, James Kingston (1776–1816).

Narrative of an expedition to explore the river Zaire, usually called the Congo, in south Africa, in 1816, under the direction of Captain J. K. Tuckey, R. N. to which is added, the journal of Professor Smith; some general observations on the country and its inhabitants; and an appendix: containing the natural history of that part of the kingdom of Congo through which the Zaire flows.

1 vol. 4° £2 2s. London: John Murray, Albemarle-Street. Printer: W. Bulmer and Co. Cleveland Row, St. James's. NSTC: 2T19494.

No revised editions.

48. 1819. Bowdich, Thomas Edward (1791?–1824).

Mission from Cape Coast Castle to Ashantee, with a statistical account of that kingdom, and geographical notices of other parts of the interior of Africa.

1 vol. 4° £3 3s. London: John Murray, Albemarle-Street. Printer: W. Bulmer and Co. Cleveland-Row, St. James's. NSTC: 2B43266.

No revised editions.

49. 1819. Burckhardt, Johann Ludwig [also known as John Lewis] (1784–1817). Note: published posthumously by the Association for Promoting the Discovery of the Interior Parts of Africa.

Travels in Nubia.

1 vol. 4° £2 8s. London: John Murray, Albemarle Street. Printer: W. Bulmer and Co. Cleveland Row, St. James's. NSTC: 2B58668.

1822: second edition (printer: C. Roworth, Bell-yard, Temple-bar).

50. 1819. Burney, James (1750–1821).

A chronological history of north-eastern voyages of discovery; and of the earliest eastern navigations of the Russians.

1 vol. 8° 12s. 6d. London: Printed for Payne and Foss, Pall-Mall; and John Murray, Albemarle Street. Printer: Luke Hansard & Sons, Lincoln's-Inn Fields. NSTC: 2B60596.

No revised editions.

51. 1819. FitzClarence, George Augustus Frederick, first earl of Munster (1794–1842).

Journal of a route across India, through Egypt, to England, in the latter end of the year 1817, and the beginning of 1818.

1 vol. 4° £2 18s. London: John Murray, Albemarle-Street. Printer: Thomas Davison, Whitefriars. NSTC: 2F7342.

No revised editions.

52. 1819. Hippisley, Gustavus (?–?).

A narrative of the expedition to the rivers Orinoco and Apuré, in South America; which sailed from England in November 1817, and joined the patriotic forces in Venezuela and Caraccas.

1 vol. 8° 15s. London: John Murray, Albemarle-Street. Printer: Thomas Davison, Whitefriars. NSTC: 2H23169.

No revised editions.

53. 1819. Lislet Geoffroy, Jean Baptiste (1755–1836). Note: "published in the original French, with an English translation."

Memoir and notice explanatory of a chart of Madagascar and the north-eastern archipelago of Mauritius; drawn up according to the latest observations, under the auspices and government of His Excellency Robert Townsend Farquhar, Governor, Commander in Chief, Captain-General of the Isle of France and Dependencies, Vice-Admiral, &c. &c. &c.

1 vol. 4° 18s. London: John Murray, Albemarle Street. Printer: C. Roworth, Bell-yard, Temple-bar. NSTC: 2L17098.

No revised editions.

54. 1819. Pritchard, John (1777–1836), Pambrun, Pierre-Chrysologue (1792–1841), and Heurter, Frederick Damien (?–?).

Narratives of John Pritchard, Pierre Chrysologue Pambrun, and Frederick Damien Heurter, respecting the aggressions of the North-West Company against the Earl of Selkirk's settlement upon the Red River.

1 vol. 8° 2s. 6d. London: John Murray, Albemarle Street. Printer: J. Brettell, Rupert Street, Haymarket. NSTC: 2D17743.

No revised editions.

55. 1819. Ross, Sir John (1777–1856).

A voyage of discovery, made under the orders of the Admiralty, in His Majesty's ships Isabella *and* Alexander, *for the purpose of exploring Baffin's Bay, and inquiring into the probability of a north-west passage.*

1 vol. 4° £3 13s. 6d. London: John Murray, Albemarle-Street. Printer: W. Clowes, Northumberland-Court, Strand. NSTC: 2R18248.

No revised editions.

56. 1819. Ross, Sir John (1777–1856).

An explanation of Captain Sabine's remarks on the late voyage of discovery to Baffin's Bay.

1 vol. 8° 2s. 6d. London: John Murray, Albemarle-Street. Printer: W. Clowes, Northumberland-Court. NSTC: 2R18253.

No revised editions.

57. 1819. Sabine, Sir Edward (1788–1883).

The North Georgia gazette, and winter chronicle.

1 vol. 4° 10s. 6d. London: John Murray, Albemarle-Street. Printer: William Clowes, Northumberland-court. NSTC: 2S913.

1822: second edition (printer: as first edition).

58. 1819. Salamé, Abraham (?–?).

A narrative of the expedition to Algiers in the year 1816, under the command of the Right Hon. Admiral Lord Viscount Exmouth.

1 vol. 8° 15s. London: John Murray, Albemarle-Street. Printer: C. Roworth, Bell-Yard, Temple-Bar. NSTC: 2S2464.

No revised editions.

59. 1820. Belzoni, Giovanni Battista (1778–1823).

Narrative of the operations and recent discoveries within the pyramids, temples, tombs, and excavations, in Egypt and Nubia; and of a journey to the coast of the Red Sea, in search of the ancient Berenice; and another to the oasis of Jupiter Ammon.

1 vol. 4° £2 2s. London: John Murray, Albemarle-Street. Printer: Thomas Davison, Whitefriars. NSTC: 2B17165. Accompanied by *Plates illustrative of the researches and operations of G. Belzoni in Egypt and Nubia*. NSTC: 2B17166.

1821: second edition (2 vols., printer: as first edition); 1822: third edition (2 vols., printer: as first edition).

60. 1820. Hamilton, Walter (bap. 1774, d. 1828).

A geographical, statistical, and historical description of Hindostan, and the adjacent countries.

2 vols. 4° £4 14s. 6d. London: John Murray, Albemarle Street. Printer: C. Roworth, Bell-yard, Temple-bar. NSTC: 2H5472.

No revised editions.

61. 1820. McLeod, John (ca. 1777–1820).

A voyage to Africa with some account of the manners and customs of the Daho-mian people.

1 vol. 8° 5s. 6d. London: John Murray. Printer: William Clowes, North-umberland-Court. NSTC: 2M7127.

No revised editions.

62. 1820. Oxley, John Joseph William Molesworth (ca. 1785–1828).

Journals of two expeditions into the interior of New South Wales, undertaken by order of the British Government in the years 1817–18.

1 vol. 4° £2 10s. London: John Murray, Albemarle-Street. Printer: Thomas Davison, Whitefriars. NSTC: 208949.

No revised editions.

63. 1820. Turner, William (1792–1867).

Journal of a tour in the Levant.

3 vols. 8° £3 3s. London: John Murray, Albemarle-Street. Printer: W. Clowes, Northumberland-Court. NSTC: 2T20727.

No revised editions.

64. 1821. Blount, Edward (1769–1843).

Notes on the Cape of Good Hope, made during an excursion in that colony in the year 1820.

1 vol. 8° 7s. 6d. London: John Murray, Albemarle-Street. Printer: Thomas Davison, Whitefriars. NSTC: 2C6473.

No revised editions.

65. 1821. Lyon, George Francis (1795–1832).

A narrative of travels in northern Africa, in the years 1818, 19, and 20; accom-panied by geographical notices of Soudan, and of the course of the Niger. With a chart of the routes, and a variety of coloured plates, illustrative of the costumes of the several natives of northern Africa.

1 vol. 4° £3 3s. London: John Murray, Albemarle Street. Printer: Thomas Davison, Whitefriars. NSTC: 2L26120.

No revised editions.

66. 1821. Parry, Sir (William) Edward (1790–1855).

Journal of a voyage for the discovery of a north-west passage from the Atlantic to the Pacific; performed in the years 1819–20, in His Majesty's ships Hecla *and* Griper, *under the orders of William Edward Parry, R.N., F.R.S., and commander of the expedition. With an appendix, containing the scientific and other observations.*

1 vol. 4° £3 13s. 6d. London: John Murray, publisher to the Admiralty, and Board of Longitude. Printer: William Clowes, Northumberland-court. NSTC: 2P5329.

1821: second edition (printer: as first edition); 1828: anthologized (see below, item 107).

67. 1821. Tulišen (1667–1741). Note: "Translated from the Chinese" by Staunton, Sir George Thomas, second baronet (1781–1859).

Narrative of the Chinese embassy to the Khan of the Tourgouth Tartars, in the years 1712, 13, 14, & 15; by the Chinese ambassador, and published, by the Emperor's authority, at Pekin. Translated from the Chinese, and accompanied by an appendix of miscellaneous translations.

1 vol. 8° 18s. London: John Murray, Albemarle Street. Printer: C. Roworth, Bell-yard, Temple-bar. NSTC: 2S37299.

No revised editions.

68. 1822. Burckhardt, Johann Ludwig [also known as John Lewis] (1784–1817). Note: published posthumously by the Association for Promoting the Discovery of the Interior Parts of Africa; edited by Leake, William Martin [known as Colonel Leake] (1777–1860).

Travels in Syria and the Holy Land.

1 vol. 4° £2 8s. London: John Murray, Albemarle Street. Printer: William Nicol, Successor to W. Bulmer & Co. Cleveland-row. NSTC: 2B58671.

No revised editions.

69. 1822. Dobrizhoffer, Martin (1717–91). Note: "From the Latin of Martin Dobrizhoffer"; translated by [Coleridge, Sara (1802–52)].

An account of the Abipones, an equestrian people of Paraguay.

3 vols. 8° £1 16s. London: John Murray, Albemarle Street. Printer: C. Roworth, Bell-yard, Temple-bar. NSTC: 2D15269.

No revised editions.

70. 1822. Edmonstone, Sir Archibald, third baronet (1795–1871).

A journey to two of the oases of upper Egypt.

1 vol. 8° 10s. 6d. London: John Murray, Albemarle Street. Printer: G. Woodfall, Angel Court, Skinner Street. NSTC: 2E4647.

No revised editions.

71. 1822. [English, George Bethune (1787–1828)]. Note: published anonymously.

A narrative of the expedition to Dongola and Sennaar, under the command of His Excellency Ismael Pasha, undertaken by order of His Highness Mehemmed Ali Pasha, Viceroy of Egypt. By an American in the service of the viceroy.

1 vol. 8° 9s. 6d. London: John Murray, Albemarle Street. Printer: C. Roworth, Bell Yard, Temple Bar. NSTC: 2I5086.

No revised editions.

72. 1822. Staunton, Sir George Thomas, second baronet (1781–1859).

Miscellaneous notices relating to China, and our commercial intercourse with that country, including a few translations from the Chinese language.

1 vol. 8° 10s. 6d. London: John Murray, Albemarle Street. Printer: H. Skelton, West-Street, Havant. NSTC: 2S37306.

1822: second edition, "enlarged" (printer: as first edition); 1850: second edition reissued, "accompanied . . . by introductory observations" (printer: as first edition).

73. 1822. Waddington, George (1793–1869), and Hanbury, Barnard (?–?).

Journal of a visit to some parts of Ethiopia.

1 vol. 8° 14s. London: John Murray, Albemarle-Street. Printer: William Clowes, Northumberland-court. NSTC: 2W750.

No revised editions.

74. 1823. Franklin, Sir John (1786–1847).

Narrative of a journey to the shores of the Polar Sea, in the years 1819, 20, 21, and 22.

1 vol. 4° £4 14s. 6d. London: John Murray, Albemarle-Street. Printer: William Clowes, Northumberland-Court. NSTC: 2F14562.

1824: second edition (2 vols., printer: as first edition); third edition (2 vols., printer: as first edition).

75. 1823. [Freygang, Frederika von (1790–1863), and Freygang, Wilhelm von (1783–1849)]. Note: published anonymously and "Translated from the French."

Letters from the Caucasus and Georgia; to which are added, the account of a journey into Persia in 1812, and an abridged history of Persia since the time of Nadir Shah.

1 vol. 8° 15s. London: John Murray, Albemarle Street. Printer: C. Roworth, Bell-Yard, Temple-Bar. NSTC: 2F16658.

No revised editions.

76. 1823. Henniker, Sir Frederick, second baronet (1793–1825).

Notes, during a visit to Egypt, Nubia, the oasis, Mount Sinai, and Jerusalem.

1 vol. 8° 12s. London: John Murray, Albemarle Street. Printer: G. Woodfall, Angel Court, Skinner Street. NSTC: 2H17283.

1824: second edition issued under the variant title *Notes, during a visit to Egypt, Nubia, the oasis Bœris, Mount Sinai, and Jerusalem* (printer: as first edition).

77. 1824. Bullock, William (ca. 1773–1849).

Six months' residence and travels in Mexico; containing remarks on the present state of New Spain, its natural productions, state of society, manufactures, trade, agriculture, and antiquities, &c.

1 vol. 8° 18s. London: John Murray, Albemarle-Street. Printer: James Bullock, Lombard Street, Whitefriars. NSTC: 2B57299.

1825: second edition (printer: as first edition).

78. 1824. Callcott [née Dundas; other married name Graham], Maria, Lady Callcott (1785–1842). Note: published when the author was Graham, Maria.

Journal of a residence in Chile, during the year 1822. And a voyage from Chile to Brazil in 1823.

1 vol. 4° £2 12s. 6d. London: Printed for Longman, Hurst, Rees, Orme, Brown, and Green, Paternoster-Row; and John Murray, Albemarle-Street. Printer: A. & R. Spottiswoode, New-Street-Square. NSTC: 2G16888.

No revised editions.

79. 1824. Callcott [née Dundas; other married name Graham], Maria, Lady Callcott (1785–1842). Note: published when the author was Graham, Maria.

Journal of a voyage to Brazil, and a residence there, during part of the years 1821, 1822, 1823.

1 vol. 4° £2 2s. London: Printed for Longman, Hurst, Rees, Orme, Brown, and Green, Paternoster-Row; and J. Murray, Albemarle-Street. Printer: A. & R. Spottiswoode, New-Street-Square. NSTC: 2G16891.

No revised editions.

80. 1824. Cochrane, John Dundas (1780–1825).

Narrative of a pedestrian journey through Russia and Siberian Tartary, from the frontiers of China to the frozen sea and Kamtchatka performed during the years 1820, 1821, 1822, and 1823.

1 vol. 8° 18s. London: John Murray, Albemarle Street. Printed: G. Woodfall, Angel Court, Skinner Street. NSTC: 2C28224.

No revised editions.

81. 1824. Cooper, Edward Joshua (1798–1863).

Views in Egypt and Nubia, executed in lithography, by Messrs. Harding and Westall, from a collection of original drawings, taken on the banks of the Nile by S. Bossi, an artist of Rome, during the winter of 1820–21, under the personal inspection and direction of Edward J. Cooper Esqr.

Published 1824–27 in 8 pts. 4° (price unrecorded). London: Published by J. Murray, Albemarle Street. Printer: C. Hullmandel's Lithographic Establishment, Marlborough Street. NSTC: 2C36655.

No revised editions.

82. 1824. Leake, William Martin [known as Colonel Leake] (1777–1860).

Journal of a tour in Asia Minor, with comparative remarks on the ancient and modern geography of that country.

1 vol. 8° 18s. London: John Murray, Albemarle-Street. Printer: Richard Taylor, Shoe-Lane. NSTC: 2L7770.

No revised editions.

83. 1824. Lyon, George Francis (1795–1832).

The private journal of Captain G. F. Lyon, of H.M.S. Hecla, *during the recent voyage of discovery under Captain Parry.*

1 vol. 8° 16s. London: John Murray, Albemarle-Street. Printer: Thomas Davison, Whitefriars. NSTC: 2L26121.

1825: "A new edition" (printer: as first edition).

84. 1824. Parry, Sir (William) Edward (1790–1855).

Journal of a second voyage for the discovery of a north-west passage from the Atlantic to the Pacific; performed in the years 1821-22-23, in His Majesty's ships Fury *and* Hecla, *under the orders of Captain William Edward Parry, R.N., F.R.S., and commander of the expedition.*

1 vol. 4° £4 14s. 6d. London: John Murray, Publisher to the Admiralty, and Board of Longitude. Printer: W. Clowes, Northumberland-court. NSTC: 2P5330.

1828: anthologized (see below, item 107).

85. 1825. Caldcleugh, Alexander (1795–1858).

Travels in South America, during the years 1819-20-21; containing an account of the present state of Brazil, Buenos Ayres, and Chile.

2 vols. 8° £1 10s. London: John Murray, Albemarle Street. Printer: C. Roworth, Bell Yard, Temple Bar. NSTC: 2C1751.

No revised editions.

86. 1825. Field, Barron (1786–1846), editor.

Geographical memoirs on New South Wales; by various hands: containing an account of the Surveyor General's late expedition to two new ports; the discovery of Moreton Bay River, with the adventures for seven months there of two shipwrecked men; a route from Bathurst to Liverpool plains; together with other papers on the Aborigines, the geology, the botany, the timber, the astronomy, and the meteorology of New South Wales and Van Diemen's Land.

1 vol. 8° 18s. London: John Murray, Albemarle-Street. Printer: Thomas Davison, Whitefriars. NSTC: 2F5261.

No revised editions.

87. 1825. Gray, William (?-?), and "the late Staff Surgeon [Duncan] Dochard."

Travels in western Africa, in the years 1818, 19, 20, and 21, from the river Gambia, through Woolli, Bondoo, Galam, Kasson, Kaarta, and Foolidoo, to the river Niger.

1 vol. 8° 18s. London: John Murray, Albemarle Street. Printer: G. Woodfall, Angel Court, Skinner Street. NSTC: 2G19140.

No revised editions.

88. 1825. Laing, Alexander Gordon (1794-1826).

Travels in the Timannee, Kooranko, and Soolima countries, in western Africa.

1 vol. 8° 16s. London: John Murray, Albemarle-Street. Printer: William Clowes, Northumberland court. NSTC: 2L1808.

No revised editions.

89. 1825. Lyon, George Francis (1795-1832).

A brief narrative of an unsuccessful attempt to reach Repulse Bay, through Sir Thomas Rowe's "Welcome," in His Majesty's ship Griper, *in the year MDCCCXXIV.*

1 vol. 8° 10s. 6d. London: John Murray, Albemarle-Street. Printer: William Clowes, Northumberland-court. NSTC: 2L26118.

No revised editions.

90. 1825. Parry, Sir (William) Edward (1790-1855).

Appendix to Captain Parry's journal of a second voyage for the discovery of a north-west passage from the Atlantic to the Pacific, performed in His Majesty's ships Fury *and* Hecla, *in the years 1821-22-23.*

1 vol. 4° £1 11s. 6d. London: John Murray, Publisher to the Admiralty and Board of Longitude. Printer: William Clowes, Northumberland-court. NSTC: unrecorded.

No revised editions.

91. 1826. [Callcott (née Dundas; other married name Graham], Maria, Lady Callcott (1785-1842), editor]. Note: published anonymously when the editor was Graham, Maria.

Voyage of H.M.S. Blonde *to the Sandwich Islands, in the years 1824-1825. Captain the Right Hon. Lord Byron, commander.*

1 vol. 4° £2 2s. London: John Murray, Albemarle-Street. Printer: ⁊ Davison, Whitefriars. NSTC: 2G160902.

No revised editions.

92. 1826. [Coleridge, Henry Nelson (1798–1843)]. Note: published anonymously.

Six months in the West Indies, in 1825.

1 vol. 8° 9s. 6d. London: John Murray, Albemarle Street. Printer: C. Roworth, Bell Yard, Temple Bar. NSTC: 2C30125.

1826: second edition, "with additions" (printer: Thomas Davison, Whitefriars); 1832: third edition, "with additions," issued as vol. 36 of Murray's Family Library series (printer: William Clowes, Stamford Street).

93. 1826. Denham, Dixon (1786–1828), Clapperton, Hugh (1788–1827), and Oudney, Walter (1790–1824).

Narrative of travels and discoveries in northern and central Africa, in the years 1822, 1823, and 1824 . . . extending across the Great Desert to the tenth degree of northern latitude, and from Kouka in Bornou, to Sackatoo, the capital of the Felatah empire.

2 vols. 4° £4 14s. 6d. London: John Murray, Albemarle-Street. Printer: Thomas Davison, Whitefriars. NSTC: 2D9035.

1826: second edition (printer: as first edition); 1828: third edition (printer: as first edition).

94. 1826. Finlayson, George (1790–1823). Note: "With a memoir of the author" by Raffles, Sir (Thomas) Stamford Bingley (1781–1826).

The mission to Siam, and Hué the capital of Cochin China, in the years 1821–2.

1 vol. 8° 15s. London: John Murray, Albemarle-Street. Printer: William Clowes, Northumberland-court. NSTC: 2F6164.

No revised editions.

95. 1826. Head, Sir Francis Bond, first baronet (1793–1875).

Rough notes taken during some rapid journeys across the Pampas and among the Andes.

1 vol. 8° 9s. 6d. London: John Murray, Albemarle-Street. Printer: William Clowes, Stamford-Street. NSTC: 2H14875.

1826: second edition (printer: as first edition); 1828: third edition (printer: as first edition); 1846: fourth edition, issued as pt. 38 (vol. 19) of Murray's Home and Colonial Library series (printer: as first edition); 1861: "New edition" (printer: as first edition).

96. 1826. Parry, Sir (William) Edward (1790–1855).

Journal of a third voyage for the discovery of a north-west passage from the Atlantic to the Pacific; performed in the years 1824–25, in His Majesty's ships Hecla *and* Fury, *under the orders of Captain William Edward Parry, R.N., F.R.S., and commander of the expedition.*

1 vol. 4° £2 10s. London: John Murray, Publisher to the Admiralty, and Board of Longitude. Printer: William Clowes, Northumberland-court. NSTC: 2P53334.

1828: anthologized (see below, item 107).

97. 1827. Andrews, Joseph (?–?).

Journey from Buenos Ayres, through the provinces of Cordova, Tucuman, and Salta, to Potosi, thence by the deserts of Caranja to Arica, and subsequently, to Santiago de Chili and Coquimbo, undertaken on behalf of the Chilean and Peruvian Mining Association, in the years 1825–26.

2 vols. 8° 18s. London: John Murray, Albemarle Street. Printer: Shackell and Baylis, Johnson's-Court. NSTC: 2A12429.

No revised editions.

98. 1827. Hamilton, John Potter (1778–1873).

Travels through the interior provinces of Columbia.

2 vols. 12° £1 1s. London: John Murray, Albemarle Street. Printer: G. Woodfall, Angel Court, Skinner Street. NSTC: 2H5280.

No revised editions.

99. 1827. King, Philip Parker (1793–1856).

Narrative of a survey of the intertropical and western coasts of Australia. Performed between the years 1818 and 1822.

2 vols. 8° £1 16s. London: John Murray, Albemarle Street. Printer: W. Clowes, Stamford-Street. NSTC: 2K5737.

No revised editions.

100. 1827. [Malcolm, Sir John (1769–1833)]. Note: published anonymously.

Sketches of Persia, from the journals of a traveller in the East.

2 vols. Post 8° 18s. London: John Murray, Albemarle Street. Printer: Thomas Davison, Whitefriars. NSTC: 2M10901.

1828: "A new edition" (printer: as first edition); 1845: Issued as pts. 17–18 (vol. 9) of Murray's Home and Colonial Library series (printer: W. Clowes & Sons, Duke Street, Stamford Street).

101. 1827. [Trent, Thomas Abercrombie (b. 1804)]. Note: published anonymously by "an officer on the staff of the Quarter-Master-General's Department."

Two years in Ava. From May 1824, to May 1826.

1 vol. 8° 16s. London: John Murray, Albemarle-Street. Printer: William Clowes, Stamford-street. NSTC: 2A19882.

No revised editions.

102. 1828. Beechey, Frederick William (1796–1856), and Beechey, Henry William (1788/89–1862).

Proceedings of the expedition to explore the northern coast of Africa, from Tripoly eastward; in MDCCCXXI. and MDCCCXXII. comprehending an account of the Greater Syrtis and Cyrenaica; and of the ancient cities comprising the Pentapolis.

1 vol. 4° £3 3s. London: John Murray, Albemarle-Street. Printer: William Clowes, Stamford-Street. NSTC: 2B14844.

No revised editions.

103. 1828. Franklin, Sir John (1786–1847).

Narrative of a second expedition to the shores of the Polar Sea, in the years 1825, 1826, and 1827.

1 vol. 4° £4 4s. London: John Murray, Albemarle-Street. Printer: William Clowes, Stamford-Street. NSTC: 2F14564.

No revised editions.

104. 1828. Heber, Reginald (1783–1826). Note: published posthumously; edited by [Heber (née Shipley), Amelia (1789–1870)].

Narrative of a journey through the upper provinces of India, from Calcutta to Bombay, 1824–1825, (with notes upon Ceylon,) an account of a journey to Madras and the southern provinces, 1826, and letters written in India.

Issued as pts. 3–6 (vols. 2 and 3) of Murray's Home and Colonial Library series. 2 vols. 4° £4 14s. 6d. London: John Murray, Albemarle Street. Printer: R. Gilbert, St. John's-Square. NSTC: 2H15682.

1828: second edition (3 vols., printer: as first edition); third edition (3 vols., printer: as first edition); 1829: fourth edition (3 vols., printer: as first edition); 1856: "New edition" (2 vols., printer: William Clowes and Sons, Stamford Street, and Charing Cross).

105. 1828. Irving, Washington (1783–1859).

A history of the life and voyages of Christopher Columbus.

4 vols. 8° 9s. 6d. London: John Murray, Albemarle-Street. Printer: W. Clowes, Stamford-Street. NSTC: 2I4654.

1830: "Abridged" and issued under the variant title *The life and voyages of Christopher Columbus* as vol. 18 of Murray's Family Library series (1 vol., printer: Thomas Davison, Whitefriars); 1849: "New and revised edition" issued under the variant title *The life and voyages of Christopher Columbus; together with the voyages of his companions* (3 vols., printer Bradbury and Evans, Printers, Whitefriars).

106. 1828. Lyon, George Francis (1795–1832).

Journal of a residence and tour in the republic of Mexico in the year 1826. With some account of the mines of that country.

2 vols. Post 8° 16s. London: John Murray, Albemarle Street. Printed: Richard Taylor, Red Lion Court, Fleet Street. NSTC: 2L26119.

No revised editions.

107. 1828. Parry, Sir (William) Edward (1790–1855).

Journals of the first, second and third voyages for the discovery of a north-west passage from the Atlantic to the Pacific, in 1819–20–21–22–23–24–25, in His Majesty's ships Hecla, Griper and Fury, under the orders of Capt. W. E. Parry, R.N. F.R.S. and commander of the expedition.

5 vols. 18° £1. London: John Murray, Albemarle Street. Printer: C. Roworth, Bell Yard, Temple Bar. NSTC: 2P5335.

No revised editions.

108. 1828. Parry, Sir (William) Edward (1790–1855).

Narrative of an attempt to reach the North Pole, in boats fitted for the purpose, and attached to His Majesty's ship Hecla, *in the year MDCCCXXVII., under the command of Captain William Edward Parry, R.N., F.R.S., and honorary member of the Imperial Academy of Sciences at St. Petersburg.*

1 vol. 4° £2 2s. London: John Murray, Publisher to the Admiralty, and Board of Longitude. Printer: William Clowes, Stamford-street. NSTC: 2P5338.

No revised editions.

109. 1829. Clapperton, Hugh (1788–1827), and Lander, Richard (1804–34).

Journal of a second expedition into the interior of Africa, from the Bight of Benin to Soccatoo. By the late commander Clapperton, of the Royal Navy. To which is added, the journal of Richard Lander from Kano to the sea-coast, partly by a more eastern route. With a portrait of Captain Clapperton, and a map of the route, chiefly laid down from actual observations for latitude and longitude.

1 vol. 4° £2 2s. London: John Murray, Albemarle-Street. Printer: Thomas Davison, Whitefriars. NSTC: 2C22897.

No revised editions.

110. 1829. Franklin, Sir John (1786–1847).

Journey to the shores of the Polar Sea, in 1819–20–21–22: with a brief account of the second journey in 1825–26–27.

4 vols. 18° 20s. London: John Murray, Albemarle Street. Printer: C. Roworth, Bell Yard, Temple Bar. NSTC: 2F14559.

No revised editions.

111. 1829. Head, Sir George (1782–1855).

Forest scenes and incidents, in the wilds of North America; being a diary of a winter's route from Halifax to the Canadas, and during four months' residence in the woods on the borders of lakes Huron and Simcoe.

1 vol. Post 8° 8s. 6d. London: John Murray, Albemarle Street. Printer: G. Woodfall, Angel Court, Skinner Street. NSTC: 2H14882.

1838: second edition (printer: William Clowes and Sons, Stamford Street).

112. 1829. Lee, Samuel (1783–1852).

The travels of Ibn Batūta; translated from the abridged Arabic manuscript copies, preserved in the public library of Cambridge. With notes, illustrative of the history, geography, botany, antiquities, &c. occurring throughout the work.

1 vol. 4° £1. London: Printed for the Oriental Translation Committee, and sold by J. Murray, Albemarle Street; Parbury, Allen, & Co., Leadenhall Street and Howell & Stewart, Holborn. Printer: J. L. Cox, Great Queen Street, Lincoln's-Inn-Fields. NSTC: 2L9322.

No revised editions.

113. 1829. Lushington [née Gascoyne], Sarah (d. 1839). Note: title page reads "By Mrs. Charles Lushington."

Narrative of a journey from Calcutta to Europe, by way of Egypt, in the years 1827 and 1828.

1 vol. Post 8° 8s. 6d. London: John Murray, Albemarle Street. Printer: William Clowes, Stamford-street. NSTC: 2L25278.

1829: second edition (printer: as first edition).

114. 1829. Maw, Henry Lister (1801–74).

Journal of a passage from the Pacific to the Atlantic, crossing the Andes in the northern provinces of Peru, and descending the river Marañon, or Amazon.

1 vol. 8° 12s. London: John Murray, Albemarle-Street. Printer: W. Clowes, Stamford-street. NSTC: 2M21114.

No revised editions.

115. 1829. Richardson, Sir John (1787–1865), "assisted by" Swainson, William (1789–1855), and Kirby, William (1759–1850).

Fauna Boreali-Americana; or the zoology of the northern parts of British America: containing descriptions of the objects of natural history collected on the late northern land expeditions, under command of Captain Sir John Franklin, R.N.

2 vols. 4° £2 12s. 6d. Published 1829-31. Vol. 1, *Part first, the quadrupeds,* dated 1829; vol. 2, *Part second, the birds,* dated 1831, authored by Swainson and Richardson alone. London: John Murray, Albemarle-Street. Printer: William Clowes, Stamford-Street. NSTC: 2R9685.

No revised editions. Vols. 3 and 4 issued subsequently under different publishers (Richard Bentley, London, and Josiah Fletcher, Norwich, respectively).

116. 1829. Thompson, George Alexander (?-?).

Narrative of an official visit to Guatemala from Mexico.

1 vol. 8° 12s. London: John Murray, Albemarle Street. Printer: G. Woodfall, Angel Court, Skinner Street. NSTC: 2T9140.

No revised editions.

117. 1830. Davy, Sir Humphry, baronet (1778-1829). Note: published posthumously and edited by [Davy, John, 1790-1868].

Consolations in travel, or the last days of a philosopher.

1 vol. 12° 6s. London: John Murray, Albemarle Street. Printer: Thomas Davison, Whitefriars. NSTC: 2D5535.

1831: "A new edition" (printer: W. Clowes, Stamford-street); 1831: third edition (printer: as 1831); 1838: fourth edition (printer: G. Woodfall, Angel Court, Skinner Street); 1851: fifth edition (printer: Bradbury and Evans, Whitefriars); 1853: sixth edition (printer: as 1851); 1869: seventh edition (printer: unrecorded).

118. 1830. Finati, Giovanni (?-?). Note: "Translated from the Italian . . . and edited by" Bankes, William John (1786-1855).

Narrative of the life and adventures of Giovanni Finati, native of Ferrara; who, under the assumed name of Mahomet, made the campaigns against the Wahabees for the recovery of Mecca and Medina; and since acted as interpreter to European travellers in some of the parts least visited of Asia and Africa.

2 vols. Foolscap 8° 14s. London: John Murray, Albemarle-Street. Printer: William Clowes, Stamford Street. NSTC: 2F5862.

No revised editions.

119. 1830. Fuller, John (?–?).

Narrative of a tour through some parts of the Turkish empire.

1 vol. 8° 15s. London: John Murray, Albemarle Street. Printer: Richard Taylor, Red Lion Court, Fleet Street. NSTC: 2F18324.

No revised editions.

120. 1830. Head, Sir Francis Bond, first baronet (1793–1875).

The life of Bruce, the African traveller.

Issued as vol. 17 of Murray's Family Library series. 1 vol. 18° 5s. London: John Murray, Albemarle Street. Printer: William Clowes, Stamford Street. NSTC: 2H14866.

1836: second edition (printer: Bradbury and Evans, Whitefriars); 1838: third edition (printer: as 1836); 1844: fourth edition (printer: as 1836).

121. 1830. Leake, William Martin (1777–1860).

Travels in the Morea.

3 vols. 8° £2 5s. London: John Murray, Albemarle Street. Printer: G. Woodfall, Angel Court, Skinner Street. NSTC: 2L7794.

No revised editions.

122. 1831. [Barrow, Sir John, first baronet (1764–1848)]. Note: published anonymously.

The eventful history of the mutiny and piratical seizure of H.M.S. Bounty: *its cause and consequences.*

Issued as vol. 25 of Murray's Family Library series. 1 vol. Small 8° 5s. London: John Murray, Albemarle-Street. Printer: William Clowes, Stamford Street. NSTC: 2B9862.

1835: second edition "sold by Thomas Tegg and Son, Cheapside" (Bradbury and Evans, Printers, Whitefriars); 1839: third edition issued in cooperation with Thomas Tegg, Cheapside (printer: as second edition); 1847: fourth edition (printer: James Nichols, Hoxton Square).

123. 1831. Parry, Sir (William) Edward (1790–1855).

Three voyages for the discovery of a north-west passage from the Atlantic to the Pacific, and narrative of an attempt to reach the North Pole.

5 vols. 12° (price unrecorded). London: John Murray, Albemarle Street. Printer: William Clowes, Stamford Street. NSTC: unrecorded.

1835: issued in four vols. (printer: Bradbury and Evans, Whitefriars. [Late T. Davison]).

124. 1832. Lander, Richard (1804–34), and Lander, John (1807–39).

Journal of an expedition to explore the course and termination of the Niger; with a narrative of a voyage down that river to its termination.

Issued as vols. 28–30 of Murray's Family Library series. 3 vols. 18° 15s. London: John Murray, Albemarle-Street. Printer: William Clowes, Stamford Street. NSTC: 2L3334.

No revised editions.

125. 1832. Mundy, Godfrey Charles (1804–60).

Pen and pencil sketches, being the journal of a tour in India.

2 vols. 8° 30s. London: John Murray, Albemarle-Street. Printer: William Clowes, Stamford-street. NSTC: 2M40852.

1833: second edition (printer: as first edition); 1858: third edition issued under the variant title *Pen and pencil sketches in India. Journal of a tour in India* (printer: as first edition).

126. 1834. Burnes, Sir Alexander (1805–41).

Travels into Bokhara; being the account of a journey from India to Cabool, Tartary, and Persia; also, narrative of a voyage on the Indus, from the sea to Lahore, with presents from the King of Great Britain; performed under the orders of the Supreme Government of India, in the years 1831, 1832, and 1833.

3 vol. 8° 18s. London: John Murray, Albemarle Street. Printer: A. Spottiswoode, New-Street-Square. NSTC: 2B60361.

1835: second edition (printer: as first edition); 1839: "New edition, with a map and illustrations" (printer: as first edition).

127. 1834. Lewis, Matthew Gregory [called Monk Lewis] (1775–1818). Note: published posthumously.

Journal of a West India proprietor, kept during a residence in the island of Jamaica.

1 vol. Post 8° 10s. 6d. London: John Murray, Albemarle Street. Printer: A. Spottiswoode, New-Street-Square. NSTC: 2L14139.

1845: issued under the variant title *Journal of a residence among the negroes in the West Indies* as pt. 16 (vol. 8) of Murray's Home and Colonial Library series (printer: William Clowes and Sons, Stamford Street); 1861: "New edition" (printer: as 1845).

128. 1835. Abdy, Edward Strutt (1791–1846).

Journal of a residence and tour in the United States of North America, from April, 1833, to October, 1834.

3 vols. Crown 8° 30s. London: John Murray, Albemarle Street. Printer: G. Woodfall, Angel Court, Skinner Street. NSTC: 2A1014.

No revised editions.

129. 1835. Irving, Washington (1783–1859).

A tour on the prairies.

1 vol. Post 8° 9s. 6d. London: John Murray, Albemarle Street. Printer: A. Spottiswoode, New-Street-Square. NSTC: 2I4724.

1835: issued as the first vol. of Irving's *Miscellanies* (printer: as first edition).

130. 1835. Kemble [married name Butler], Frances Anne [Fanny] (1809–93).

Journal.

2 vols. Post 8° 18s. London: John Murray, Albemarle Street. Printer: A. Spottiswoode, New-Street-Square. NSTC: 2K2468.

No revised editions.

131. 1835. Wilkinson, Sir John Gardner (1797–1875).

Topography of Thebes, and general view of Egypt. Being a short account of the principal objects worthy of notice in the valley of the Nile, to the second cataract and Wadee Samneh, with the Fyoom, oases, and eastern desert, from Sooz to Berenice; with remarks on the manners and customs of the ancient Egyptians and the productions of the country, &c. &c.

1 vol. 8° 30s. London: John Murray, Albemarle Street. Printer: William Clowes, Duke Street, Lambeth. NSTC: 2W21000.

No revised editions.

132. 1836. Back, Sir George (1796–1878).

Narrative of the Arctic land expedition to the mouth of the Great Fish River, and along the shores of the Arctic Ocean, in the years 1833, 1834, and 1835.

1 vol. 8° 30s. London: John Murray, Albemarle Street. Printer: A. Spottiswoode, New-Street-Square. NSTC: 2B1451.

No revised editions.

133. 1836. Laborde, Léon, marquis de (1807–69).

Journey through Arabia Petræa, to Mount Sinai, and the excavated city of Petra, the Edom of the prophecies.

1 vol. 8° 18s. London: John Murray, Albemarle Street. Printer: A. Spottiswoode, New-Street-Square. NSTC: 2L569.

1838: second edition (printer: as first edition).

134. 1836. Smyth, William Henry (1788–1865), and Lowe, Frederick (1811–47).

Narrative of a journey from Lima to Para, across the Andes and down the Amazon: undertaken with a view of ascertaining the practicability of a navigable communication with the Atlantic, by the rivers Pachitea, Ucayali, and Amazon.

1 vol. 8° 12s. London: John Murray, Albemarle-Street. Printer: William Clowes and Sons, Stamford Street. NSTC: 2S29488.

No revised editions.

135. 1837. Taitbout de Marigny, Edouard (?–?).

Three voyages in the Black Sea to the coast of Circassia: including descriptions of the ports, and the importance of their trade: with sketches of the manners, customs, religion, &c. &c., of the Circassians.

1 vol. 8° 6s. 10d. London: John Murray, Albemarle Street. Printer: W. Clowes and Sons, Stamford Street. NSTC: 2T1025.

No revised editions.

136. 1838. Back, Sir George (1796–1878).

Narrative of an expedition in H.M.S. Terror, undertaken with a view to geographical discovery on the Arctic shores, in the years 1836–7.

1 vol. 8° 12s. London: John Murray, Albemarle Street. Printer: A. Spottis-woode, New-Street-Square. NSTC: 2B1450.

No revised editions.

137. 1838. Parish, Sir Woodbine (1796–1882).

Buenos Ayres, and the provinces of the Rio de la Plata: their present state, trade, and debt; with some account from original documents of the progress of geo-graphical discovery in those parts of South America during the last sixty years.

1 vol. 8° 18s. London: John Murray, Albemarle Street. Printer: William Clowes and Sons, Stamford Street. NSTC: 2P3375.

1852: second edition, "enlarged, with a new map and illustrations" issued under the variant title *Buenos Ayres and the provinces of the Rio de la Plata: from their discovery and conquest by the Spaniards to the establishment of their political independence. With some account of their present state, trade, debt, etc.; an appendix of historical and statistical documents; and a description of the geol-ogy and fossil monsters of the Pampas* (printer: as first edition).

138. 1838. Robertson, John Parish (1792–1843), and Robertson, William Parish (?–?).

Letters on Paraguay: comprising an account of a four years' residence in that republic, under the government of the dictator Francia.

3 vols. Post 8° 28s. 6d. Published 1838–39. Vol. 3 dated 1839 (see below, item 143). London: John Murray, Albemarle Street. Printer: W. Clowes and Sons, 14, Charing Cross. NSTC: 2R13195.

1839: second edition (printer: as first edition).

139. 1838. Wellsted, James Raymond (1805–42).

Travels in Arabia.

2 vols. 8° 24s. London: John Murray, Albemarle-Street. Printer: William Clowes and Sons, Stamford Street. NSTC: 2W12622.

No revised editions.

140. 1839. Fellows, Sir Charles (1799–1860).

A journal written during an excursion in Asia Minor.

1 vol. Imperial 8° 42s. London: John Murray, Albemarle Street. Printer: Richard and John E. Taylor, Red Lion Court, Fleet Street. NSTC: 2F3665.

1852: abridged under the variant title *Travels and researches in Asia Minor,* *more particularly in the province of Lycia* (printer: John Edward Taylor, Little Queen Street, Lincoln's Inn Fields).

141. 1839. Harris, Sir William Cornwallis (bap. 1807, d. 1848).

The wild sports of Southern Africa; being the narrative of an expedition from the Cape of Good Hope, through the territories of the chief Moselekatse, to the Tropic of Capricorn.

1 vol. 8° 10s. 6d. London: John Murray, Albemarle Street. Printer: W. Clowes and Sons, 14, Charing Cross. NSTC: 2H9549.

No revised editions.

142. 1839. Newbold, Thomas John (1807-50).

Political and statistical account of the British settlements in the Straits of Malacca, viz. Pinang, Malacca, and Singapore; with a history of the Malayan states of the peninsula of Malacca.

2 vols. 8° 26s. London: John Murray, Albemarle Street. Printer: Stewart and Murray, Old Bailey. NSTC: 2N538.

No revised editions.

143. 1839. Robertson, John Parish (1792-1843), and Robertson, William Parish (?-?).

Francia's reign of terror, being the continuation of Letters on Paraguay.

Vol. 3 of their *Letters on Paraguay* (1838). 1 vol. Post 8° 10s. 6d. London: John Murray, Albemarle Street. Printer: W. Clowes and Sons, 14, Charing Cross. NSTC: 2R13195.

No revised editions.

144. 1839. Venables, Richard Lister (1809-94).

Domestic scenes in Russia: in a series of letters describing a year's residence in that country, chiefly in the interior.

1 vol. Post 8° 9s. 6d. London: John Murray, Albemarle Street. Printer: Stewart and Murray, Old Bailey. NSTC: 2V2063.

1856: second edition, "revised" (printer: W. Clowes and Sons, Stamford Street, and Charing Cross).

145. 1839. Wilbraham, Sir Richard (1811–1900).

Travels in the Trans-Caucasian provinces of Russia, and along the southern shore of the lakes of Van and Urumiah, in the autumn and winter of 1837.

1 vol. 8° 18s. London: John Murray, Albemarle Street. Printer: William Clowes and Sons, Stamford Street. NSTC: 2W20146.

No revised editions.

146. 1840. Allen, William (1792–1864).

Picturesque views on the river Niger, sketched during Lander's last visit in 1832–33.

1 vol. Oblong 4° 25s. London: John Murray, Albemarle Street; Hodgson & Graves, Pall Mall; and Ackerman, Strand. Printer: W. Clowes and Sons, Charing Cross. NSTC: 2A9289.

No revised editions.

147. 1840. Gurney, Joseph John (1788–1847).

A winter in the West Indies, described in familiar letters to Henry Clay, of Kentucky.

1 vol. 8° 5s. London: John Murray; Norwich: Josiah Fletcher. Printer: Josiah Fletcher, Upper Haymarket, Norwich. NSTC: 2G25716.

1840: second edition (printer: as first edition); 1841: third edition; fourth edition (printer: as first edition).

148. 1841. [Busk (née Blair), Mary Margaret (1779–1863), editor]. Note: published anonymously.

Manners and customs of the Japanese, in the nineteenth century. From recent Dutch visitors of Japan, and the German of Dr. Ph. Fr. von Siebold.

1 vol. Post 8° 9s. 6d. London: John Murray, Albemarle Street. Printer: Stewart and Murray, Old Bailey. NSTC: 2B62410.

1852: "New and cheaper edition" (printer: as first edition).

149. 1841. Grant, Asahel (1807–44).

The Nestorians; or, the Lost Tribes: containing evidence of their identity; an account of their manners, customs, and ceremonies; together with sketches of travel in ancient Assyria, Armenia, Media, and Mesopotamia; and illustrations of scripture prophecy.

1 vol. 8° 8s. 6d. London: John Murray, Albemarle Street. Printer: William Clowes and Sons, Stamford-street. NSTC: 2G17475.

1843: second edition (printer: as first edition); 1844: third edition (printer: A. Spottiswoode, New-Street-Square).

150. 1841. Jocelyn, Robert Jocelyn, Viscount (1816–54).

Six months with the Chinese expedition; or, leaves from a soldier's note-book.

1 vol. 12° 5s. 6d. London: John Murray, Albemarle Street. Printer: William Clowes and Sons, Stamford Street. NSTC: 2J7540.

1841: second edition (printer: as first edition).

151. 1841. Kinnear, John Gardiner (1800–1866).

Cairo, Petra, and Damascus, in 1839. With the remarks on the government of Mehemet Ali, and on the present prospects of Syria.

1 vol. Post 8° 9s. 6d. London: John Murray, Albemarle Street. Printer: Bradbury and Evans, Whitefriars. NSTC: 2K6488.

No revised editions.

152. 1841. Moorcroft, William (bap. 1767, d. 1825), and Trebeck, George (1800–1825). Note: "Prepared for the press, from original journals and correspondence" by Wilson, Horace Hayman (1786–1860).

Travels in the Himalayan provinces of Hindustan and the Panjab; in Ladakh and Kashmir; in Peshawar, Kabul, Kunduz, and Bokhara.

2 vols. 8° 30s. John Murray, Albemarle Street. Printer: William Clowes and Sons, Stamford Street. NSTC: 2M34498.

No revised editions.

153. 1841. Robinson, Edward (1794–1863), and Smith, Eli (1801–1857). Note: "Drawn up from the original diaries, with historical illustrations" by Robinson.

Biblical researches in Palestine, Mount Sinai and Arabia Petræa. A journal of travels in the years 1838.

3 vols. 8° 45s. London: John Murray, Albemarle Street. Printer: A. Spottiswoode, New-Street-Square. NSTC: 2R13678.

1856: second edition, "with new maps and plans," issued under the variant title *Biblical researches in Palestine and the adjacent regions: a journal of travels in the years 1838 & 1852* (original condensed as 2 vols. plus new supplementary vol., printer: as first edition).

154. 1841. Stephens, John Lloyd (1805-52).

Incidents of travel in Central America, Chiapas, and Yucatan.

2 vols. 8° 32s. London: John Murray, Albemarle Street. Printer: W. Clowes and Sons, 14, Charing Cross. NSTC: 2S38590.

1842: "New edition" (printer: as first edition).

155. 1841. Wood, John (1811-71).

A personal narrative of a journey to the source of the river Oxus, by the route of the Indus, Kabul, and Badakhshan, performed under the sanction of the Supreme Government of India, in the years 1836, 1837, and 1838.

1 vol. 8° 14s. London: John Murray, Albemarle Street. Printer: William Clowes and Sons, Stamford Street. NSTC: 2W29622.

No revised editions.

156. 1842. Burnes, Sir Alexander (1805-41).

Cabool: being a personal narrative of a journey to, and residence in that city, in the years 1836, 7, and 8.

1 vol. 8° 18s. London: John Murray, Albemarle Street. London: Printed by William Clowes and Sons, Stamford Street. NSTC: 2B60359.

1843: second edition (printer: as first edition).

157. 1842. Hamilton, William John (1805-67).

Researches in Asia Minor, Pontus, and Armenia; with some account of their antiquities and geology.

2 vols. 8° 38s. London: John Murray, Albemarle Street. Printer: William Clowes and Sons, Stamford Street. NSTC: 2H5627.

No revised editions.

158. 1842. Jukes, (Joseph) Beete (1811-69).

Excursions in and about Newfoundland, during the years 1839 and 1840.

2 vols. Post 8° 21s. London: John Murray, Albemarle Street. Printer: William Clowes and Sons, Stamford Street. NSTC: 2J13156.

No revised editions.

159. 1843. Dieffenbach, Johann Karl Ernst (1811-55).

Travels in New Zealand; with contributions to the geography, geology, botany, and natural history of that country.

2 vols. 8° 24s. London: John Murray, Albemarle Street. Printer: William Clowes and Sons, Stamford Street. NSTC: 2D13335.

No revised editions.

160. 1843. Hood, John (?-?).

Australia and the East: being a journal narrative of a voyage to New South Wales in an emigrant ship: with a residence of some months in Sydney and the bush, and the route home by way of India and Egypt, in the years 1841 and 1842.

1 vol. 8° 14s. London: John Murray, Albemarle Street. Printer: Bradbury and Evans, Whitefriars. NSTC: 2H28630.

No revised editions.

161. 1843. [Maitland (née Barrett; other married name Thomas), Julia Charlotte (1808-64)]. Note: published anonymously, "By a lady."

Letters from Madras, during the years 1836-1839.

1 vol. Crown 8° 9s. 6d. London: John Murray, Albemarle Street. Printer: William Clowes and Sons, Stamford Street. NSTC: 2M10486.

1846: issued as pt. 35 (vol. 17) of Murray's Home and Colonial Library (printer: as first edition); 1861: "New edition" (printer: as first edition).

162. 1843. Robertson, John Parish (1792-1843), and Robertson, William Parish (?-?).

Letters on South America; comprising travels on the banks of the Paraná and Rio de la Plata.

3 vols. Post 8° 28s. 6d. London: John Murray, Albemarle Street. Printer: William Clowes and Sons, Stamford Street. NSTC: 2R13196.

No revised editions.

163. 1843. Sale [née Wynch], Florentia (1790-1853).

A journal of the disasters in Affghanistan, 1841-2.

1 vol. Post 8° 12s. London: John Murray, Albemarle Street. Printer: A. Spottiswoode, New-Street-Square. NSTC: 2S2485.

No revised editions.

164. 1843. Stephens, John Lloyd (1805–52).

Incidents of travel in Yucatan.

2 vols. 8° 42s. London: John Murray, Albemarle Street. Printer: W. Clowes and Sons, 14, Charing Cross. NSTC: 2S38599.

No revised editions.

165. 1843. Wilkinson, Sir John Gardner (1797–1875).

Modern Egypt and Thebes: being a description of Egypt; including the information required for travellers in that country.

2 vols. Post 8° 15s. London: John Murray, Albemarle Street. Printer: A. Spottiswoode, New-Street-Square. NSTC: 2W20992.

1847: "A new edition, corrected and condensed," issued under the variant title *Hand-book for travellers in Egypt; including descriptions of the course of the Nile to the second cataract, Alexandria, Cairo, the Pyramids, and Thebes, the overland transit to India, the peninsula of Mount Sinai, the oases, &c.* (imprint: London: John Murray, Albemarle Street; Paris, Galignani; Stassin & Xavier; Malta, Muir. Printer: Spottiswoode and Shaw, New-Street-Square); 1858: "A new edition, with corrections and addition" issued under the variant title *A handbook for travellers in Egypt* . . . (printer: W. Clowes and Sons, Stamford Street, and Charing Cross); 1867: "A new edition, with corrections and additions" (printer: as 1858); 1873: fourth edition, "revised on the spot" (printer: as 1858); 1875: fifth edition, "revised on the spot" (printer: as 1858); 1880: sixth edition, "revised on the spot," issued, in 2 pts., under the variant title *A handbook for travellers in lower and upper Egypt* . . . (printer: as 1858); 1888: seventh edition, issued in 2 pts. (printer: as 1858); 1891: eighth edition, "revised" (printer: as 1858); 1896: ninth edition, "rewritten" (printer: 1858); 1900: tenth edition, "revised" (printer: as 1858).

166. 1844. Featherstonehaugh, George William (1780–1866). Note: title page renders author's name as Featherstonhaugh.

Excursion through the slave states, from Washington on the Potomac to the frontiers of Mexico; with sketches of popular manners and geological notices.

2 vols. 8° 26s. London: John Murray, Albemarle Street. Printer: William Clowes and Sons, Stamford Street. NSTC: 2F3390.

No revised editions.

167. 1844. Godley, John Robert (1814–61).

Letters from America.

2 vols. Post 8° 16s. London: John Murray, Albemarle Street. Printer: A. Spottiswoode, New-Street-Square. NSTC: 2G11342.

No revised editions.

168. 1844. Greenhow, Robert (1800–1854).

The history of Oregon and California, and the other territories on the north-west coast of North America; accompanied by a geographical view and map of those countries, and a number of documents as proofs and illustrations of the history.

1 vol. 8° 16s. London: John Murray, Albemarle Street. Printer: W. Clowes and Sons, 14, Charing Cross. NSTC: 2G20638.

No revised editions.

169. 1844. Hay, Sir John Hay Drummond- (1816–93).

Western Barbary: its wild tribes and savage animals.

Issued as pt. 9 (vol. 5) of Murray's Home and Colonial Library series. 1 vol. 12° 2s. 9d. London: John Murray, Albemarle Street. Printer: William Clowes and Sons, Stamford Street. NSTC: 2H13686.

1861: "New edition" (printer: as first edition).

170. 1844. Houstoun [née Jesse; other married name Fraser], Matilda Charlotte (1815–92).

Texas and the Gulf of Mexico; or yachting in the New World.

2 vols. Post 8° 21s. London: John Murray, Albemarle Street. Printer: W. Nicol, 60, Pall Mall. NSTC: 2H32481.

No revised editions.

171. 1844. Irby, Charles Leonard (1789–1845), and Mangles, James (1786–1867).

Travels in Egypt and Nubia, Syria, and the Holy Land; including a journey round the Dead Sea, and through the country east of the Jordan.

Issued as pt. 7 (vol. 4) of Murray's Home and Colonial Library series. 1 vol. Post 8° 2s. 6d. London: John Murray, Albemarle Street. Printer: Bradbury and Evans, Whitefriars. NSTC: 2I3730.

1852: "New edition" (printer: W. Clowes and Sons, Stamford Street); 1868: "New edition" (printer: as 1852).

172. 1844. Meredith [née Twamley], Louisa Anne (1812–95). Note: title page reads "By Mrs. Charles Meredith."

Notes and sketches of New South Wales, during a residence in that colony from 1839 to 1844.

Issued as pt. 13 (vol. 7) of Murray's Home and Colonial Library series. 1 vol. 12° 2s. 6d. London: John Murray, Albemarle Street. Printer: William Clowes and Sons, Stamford Street. NSTC: 2T21095.

No revised editions.

173. 1844. Ripa, Matteo (1682–1746). Notes: "Selected and translated from the Italian" by Prandi, Fortunato (?–?).

Memoirs of Father Ripa, during thirteen years' residence at the court of Peking in the service of the Emperor of China; with an account of the foundation of the College for the Education of Young Chinese at Naples.

Issued as pt. 15 (vol. 8) of Murray's Home and Colonial Library series. 1 vol. Post 8° 2s. 6d. London: John Murray, Albemarle Street. Printer: W. Clowes and Sons, Stamford Street. NSTC: 2R11334.

1855: "New edition" (printer: as first edition); 1861: "New edition" (printer: as first edition).

174. 1845. Darwin, Charles Robert (1809–82).

Journal of researches into the natural history and geology of the countries visited during the voyage of H.M.S. Beagle *round the world, under the command of Capt. FitzRoy, R.N.*

Issued as pts. 22–24 (vol. 12) of Murray's Home and Colonial Library series. 1 vol. Post 8° 2s. 6d. London: John Murray, Albemarle Street. Printer: William Clowes and Sons, Stamford Street. NSTC: unrecorded.

1852: "New edition" (printer: as first edition); 1870: "New edition" (printer: as first edition); 1890: "New edition, with portrait" (printer: as first edition); "New edition with illustrations" (printer: R. & R. Clark, Edinburgh).

175. 1845. Lyell, Sir Charles, first baronet (1797–1875).

Travels in North America; with geological observations on the United States, Canada, and Nova Scotia.

2 vols. Post 8° 21s. London: John Murray, Albemarle Street. Printer: A. Spot-tiswoode, New-Street-Square. NSTC: 2L25694.

1855: second edition, issued under the variant title *Travels in North America, Canada, and Nova Scotia. With geological observations* (printer: as first edition).

176. 1845. Wakefield, Edward Jerningham (1820–79).

Adventure in New Zealand, from 1839 to 1844; with some account of the begin-nings of the British colonization of the islands.

2 vols. 8° 28s. London: John Murray, Albemarle Street. Printer: William Clowes and Sons, Stamford Street. NSTC: 2W1392.

No revised editions.

177. 1846. Abbott, Joseph (bap. 1790, d. 1862). Note: Abbott is listed as edi-tor, although he was the author. Philip Musgrave was a fictional character.

Philip Musgrave; or memoirs of a Church of England missionary in the North American colonies.

Issued as pt. 33 (vol. 16) of Murray's Home and Colonial Library series. 1 vol. Post 8° 2s. 6d. London: John Murray, Albemarle Street. Printer: Wil-liam Clowes and Sons, Stamford Street. NSTC: 2A880.

1850: "New edition" (printer: as first edition).

178. 1846. Barrow, Sir John, first baronet (1764–1848).

Voyages of discovery and research within the Arctic regions, from the year 1818 to the present time: under the command of several naval officers employed by sea and land in search of a north-west passage from the Atlantic to the Pacific; with two attempts to reach the North Pole.

1 vol. 8° 15s. London: John Murray, Albemarle Street. Printer: William Clowes and Sons, Stamford Street. NSTC: 2B9868.

No revised editions.

179. 1846. Browne, John Ross (1821–75).

Etchings of a whaling cruise, with notes of a sojourn on the island of Zanzibar; and a brief history of the whale fishery, in its past and present condition.

1 vol. 8° 18s. London: John Murray, Albemarle Street. Printer: W. Clowes and Sons, 14, Charing Cross. NSTC: 2B53753.

No revised editions.

180. 1846. Melville, Herman (1819–91).

Narrative of a four months' residence among the natives of a valley of the Marquesas Islands; or, a peep at Polynesian life.

Issued as pts. 30–31 (vol. 15) of Murray's Home and Colonial Library series. 1 vol. Post 8° 6s. London: John Murray, Albemarle Street. Printer: William Clowes and Sons, Stamford Street. NSTC: 2M24032.

1847: issued under the variant title *Typee; or, a narrative of a four months' residence among the natives of a valley of the Marquesas Islands; or, a peep at Polynesian life* (printer: as first edition); 1893: "New edition" (printer: as first edition).

181. 1847. Acland, Charles (d. 1845).

A popular account of the manners and customs of India.

Issued as pt. 50 (vol. 25) of Murray's Home and Colonial Library series. 1 vol. Square 12° 6s. London: John Murray, Albemarle Street. Printer: W. Clowes and Sons, Stamford Street. NSTC: 2A2092.

No revised editions.

182. 1847. Ali, Shahamat (?–?).

The Sikhs and Afghans, in connection with India and Persia, immediately before and after the death of Ranjeet Singh: from the journal of an expedition to Kabul, through the Panjab and the Khaibar Pass.

1 vol. Post 8° 12s. London: John Murray, Albemarle Street. W. Clowes and Sons, Stamford Street. NSTC: 2S14854.

No revised editions.

183. 1847. Barrow, Sir John, first baronet (1764–1848).

An auto-biographical memoir of Sir John Barrow, Bart., late of the Admiralty; including reflections, observations, and reminiscences at home and abroad, from early life to advanced age.

1 vol. 8° 16s. London: John Murray, Albemarle Street. Printer: W. Clowes and Sons, Stamford Street. NSTC: 2B9856.

No revised editions.

184. 1847. Edwards, William Henry (1822–1909).

A voyage up the river Amazon, including a residence at Pará.

Issued as pt. 48 (vol. 24) of Murray's Home and Colonial Library series. 1 vol. Post 8° 2s. 6d. London: John Murray, Albemarle Street. Printer: William Clowes and Sons, Stamford Street. NSTC: 2E5745.

No revised editions.

185. 1847. Fortune, Robert (1812–80).

Three years' wanderings in the northern provinces of China, including a visit to the tea, silk, and cotton countries: with an account of the agriculture of the Chinese, new plants, etc.

1 vol. 8° 15s. London: John Murray, Albemarle Street. Printer: Spottiswoode and Shaw, New-Street-Square. NSTC: 2F11549.

1847: second edition (printer: as first edition).

186. 1847. Melville, Herman (1819–91).

Omoo: a narrative of adventures in the South Seas; being a sequel to the "Residence in the Marquesas Islands."

Issued as pts. 43–44 (vol. 22) of Murray's Home and Colonial Library series. 1 vol. Post 8° 6s. London: John Murray, Albemarle Street. Printer: Spottiswoode and Shaw, New-Street-Square. NSTC: 2M24033.

1861: "New edition" (printer: as first edition); 1893: "New edition, with a memoir of the author and illustrations" (printer: as first edition).

187. 1847. Ross, Sir James Clark (1800–1862).

A voyage of discovery and research in the southern and Antarctic regions, during the years 1839–43.

2 vols. 8° 36s. London: John Murray, Albemarle Street. Printer: Spottiswoode and Shaw, New-Street-Square. NSTC: 2R18176.

No revised editions.

188. 1847. Ruxton, George Frederick Augustus (1820–48).

Adventures in Mexico and the Rocky Mountains.

Issued as pts. 52–53 (vo. 26) of Murray's Home and Colonial Library series. 1 vol. Post 8° 6s. London: John Murray, Albemarle Street. Printer: William Clowes and Sons, Stamford Street. NSTC: 2R22195.

1861: "New edition" (printer: as first edition).

189. 1848. Brooke, Sir James (1803–68), and Mundy, Sir George Rodney (1805–84).

Narrative of events in Borneo and Celebes, down to the occupation of Labuan: from the journals of James Brooke, Esq. Rajah of Sarāwak, and Governor of Labuan. Together with a narrative of the operations of H.M.S. Iris. *By Captain Rodney Mundy, R.N.*

2 vols. 8° 32s. London: John Murray, Albemarle Street. Printer: Spottiswoode and Shaw, New-street-Square. NSTC: 2B50344.

No revised editions.

190. 1848. Bunbury, Sir Charles James Fox, eighth baronet (1809–86).

Journal of a residence at the Cape of Good Hope; with excursions into the interior, and notes on the natural history, and the native tribes.

1 vol. Post 8° 9s. London: John Murray, Albemarle Street. Printer: Spottiswoode and Shaw, New-street-Square. NSTC: 2B57635.

No revised editions.

191. 1848. Haygarth, Henry William (?–?).

Recollections of bush life in Australia, during a residence of eight years in the interior.

Issued as pt. 58 (vol. 29) of Murray's Home and Colonial Library series. 1 vol. 12° 2s. 6d. London: John Murray, Albemarle Street. Printer: William Clowes and Sons, Stamford Street. NSTC: 2H14066.

1861: "New edition" (printer: as first edition).

192. 1848. Wilkinson, George Blakiston (1817–88).

South Australia; its advantages and its resources. Being a description of that colony, and a manual of information for emigrants.

1 vol. Post 8° 10s. 6d. London: John Murray, Albemarle Street. Printer: George Woodfall and Son, Angel Court, Skinner Street. NSTC: 2W20859.

No revised editions.

193. 1849. Curzon, Robert, fourteenth Baron Zouche of Harringworth (1810–73).

Visits to monasteries in the Levant.

1 vol. Post 8° 15s. London: John Murray, Albemarle Street. Printer: W. Clowes and Sons, Stamford Street. NSTC: 2C47841.

1849: second edition (printer: as first edition); 1850: third edition (printer: as first edition); 1851 fourth edition (printer: as first edition); 1865: fifth edition (printer: R. Clark, Edinburgh); 1881: sixth edition (printer: Henry Hansard and Son, Lincoln's Inn Fields).

194. 1849. Herschel, Sir John Frederick William, first baronet (1792–1871). Note: "Edited by."

A manual of scientific enquiry; prepared for the use of Her Majesty's navy: and adapted for travellers in general.

1 vol. Post 8° 10s. 6d. London: John Murray, Albemarle Street, Publisher to the Admiralty. Printer: William Clowes and Sons, Stamford Street.

1851: second edition (printer: Bradbury and Evans, Printers, Whitefriars); 1859: third edition, "Superintended by the Rev. Robert Main" (printer: as first edition); 1871: fourth edition, "Superintended by Rev. Robert Main" and "Printed for Her Majesty's Stationary Office, and sold by John Murray" (printer: as first edition).

195. 1849. Humboldt, Alexander von (1769–1859). Note: "Translated by Mrs. Sabine," Sabine [née Leeves], Elizabeth Juliana (1807–79).

Aspects of nature, in different lands and different climates; with scientific elucidations.

2 vols. Square 12° 6s. London: Printed for Longman, Brown, Green and Longmans; Paternoster Row; and John Murray, Albemarle Street. Printer: Wilson and Ogilvy, 57, Skinner Street, Snowhill. NSTC: 2H36371.

No revised editions.

196. 1849. Layard, Sir Austen Henry (1817–94).

Nineveh and its remains: with an account of a visit to the Chaldæan Christians of Kurdistan, and the Yezidis, or devil-worshippers; and an enquiry into the manners and arts of the ancient Assyrians.

2 vols. 8° 36s. London: John Murray, Albemarle Street. Printer: Spottiswoodes [*sic*] and Shaw, New-street-Square. NSTC: 2L6960.

1849: second edition (printer: as first edition); 1849: third edition (printer: as first edition); 1849: fourth edition (printer: as first edition); 1850: fifth edition (printer: as first edition); 1851: "Abridged by him" and issued under the variant title *A popular account of discoveries at Nineveh* (printer: as first edition); 1852: "New edition" of *Popular account* (printer: as first edition); 1854: sixth edition of *Nineveh and its remains* (printer: as first edition); 1867: "Abridged by the author" and issued under the variant title *Nineveh and its remains: a narrative of an expedition to Assyria during the years 1845, 1846, & 1847* (printer: as first edition). Note: no fourth edition identified.

197. 1849. Layard, Sir Austen Henry (1817–94).

The monuments of Nineveh. From drawings made on the spot.

1 vol. 2° £10 10s. London: John Murray, Albemarle Street. Printer: Bradbury and Evans, Whitefriars. NSTC: 2L6958.

1853: second volume issued under the title *A second series of the monuments of Nineveh; including bas-reliefs from the Palace of Sennacherib and bronzes from the ruins of Nimroud. From drawings made on the spot during a second expedition to Assyria* (printer: as first volume, although here described as "printers extraordinary to the Queen").

198. 1849. Lyell, Sir Charles, first baronet (1797–1875).

A second visit to the United States of North America.

2 vols. Post 8° 18s. London: John Murray, Albemarle Street. Printer: Spottiswoodes [*sic*] and Shaw, New-street-Square. NSTC: 2L25695.

1850: second edition, "revised and corrected" (printer: as first edition); 1855: third edition (printer: as first edition).

199. 1849. [Melville, Elizabeth Helen (?–?)]. Note: published anonymously, "By a lady," and "Edited by the Hon. Mrs. Norton," Norton [née Sheridan], Caroline Elizabeth Sarah [other married name Caroline Elizabeth Sarah Stirling Maxwell, Lady Stirling Maxwell] (1808–77)].

A residence at Sierra Leone. Described from a journal kept on the spot, and from letters written to friends at home.

Issued as pts. 68–69 (vol. 34) of Murray's Home and Colonial Library series. 1 vol. 12° 6s. London: John Murray, Albemarle Street. Printer: William Clowes and Sons, Stamford Street. NSTC: 2L1373.

No revised editions.

200. 1849. St John, Bayle Frederick (1822–59).

Adventures in the Libyan Desert and the oäsis of Jupiter Ammon.

Issued as pt. 67 (vol. 33) of Murray's Home and Colonial Library series. 1 vol. Post 8° 2s. 6d. London: John Murray, Albemarle Street. Printer: William Clowes and Sons, Stamford Street. NSTC: 2S2095.

No revised editions.

201. 1849. Wilkinson, George Blakiston (1817–88).

The working man's handbook to South Australia. With advice to the farmer, and detailed information for the several classes of labourers and artizans.

1 vol. 18° 1s. 6d. London: John Murray, Albemarle Street; Trelawney Saunders, Charing Cross. Printer: G. Woodfall and Son, Angel Court, Skinner Street. NSTC: 2W20860.

No revised editions.

202. 1850. Cumming, Roualeyn George Gordon- (1820–66).

Five years of a hunter's life in the far interior of South Africa. With notices of the native tribes, and anecdotes of the chase of the lion, elephant, hippopotamus, giraffe, rhinoceros, &c.

2 vols. Post 8° 24s. London: John Murray, Albemarle Street. Printer: W. Clowes and Sons, Stamford Street. NSTC: 2C46471.

1850: second edition (printer: as first edition); 1855: "New edition" (printer: as first edition); 1856: "New edition, revised and condensed" issued under the variant title *Five years' adventures in the far interior of South Africa; with notices of the native tribes and savage animals* (printer: as first edition); 1857: "New edition, revised and condensed" (printer: as first edition); 1863: "New edition" (printer: as first edition).

203. 1850. MacFarlane, Charles (1799–1858).

Turkey and its destiny: the result of journeys made in 1847 and 1848 to examine into the state of that country.

2 vols. 8° 28s. London: John Murray, Albemarle Street. Printer: W. Clowes and Sons, Stamford Street. NSTC: 2M3815.

No revised editions.

204. 1850. Spencer, Jesse Ames (1816–98).

The East: sketches of travels in Egypt and the Holy Land.

1 vol. 8° 21s. New York: George P. Putnam, 163 Broadway; London: John Murray. Printer: Stereotyped by T. B. Smith, 216 William Street. NSTC: 2S33809.

No revised editions.

205. 1850. Weld, Charles Richard (1813–69).

Arctic expeditions: a lecture delivered at the London Institution, February 6, 1850.

1 vol. Post 8° 1s. London: John Murray, Albemarle Street. Printer: G. Woodfall and Son, Printers, Angel Court, Skinner Street, London. NSTC: 2W11713.

No revised editions.

206. 1851. [Blenkinsop, Adam (?–?)]. Note: published anonymously.

A transport voyage to the Mauritius and back; touching at the Cape of Good Hope and St. Helena.

1 vol. Post 8° 9s. 6d. London: John Murray, Albemarle Street. Printer: W. Clowes and Sons, Stamford Street. NSTC: 2G21543.

No revised editions.

207. 1851. Harcourt, Edward William (1825–91).

A sketch of Madeira; containing information for the traveller, or invalid visitor.

1 vol. Post 8° 8s. 6d. London: John Murray, Albemarle Street. Printer: George Woodfall and Son, Angel Court, Skinner Street. NSTC: 2H7317.

No revised editions.

208. 1852. Egerton [formerly Leveson-Gower], Francis, first earl of Ellesmere (1800–1857).

Journal of a winter's tour in India: with a visit to the court of Nepaul.

2 vols. Post 8° 18s. London: John Murray, Albemarle Street. Printer: Bradbury and Evans, Printers, Whitefriars. NSTC: 2E5956.

No revised editions.

209. 1852. Fortune, Robert (1812–80).

A journey to the tea countries of China; including Sung-Lo and the Bohea hills; with a short notice of the East India Company's tea plantations in the Himalaya Mountains.

1 vol. Post 8° 18s. London: John Murray, Albemarle Street. Printer: W. Clowes and Sons, Stamford Street. NSTC: 2F11546.

1853: third edition, issued under the variant title *Two visits to the tea countries of China and the British tea plantations in the Himalaya; with a narrative of adventures, and a full description of the culture of the tea plant, the agriculture, horticulture, and botany of China* (printer: as first edition). Note: no second edition identified.

210. 1852. Meredith [née Twamley], Louisa Anne (1812–95). Note: title page reads "By Mrs. Charles Meredith."

My home in Tasmania, during a residence of nine years.

2 vols. Post 8° 2s. 6d. London: John Murray, Albemarle Street. Printer: G. Woodfall and Son, Angel Court, Skinner Street, London. NSTC: 2T21094.

No revised editions.

211. 1852. Oliphant, Laurence (1829–88).

A journey to Katmandu (the capital of Nepaul), with the camp of Jung Bahadoor; including a sketch of the Nepaulese ambassador at home.

Issued as part of Murray's Railway Reading series. 1 vol. 12° 6s. 6d. London: John Murray, Albemarle Street. Printer: W. Clowes and Sons, Stamford Street. NSTC: 203054.

No revised editions.

212. 1852. Tremenheere, Hugh Seymour (1804–93).

Notes on public subjects, made during a tour in the United States and in Canada.

1 vol. Post 8° 10s. 6d. John Murray, Albemarle Street. Printer: W. Clowes and Sons, Stamford Street. NSTC: 2T17102.

No revised editions.

213. 1853. Erskine, John Elphinstone (1806–87).

Journal of a cruise among the islands of the western Pacific, including the Feejees and others inhabited by the Polynesian Negro races. In her Majesty's ship Havannah.

1 vol. 8° 16s. London: John Murray, Albemarle Street. Printer: William Clowes and Sons, Stamford Street. NSTC: 2E11476.

No revised editions.

214. 1853. Galton, Sir Francis (1822–1911).

The narrative of an explorer in tropical South Africa.

1 vol. Post 8° 12s. London: John Murray, Albemarle Street. Printer: Bradbury and Evans, Printers, Whitefriars. NSTC: 2G1428.

No revised editions.

215. 1853. Hooper, William Hulme (1827–54).

Ten months among the tents of the Tuski, with incidents of an Arctic boat expedition in search of Sir John Franklin, as far as the Mackenzie River, and Cape Bathurst.

1 vol. 8° 14s. London: John Murray, Albemarle Street. Printer: Bradbury and Evans, Whitefriars. NSTC: 2H29609.

No revised editions.

216. 1853. Layard, Sir Austen Henry (1817–94).

Discoveries in the ruins of Nineveh and Babylon; with travels in Armenia, Kurdistan and the desert: being the result of a second expedition undertaken for the Trustees of the British Museum.

1 vol. 8° 21s. London: John Murray, Albemarle Street. Printer: Spottiswoodes [sic] and Shaw, New-Street-Square. NSTC: 2L6949.

No revised editions.

217. 1853. Palliser, John (1817–87).

Solitary rambles and adventures of a hunter in the prairies.

1 vol. Post 8° 10s. 6d. London: John Murray, Albemarle Street. Printer: Bradbury and Evans, Whitefriars. NSTC: 2P2065.

No revised editions.

218. 1853. Parkyns, Mansfield (1823–94).

Life in Abyssinia: being notes collected during three years' residence and travels in that country.

2 vols. 8° 30s. London: John Murray, Albemarle Street. Printer: William Clowes and Sons, Stamford Street. NSTC: 2P4563.

1868: second edition, "with a new introduction" (printer: as first edition).

219. 1854. Curzon, Robert, fourteenth Baron Zouche of Harringworth (1810–73).

Armenia: a year at Erzeroom, and on the frontiers of Russia, Turkey, and Persia.

1 vol. Post 8° 7s. 6d. London: John Murray, Albemarle Street. Printer: W. Clowes and Sons, Stamford Street, and Charing Cross. NSTC: 2C47835.

1854: third edition (printer: as first edition). Note: second edition not identified.

220. 1854. Hooker, Sir Joseph Dalton (1817–1911).

Himalayan journals; or, notes of a naturalist in Bengal, the Sikkim and Nepal Himalayas, the Khasia Mountains, &c.

2 vols. 8° 36s. London: John Murray, Albemarle Street. Printer: Bradbury and Evans, Whitefriars. NSTC: 2H29205.

1855: "A new edition, carefully revised and condensed" (printer: as first edition).

221. 1855. Galton, Sir Francis (1822–1911).

The art of travel; or, shifts and contrivances available in wild countries.

1 vol. 12° 7s. 6d. London: John Murray, Albemarle Street. Printer: Woodfall and Kinder, Angel Court, Skinner Street. NSTC: 2G1422.

1856: second edition, "revised and enlarged, with many additional woodcuts" (printer: as first edition); 1860: third edition, "revised and enlarged" (printer: William Clowes and Sons, Duke Street, Stamford Street, and Charing Cross); 1867: fourth edition, "entirely re-cast and much enlarged" (printer: as 1860); 1872: fifth edition (printer: as 1860); 1876: sixth edition (printer: as 1860); 1883: seventh edition (printer: as 1860); 1893: eighth edition (printer: as 1860).

222. 1855. Perry, Sir Thomas Erskine (1806–82).

A bird's-eye view of India, with extracts from a journal kept in the provinces, Nepal, &c.

1 vol. 12° 5s. London: John Murray, Albemarle Street. Printer: Bradbury and Evans, Whitefriars. NSTC: 2P11896.

No revised editions.

223. 1855. Porter, Josias Leslie (1823–89).

Five years in Damascus: including an account of the history, topography, and antiquities of the city; with travels and researches in Palmyra, Lebanon, and the Hauran.

2 vols. Post 8° 21s. London: John Murray, Albemarle Street. Printer: W. Clowes and Sons, Stamford Street, and Charing Cross. NSTC: 2P22494.

1870: second edition, "revised" (printer: as first edition).

224. 1856. Baikie, William Balfour (1825–64).

Narrative of an exploring voyage up the rivers Kwóra and Bínue (commonly known as the Niger and Tsádda) in 1854.

1 vol. 8° 16s. London: John Murray, Albemarle Street. Printer: Bradbury and Evans, Whitefriars. NSTC: 2B2301.

No revised editions.

225. 1856. [Bishop (née Bird), Isabella Lucy (1831–1904)]. Note: published anonymously.

The Englishwoman in America.

1 vol. Post 8° 10s. 6d. London: John Murray, Albemarle Street. Printer: W. Clowes and Sons, Stamford Street, and Charing Cross. NSTC: 2E10620.

No revised editions.

226. 1856. Ferrier, Joseph Pierre (?–?). Note: "Translated from the original unpublished manuscript" by Jesse, William (1809–71), and edited by Seymour, Henry Danby (1820–77).

Caravan journeys and wanderings in Persia, Afghanistan, Turkistan, and Beloochistan; with historical notices of the countries lying between Russia and India.

1 vol. 8° 21s. London: John Murray, Albemarle Street. Printer: William Clowes and Sons, Stamford Street and Charing Cross. NSTC: 2F4861.

1857: second edition (printer: as first edition).

227. 1856. Hamilton, James (?-?).

Wanderings in North Africa.

1 vol. Post 8° 12s. London: John Murray, Albemarle Street. Printer: Woodfall and Kinder, Angel Court, Skinner Street. NSTC: 2H5201.

No revised editions.

228. 1856. Robinson, Edward (1794-1863), Smith, Eli (1801-1857), "and others."

Later biblical researches in Palestine and in the adjacent regions: a journal of travels in the year 1852.

1 vol. 8° 15s. John Murray, Albemarle Street. Printer: Spottiswoode & Co., New Street-Square. NSTC: 2R13699.

No revised editions.

229. 1856. Sandwith, Humphry (1822-81).

A narrative of the siege of Kars and of the six months' resistance by the Turkish garrison under General Williams to the Russian army: together with a narrative of travels and adventures in Armenia and Lazistan; with remarks on the present state of Turkey.

1 vol. Post 8° 10s. 6d. London: John Murray, Albemarle Street. Printer: William Clowes and Sons, Stamford Street, and Charing Cross. NSTC: 2S4194.

1856: second edition (printer: as first edition); 1856: "A new edition, abridged" (printer: as first edition).

230. 1856. Sheil, Mary Leonora Woulfe, Lady (?-?).

Glimpses of life and manners in Persia. By Lady Sheil. With notes on Russia, Koords, Toorkomans, Nestorians, Khiva, and Persia.

1 vol. Post 8° 12s. London: John Murray, Albemarle Street. Printer: William Clowes and Sons, Stamford Street, and Charing Cross. NSTC: 2S18308.

No revised editions.

231. 1856. Stanley, Arthur Penrhyn (1815-81).

Sinai and Palestine in connection with their history.

1 vol. 8° 16s. London: John Murray, Albemarle Street. Printer: Bradbury and Evans, Whitefriars. NSTC: 2S36323.

1856: second edition (printer: as first edition); 1856: third edition (printer: as first edition); 1857: fourth edition (printer: as first edition); 1858: fifth edition (printer: as first edition); 1866, 1868, 1871, 1873, 1875, 1877, 1881, 1883, 1887, 1889, 1896: "New edition[s]" (printer: as first edition).

232. 1857. Barrow, Sir George, second baronet (1806–76).

Ceylon: past and present.

1 vol. Post 8° 6s. London: John Murray, Albemarle Street. Printer: W. Clowes and Sons, Stamford Street, and Charing Cross. NSTC: 2B9787.

No revised editions.

233. 1857. Davis, Sir John Francis, first baronet (1795–1890).

China: a general description of that empire and its inhabitants; with the history of foreign intercourse down to the events which produced the dissolution of 1857.

2 vols. Post 8° 14s. London: John Murray, Albemarle Street. Printed: W. Clowes and Sons, Stamford Street, and Charing Cross. NSTC: 2D5134.

No revised editions.

234. 1857. Fortune, Robert (1812–80).

A residence among the Chinese: inland, on the coast, and at sea. Being a narrative of scenes and adventures during a third visit to China, from 1853 to 1856. Including notices of many natural productions and works of art, the culture of silk, &c.; with suggestions on the present war.

1 vol. 8° 16s. London: John Murray, Albemarle Street. Printer: W. Clowes and Sons, Stamford Street, and Charing Cross. NSTC: 2F11548.

No revised editions.

235. 1857. Livingstone, David (1813–73).

Missionary travels and researches in South Africa including a sketch of sixteen years' residence in the interior of Africa, and a journey from the Cape of Good Hope to Loanda on the west coast; thence across the continent, down the river Zambesi, to the eastern ocean.

1 vol. 8° 21s. London: John Murray, Albemarle Street. Printer: W. Clowes and Sons, Stamford Street, and Charing Cross. NSTC: 2L18333.

1861: issued under the variant title *A popular account of missionary travels and researches in South Africa* (printer: as first edition); 1875: "New edition" of *A popular account* (printer: as first edition); 1899: "New edition" of *Missionary travels* (printer: W. Brendon & Son, Plymouth).

236. 1858. Ellis, William (1794–1872).

Three visits to Madagascar during the years 1853–1854–1856. Including a journey to the capital. With notices of the natural history of the country and of the present civilisation of the people.

1 vol. 8° 16s. London: John Murray, Albemarle Street. Printer: Spottiswoode and Co. New-Street-Square. NSTC: 2E8165.

No revised editions.

237. 1859. [Kinglake, Alexander William (1809–91)]. Note: published anonymously.

Eothen.

1 vol. Post 8° 7s. 6d. London: John Murray, Albemarle Street. Printer: R. and R. Clark, Printers, Edinburgh. ESTC: 2K6020.

No revised editions.

238. 1859. McClintock, Sir (Francis) Leopold (1819–1907).

The voyage of the 'Fox' in the Arctic seas. A narrative of the discovery of the fate of Sir John Franklin and his companions.

1 vol. 8° 16s. London: John Murray, Albemarle Street, Publisher to the Admiralty. Printer: W. Clowes and Sons, Stamford Street, and Charing Cross. NSTC: 2F14529.

No revised editions.

239. 1859. Thomson, Arthur Saunders (1816–60).

The story of New Zealand: past and present — savage and civilized.

2 vols. Post 8° 24s. London: John Murray, Albemarle Street. Printer: Spottiswoode and Co. New-Street Square. NSTC: 2T9947.

No revised editions.

Notes

CHAPTER ONE

1. [Harriet Martineau], "Travel during the Last Half Century," review of *The Principal Navigations, Voyages, Traffiques and Discoveries of the English Nation*, by Richard Hakluyt and *The English Cyclopædia*, by Charles Knight, *Westminster Review* 70, no. 138 (October 1858): 236.

2. Richard Holmes, *The Age of Wonder: How the Romantic Generation Discovered the Beauty and Terror of Science* (London: Harper Press, 2008).

3. Marie-Noëlle Bourguet, "The Explorer," in *Enlightenment Portraits*, ed. Michel Vovelle (Chicago: University of Chicago Press, 1997), 296.

4. Ian S. MacLaren, "In Consideration of the Evolution of Explorers and Travellers into Authors: A Model," *Studies in Travel Writing* 15, no. 3 (2011): 223.

5. Steven Shapin, "Rarely Pure and Never Simple: Talking about Truth," *Configurations*, vol. 7, no. 1 (1999).

6. One volume to address these questions is Tim Fulford, Debbie Lee, and Peter J. Kitson, *Literature, Science and Exploration in the Romantic Era: Bodies of Knowledge* (Cambridge: Cambridge University Press, 2004). See also Andrew Cunningham and Nicholas Jardine, eds., *Romanticism and the Sciences* (Cambridge: Cambridge University Press, 1990); Holmes, *Age of Wonder*. On the Enlightenment, see Charles W. J. Withers, *Placing the Enlightenment: Thinking Geographically about the Age of Reason* (Chicago: University of Chicago Press, 2007).

7. See, e.g., John Gascoigne, *Science in the Service of Empire: Joseph Banks, the British State and the Uses of Science in the Age of Revolution* (Cambridge: Cambridge University Press, 1998); John Gascoigne, *The Enlightenment and the Origins of European Australia* (Cambridge: Cambridge University Press, 2002); Larry Stewart, "Global Pillage: Science, Commerce, and Empire," in *The Cambridge History of Science*, vol. 4, *Eighteenth-Century Science*, ed. Roy Porter (Cambridge: Cambridge University Press, 2003).

8. Kapil Raj, *Relocating Modern Science: Circulation and the Construction of Knowledge in South Asia and Europe, 1650–1900* (Basingstoke: Palgrave Macmillan, 2006); see also Roy Bridges, "Exploration and Travel Outside Europe (1720–1914)," in *The Cambridge Companion to Travel Writing*, ed. Peter Hulme and Tim Youngs (Cambridge: Cambridge University Press, 2002); see also the essays in Dane Kennedy, ed., *Reinterpreting Exploration: The West in the World* (New York: Oxford University Press, 2014).

9. On travel, see Jaś Elsner and Joan-Pau Rubiés, eds., *Voyages and Visions: Towards a Cultural History of Travel* (London: Reaktion, 1999); and Justin Stagl, *A History of Curiosity: Theory of*

Travel, 1550–1800 (Chur: Harwood Academic, 1995). On the huge field that is travel writing, see, e.g., Glenn Hooper and Tim Youngs, eds., *Perspectives on Travel Writing* (Aldershot: Ashgate, 2004); Hulme and Youngs, eds., *Cambridge Companion to Travel Writing*; Chris Rojek and John Urry, eds., *Touring Cultures: Transformations of Travel and Theory* (London: Routledge, 1997); Mary Louise Pratt, *Imperial Eyes: Travel Writing and Transculturation*, 2nd ed. (New York: Routledge, 2008); Nigel Leask, *Curiosity and the Aesthetics of Travel Writing, 1770–1840* (Oxford: Oxford University Press, 2002); Julia Kuehn and Paul Smethurst, eds., *Travel Writing, Form, and Empire: The Poetics and Politics of Mobility* (New York: Routledge, 2009).

10. On explorers' notebooks, see Marie-Noëlle Bourguet, "A Portable World: The Notebooks of European Travellers (Eighteenth to Nineteenth Centuries)," *Intellectual History Review*, vol. 20, no. 3 (2010). The phrase "instabilities of print" is from David McKitterick, *Print, Manuscript and the Search for Order, 1450–1830* (Cambridge: Cambridge University Press, 2003), 217. On book history, see Adrian Johns, *The Nature of the Book: Print and Knowledge in the Making* (Chicago: University of Chicago Press, 1998). The importance of books and print in stimulating knowledge formation as communicative action is stressed in James A. Secord, "Knowledge in Transit," *Isis*, vol. 95, no. 4 (2004).

11. Janice Cavell, *Tracing the Connected Narrative: Arctic Exploration in British Print Culture, 1818–1860* (Toronto: University of Toronto Press, 2008). On media and cultural interest in polar exploration, see Adriana Craciun, "Writing the Disaster: Franklin and Frankenstein," *Nineteenth-Century Literature*, vol. 65, no. 4 (2011).

12. On blank spaces and travel narration, see Dane Kennedy, *The Last Blank Spaces: Exploring Africa and Australia* (Cambridge, MA: Harvard University Press, 2013).

13. John Scott, *A Visit to Paris in 1814*, 2nd ed. (London: Longman, Hurst, Rees, Orme, & Brown, 1815), 3.

14. Julian Jackson, "On Picturesque Description in Books of Travels," *Journal of the Royal Geographical Society of London* 5 (1835): 387.

15. Benjamin Colbert, "Bibliography of British Travel Writing," *Cardiff Corvey*, vol. 13 (2004), http://www.cardiff.ac.uk/encap/journals/corvey/articles/cc13_n01.html; Simon Eliot, *Some Patterns and Trends in British Publishing, 1800–1819* (London: Bibliographical Society, 1994).

16. Eliot, *Some Patterns*, 44; Simon Eliot, "Some Trends in British Book Production, 1800–1919," in *Literature in the Marketplace: Nineteenth-Century British Publishing and Reading Practices*, ed. John O. Jordan and Robert L. Patten (Cambridge: Cambridge University Press, 1995); Leslie Howsam, "The History of the Book in Britain, 1801–1914," in *The Oxford Companion to the Book*, ed. Michael F. Suarez and H. R. Woudhuysen (Oxford: Oxford University Press, 2010), vol. 1.

17. Bill Bell, "The Secret History of Smith and Elder: *The Publisher's Circular* as a Source for Publishing History," in *A Genius for Letters: Booksellers and Bookselling from the 16th to the 20th Century*, ed. Robin Myers and Michael Harris (Winchester: St Paul's Bibliographies, 1995).

18. Humphrey Carpenter, *The Seven Lives of John Murray: The Story of a Publishing Dynasty, 1768–2002* (London: John Murray, 2008), 124. On John Murray I, see William Zachs, *The First John Murray and the Late Eighteenth-Century London Book Trade: With a Checklist of His Publications* (Oxford: Oxford University Press for the British Academy, 1999).

19. On Barrow's role as an organizing influence behind British voyages of exploration in the early nineteenth century, see Fergus Fleming, *Barrow's Boys: A Stirring Story of Daring, Fortitude and Outright Lunacy* (London: Granta, 1998); Kim Wheatley, "The Arctic in the *Quarterly Review*," *European Romantic Review*, vol. 20, no. 4 (2009); Adriana Craciun, "What Is an Explorer?" *Eighteenth-Century Studies*, vol. 45, no. 1 (2011).

20. Jan Borm, "Defining Travel: On the Travel Book, Travel Writing and Terminology," in *Perspectives on Travel Writing*, ed. Glen Hooper and Tim Youngs (Aldershot: Ashgate, 2004), 13.

21. Borm, "Defining Travel," 19.

22. On an example for one notable explorer and John Murray author, see Lawrence Dritsas, *Zambesi: David Livingstone and Expeditionary Science in Africa* (London: I. B. Tauris, 2010).

23. Rob Iliffe, "Science and Voyages of Discovery," in *Cambridge History of Science*, vol. 4, *Eighteenth-Century Science*, ed. Porter, 618–45; Harry Liebersohn, "Scientific Ethnography and Travel," in *The Cambridge History of Science*, vol. 7, *The Modern Social Sciences*, ed. Theodore M. Porter and Dorothy Ross (Cambridge: Cambridge University Press, 2003); David Philip Miller and Peter Hanns Reill, eds., *Visions of Empire: Voyages, Botany, and Representations of Nature* (Cambridge: Cambridge University Press, 1996).

24. Barbara Maria Stafford, *Voyage into Substance: Art, Science, Nature, and the Illustrated Travel Account, 1760–1840* (Cambridge, MA: MIT Press, 1984).

25. For a brilliant exposition of the problems of transience in restricting truth, and sedentary science as a basis for measured reflection, see Dorinda Outram, "On Being Perseus: New Knowledge, Dislocation, and Enlightenment Exploration," in *Geography and Enlightenment*, ed. David N. Livingstone and Charles W. J. Withers (Chicago: University of Chicago Press, 1999). On writing about the body, on the body, and about bodies of knowledge in the late Enlightenment each understood as forms of embodied inscription, see Simon Schaffer, "'On Seeing Me Write': Inscription Devices in the South Seas," *Representations*, vol. 97, no. 1 (2007). For discussions of these issues with regard to the visual depiction of Africa, see Leila Koivunen, *Visualizing Africa in Nineteenth-Century British Travel Accounts* (New York: Routledge, 2009).

26. Stafford, *Voyage into Substance*; Brian J. Ford, *Images of Science: A History of Scientific Illustration* (New York: Oxford University Press, 1982); Charlotte Klonk, "Science, Art, and the Representation of the Natural World," in *Cambridge History of Science*, vol. 4, *Eighteenth-Century Science*, ed. Porter; Elisabeth A. Fraser, "Books, Prints, and Travel: Reading in the Gaps of the Orientalist Archive," *Art History*, vol. 31, no. 3 (2008); Jonathan Crary, *Techniques of the Observer: On Vision and Modernity in the Nineteenth Century* (Cambridge, MA: MIT Press, 1990); Martin Kemp, "Taking It on Trust: Form and Meaning in Naturalistic Representation," *Archives of Natural History*, vol. 17 (1990); Martin J. S. Rudwick, "The Emergence of a Visual Language for Geological Science, 1760–1840," *History of Science*, vol. 14 (1976). The importance of visual representation in the earth sciences—and to earth scientists' depiction as quintessential field agents—is stressed in Martin J. S. Rudwick, *Bursting the Limits of Time: The Reconstruction of Geohistory in the Age of Revolution* (Chicago: University of Chicago Press, 2005). The point about the portrait of the author helping establish the credibility of book and the author is made of Enlightenment texts by Richard B. Sher, *The Enlightenment and the Book: Scottish Authors and Their Publishers in Eighteenth-Century Britain, Ireland, and America* (Chicago: University of Chicago Press, 2006). On the complex emergence of "objectivity" and its criteria in illustration, see Lorraine Daston and Peter Galison, *Objectivity*, 2nd ed. (New York: Zone Books, 2010).

27. In a large literature on this topic, see Christian Jacob, *The Sovereign Map: Theoretical Approaches in Cartography throughout History*, trans. Tom Conley, ed. Edward H. Dahl (Chicago: University of Chicago Press, 2006); Matthew Edney, "Cartography: Disciplinary History," in *Sciences of the Earth: An Encyclopedia of Events, People, and Phenomena*, ed. Gregory A. Good, (New York: Garland, 1998); J. B. Harley, *The New Nature of Maps: Essays in the History of Cartography*, ed. Paul Laxton (Baltimore: Johns Hopkins University Press, 2001). For examples of mapping, exploration, and maps as forms of territorial inscription, see Matthew Edney, *Mapping an Empire: The Geographical Construction of British India, 1765–1843* (Chicago: University of Chicago Press, 1998); D. Graham Burnett, *Masters of All They Surveyed: Exploration, Geography, and a British El Dorado* (Chicago: University of Chicago Press, 2000)—on British Guiana in the 1830s; Anne Marie Claire Godlewska, *Geography Unbound: French Geographic Science from Cassini to Humboldt* (Chicago: University of Chicago Press, 1999); Haim Goren, *Dead Sea Level: Science, Exploration and Imperial Interests in the Near East* (London: I. B. Tauris, 2011).

28. For example, James Duncan and Derek Gregory, eds., *Writes of Passage: Reading Travel Writing* (London: Routledge, 1999); Gerry Kearns, "The Imperial Subject: Geography and Travel in the Work of Mary Kingsley and Halford Mackinder," *Transactions of the Institute of British Geographers*, vol. 22, no. 4 (1997); Leonard Guelke and Jeanne Kay Guelke, "Imperial Eyes on South

Africa: Reassessing Travel Narratives," *Journal of Historical Geography*, vol. 30, no. 1 (2004); Alison Blunt, *Travel, Gender, and Imperialism: Mary Kingsley and West Africa* (New York: Guilford Press, 1994); Cheryl McEwan, *Gender, Geography and Empire: Victorian Women Travellers in West Africa* (Aldershot: Ashgate, 2000).

29. For example, James Clifford and George E. Marcus, eds., *Writing Culture: The Poetics and Politics of Ethnography* (Berkeley: University of California Press, 1986); James Clifford, *Routes: Travel and Translation in the Late Twentieth Century* (Cambridge MA: Harvard University Press, 1997).

30. Robert J. Mayhew, "Materialist Hermeneutics, Textuality and the History of Geography: Print Spaces in British Geography, c.1500–1900," *Journal of Historical Geography*, vol. 33, no. 3 (2007); Robert J. Mayhew, "Printing Posterity: Editing Varenius and the Construction of Geography's History," in *Geographies of the Book*, ed. Miles Ogborn and Charles W. J. Withers (Farnham: Ashgate, 2010); Miles Ogborn, "*Geographia*'s Pen: Writing, Geography and the Arts of Commerce, 1660–1760," *Journal of Historical Geography*, vol. 30, no. 2 (2004); Miles Ogborn, *Indian Ink: Script and Print in the Making of the English East India Company* (Chicago: University of Chicago Press, 2007); Innes M. Keighren and Charles W. J. Withers, "Questions of Inscription and Epistemology in British Travelers' Accounts of Early Nineteenth-Century South America," *Annals of the Association of American Geographers*, vol. 101, no. 6 (2011); Charles W. J. Withers, "Writing in Geography's History: *Caledonia*, Networks of Correspondence and Geographical Knowledge in the Late Enlightenment," *Scottish Geographical Journal*, vol. 120, nos. 1–2 (2004); Charles W. J. Withers, "Geography, Enlightenment and the Book: Authorship and Audience in Mungo Park's African Texts," in *Geographies of the Book*, ed. Miles Ogborn and Charles W. J. Withers (Farnham: Ashgate, 2010); Charles W. J. Withers and Innes M. Keighren, "Travels into Print: Authoring, Editing and Narratives of Travel and Exploration, c.1815–c.1857," *Transactions of the Institute of British Geographers*, vol. 36, no. 4 (2011).

31. Elizabeth Eisenstein, *The Printing Press as an Agent of Change: Communications and Cultural Transformations in Early-Modern Europe*, 2 vols. (Cambridge: Cambridge University Press, 1979).

32. Walter J. Ong, *Ramus Method, and the Decay of Dialogue: From the Art of Discourse to the Art of Reason* (Cambridge, MA: Harvard University Press, 1958).

33. Robert Darnton, "What Is the History of Books?" *Daedalus* 111, no. 3 (1982): 69. On the interdisciplinarity of book history, see also Cyndia Susan Clegg, "History of the Book: An Undisciplined Discipline?" *Renaissance Quarterly*, vol. 54, no. 1 (2001).

34. Thomas R. Adams and Nicholas Barker, "A New Model for the History of the Book," in *A Potencie of Life: Books in Society*, ed. Nicholas Barker (London: British Library, 1993), 5–43.

35. Leslie Howsam, *Old Books and New Histories: An Orientation to Studies in Book and Print Culture* (Toronto: University of Toronto Press, 2006), vii. For a related discussion, see Miles Ogborn and Charles W. J. Withers, "Introduction: Book Geography, Book History," in *Geographies of the Book*, ed. Miles Ogborn and Charles W. J. Withers (Farnham: Ashgate, 2010), 1–25.

36. The phrase "cross-border approaches" we take from Tim Youngs, "Where Are We Going? Cross-Border Approaches to Travel Writing," in *Perspectives on Travel Writing*, ed. Glenn Hooper and Tim Youngs (Aldershot: Ashgate, 2004). On Said's "traveling theory," see Edward Said, *The World, the Text, and the Critic* (Cambridge MA: Harvard University Press, 1982), 226–47. The significance of Said's "traveling theory" to the notion of "traveling concepts" in English studies (albeit less to their applicability in studies of travel and travel writing in the English language) is addressed in Birgit Neumann and Frederik Tygstrup, "Travelling Concepts in English Studies," *European Journal of English Studies*, vol. 13, no. 1 (2009), and in the articles to which their essay is an introduction; see, particularly, in this context, Michael C. Frank, "Imaginative Geography as a Travelling Concept: Foucault, Said and the Spatial Turn," *European Journal of English Studies*, vol. 13, no. 1 (2009).

37. Bourguet, "The Explorer" (n. 3 above, this chap.), 257–58.

38. Percy G. Adams, *Travelers and Travel Liars, 1660–1800* (Berkeley: University of Califor-

nia Press, 1962). On attempts to regulate travel facts in earlier periods, see the essays in Judy A. Hayden, ed., *Travel Narratives, the New Science, and Literary Discourse, 1569–1750* (Farnham: Ashgate, 2012). For a recent study of these issues, see Benjamin Breen, "No Man Is an Island: Early Modern Globalization, Knowledge Networks, and George Psalmanazar's Formosa," *Journal of Early Modern History* 17, no. 4 (2013): 391–417.

39. Martyn Lyons, *Reading Culture and Writing Practices in Nineteenth-Century France* (Toronto: University of Toronto Press, 2008), and *The Writing Culture of Ordinary People in Europe, c. 1860– 1920* (Cambridge: Cambridge University Press, 2013). On correspondence as the basis of scholarly networks, see Lorraine Daston, "The Ideal and the Reality of the Republic of Letters in the Enlightenment," *Science in Context*, vol. 4, no. 2 (1992); Anne Goldgar, *Impolite Learning: Conduct and Community in the Republic of Letters, 1680–1850* (New Haven, CT: Yale University Press, 1995); Steven Shapin, "The Image of the Man of Science," in *The Cambridge History of Science*, vol. 4, *Eighteenth-Century Science*, ed. Porter, and "The Man of Science," in *The Cambridge History of Science*, vol. 3, *Early Modern Science*, ed. Katharine Park and Lorraine Daston (Cambridge: Cambridge University Press, 2006).

40. Amy Elizabeth Smith, "Travel Narratives and the Familiar Letter Form in the Mid-Eighteenth Century," *Studies in Philology*, vol. 95, no. 1 (1998).

41. Kieron O'Hara, *Trust: From Socrates to Spin* (London: Icon Books, 2004); Bernard Williams, *Truth and Truthfulness: An Essay in Genealogy* (Princeton, NJ: Princeton University Press, 2002), 88–92; Michael Williams, *Problems of Knowledge: A Critical Introduction to Epistemology* (Oxford: Oxford University Press, 2001), 138–45. The point about "metonymics," a "standing-for" relationship, we take from Steven Shapin, *Never Pure: Historical Studies of Science as If It Was Produced by People with Bodies, Situated in Time, Space, Culture, and Society, and Struggling for Credibility and Authority* (Baltimore: Johns Hopkins University Press, 2010), 22–23.

42. Frederic Regard, "Introduction: Articulating Empire's Unstable Zones," in *British Narratives of Exploration*, ed. Frederic Regard (London: Pickering & Chatto, 2009), 10.

43. Hayden White, *Tropics of Discourse: Essays in Cultural Criticism* (Baltimore: Johns Hopkins University Press, 1985); Hayden White, *Content of the Form: Narrative Discourse and Historical Representation* (Baltimore: Johns Hopkins University Press, 1987).

44. Adriana Craciun, "Oceanic Voyages, Maritime Books, and Eccentric Inscriptions," *Atlantic Studies* 10, no. 2 (2013): 172.

45. Steven Shapin, "'A Scholar and a Gentleman': The Problematic Identity of the Scientific Practitioner in Early Modern England," *History of Science*, vol. 29 (1991), and *A Social History of Truth: Civility and Science in Seventeenth-Century England* (Chicago: University of Chicago Press, 1994); Mario Biagioli and Peter Galison, eds., *Scientific Authorship: Credit and Intellectual Property in Science* (New York: Routledge, 2003).

46. In his analysis of "epistemological decorum," Shapin discerns "seven maxims for the evaluation of testimony canvassed in the seventeenth-century literature: (i) assent to testimony which is plausible; (ii) assent to testimony which is multiple; (iii) assent to testimony which is consistent; (iv) assent to testimony which is immediate; (v) assent to testimony from knowledgeable or skilled sources; (vi) assent to testimony given in a manner which inspires a just confidence; and (vii) assent to testimony from sources of acknowledged integrity and disinterestedness" (*A Social History of Truth*, 212).

47. Adrian Johns, "Science and the Book in Modern Cultural Historiography," *Studies in the History and Philosophy of Science* 29, no. 2 (1998): 194.

48. On the Tibet and Himalayan example, see Kapil Raj, "When Human Travellers Become Instruments: The Indo-British Exploration of Central Asia in the Nineteenth Century," in *Instruments, Travel and Science: Itineraries of Precision from the Seventeenth to the Twentieth Century*, ed. Marie-Noëlle Bourguet, Christian Licoppe, and H. Otto Sibum (London: Routledge, 2002). On Timbuktu, see Michael Heffernan, "'A Dream as Frail as Those of Ancient Time': The In-Credible Geographies of Timbuctoo," *Environment and Planning D: Society and Space*, vol. 19, no. 2 (2001).

49. On these points, see Johannes Fabian, *Out of Our Minds: Reason and Madness in the Exploration of Central Africa* (Berkeley: University of California Press, 2000); Clive Barnett, "Impure and Worldly Geography: The Africanist Discourse of the Royal Geographical Society, 1831-73," *Transactions of the Institute of British Geography*, vol. 23, no. 19 (1998); and the essays in Simon Schaffer, Lissa Roberts, Kapil Raj, and James Delbourgo, eds., *The Brokered World: Go-Betweens and Global Intelligence, 1770-1820* (Sagamore Beach, MA: Science History Publications, 2009); and in Kristian H. Nielsen, Michael Harbsmeier and Christopher J. Ries, eds., *Scientists and Scholars in the Field: Studies in the History of Fieldwork and Expeditions* (Aarhus: Aarhus University Press, 2012).

50. Michael Bravo, "Ethnographic Navigation and the Geographical Gift," in *Geography and Enlightenment*, ed. Livingstone and Withers (n. 25 above, this chap.).

51. For discussion of this topic, see Steven Shapin, *The Scientific Revolution* (Chicago: University of Chicago Press, 1996); John Henry, *The Scientific Revolution and the Origins of Modern Science* (Basingstoke: Macmillan, 1997). On the early history of instructional travel guides, see Joan-Pau Rubiés, "Instructions for Travellers: Teaching the Eye to See," *History and Anthropology*, vol. 9, no. 2-3 (1996); this essay, with other relevant material on this question, also appears in Joan-Pau Rubiés, *Travellers and Cosmographers: Studies in the History of Early Modern Travel and Ethnology* (Aldershot: Ashgate, 2007).

52. Bourguet, "Portable World" (n. 10 above, this chap.); Richard Yeo, "Between Memory and Paperbooks: Baconianism and Natural History in Seventeenth-Century England," *History of Science*, vol. 45, no. 1 (2007).

53. Bourguet, "Portable World," 381.

54. On these issues in different periods and regions, see Neil Safier, "'Every Day That I Travel . . . is a Page that I Turn': Reading and Observing in Eighteenth-Century Amazonia," *Huntington Library Quarterly*, vol. 70, no. 1 (2007); Anne Secord, "Pressed into Service: Specimens, Space, and Seeing in Botanical Practice," in *Geographies of Nineteenth-Century Science*, ed. David N. Livingstone and Charles W. J. Withers (Chicago: University of Chicago Press, 2011).

55. Felix Driver, *Geography Militant: Cultures of Exploration and Empire* (Oxford: Blackwell, 2001), 8.

56. William H. Sherman, "Stirrings and Searchings (1500-1720)," in *Cambridge Companion to Travel Writing*, ed. Hulme and Youngs (n. 8 above, this chap.).

57. Ian S. MacLaren, "Exploration/Travel Literature and the Evolution of the Author," *International Journal of Canadian Studies* 5 (1992): 41, 42.

58. MacLaren, "In Consideration" (n. 4 above, this chap.), 227.

59. Craciun, "Oceanic Voyages" (n. 44 above, this chap.), 172.

60. MacLaren, "In Consideration," 229.

61. Rob Iliffe, "Author-Mongering: The 'Editor' between Producer and Consumer," in *The Consumption of Culture, 1600-1800: Image, Object, Text*, ed. Ann Bermingham and John Brewer (Abingdon: Routledge, 1995).

62. Jerome J. McGann, *The Textual Condition* (Princeton, NJ: Princeton University Press, 1991).

63. Ian S. MacLaren, "From Exploration to Publication: The Evolution of a 19th-Century Arctic Narrative," *Arctic* 47, no. 1 (1992): 51; on alterations to the texts resulting from the Canadian exploration of Samuel Hearne in the years 1769-72, see Ian S. MacLaren, "Exploration/Travel Literature," "Samuel Hearne's Accounts of the Massacre at Bloody Fall, 17 July 1771," *Ariel*, vol. 22, no. 1 (1991), and "Explorers' and Travelers' Narratives: A Peregrination through Different Editions," *History in Africa*, vol. 30 (2003).

64. Janice Cavell, "Representing Akaitcho: European Vision and Revision in the Writing of John Franklin's *Narrative of a Journey to the Shores of the Polar Sea . . .*," *Polar Record* 44, no. 1 (2008): 26. For discussion of the supposed amendment of parts of the narrative of Franklin's 1819-22 expedition by one of its members, John Richardson, and the historiographical assessment

of Franklin and his companions by later historians of Polar exploration, see Janice Cavell, "The Hidden Crime of Dr Richardson," *Polar Record,* vol. 43, no. 2 (2007).

65. Jamie Bruce Lockhart, "In the Raw: Some Reflections on Transcribing and Editing Lieutenant Hugh Clapperton's Writings on the Borno Mission of 1822–25," *History in Africa,* vol. 26 (1999). Despite its intriguing title, the focus of David Henige, "'Twixt the Cup and the Lip:' Field Notes on the Way to Print," *History in Africa,* vol. 25 (1998), is with the different interpretations put on the field notes of the leading anthropologist Margaret Mead in her Samoan work in the 1920s.

66. David Finkelstein, "Unraveling Speke: The Unknown Revision of an African Exploration Classic," *History in Africa,* vol. 30 (2003). On the Blackwood publishing firm in this period, see David Finkelstein, *The House of Blackwood: Author-Publisher Relations in the Victorian Era* (University Park: Pennsylvania State University Press, 2002).

67. Alison E. Martin, "'These Changes and Accessions of Knowledge': Translation, Scientific Travel Writing and Modernity—Alexander von Humboldt's *Personal Narrative,*" *Studies in Travel Writing,* vol. 15, no. 1 (2011). As Nicolaas Rupke has shown, Humboldt's publications were reviewed and received differently in different European periodicals, with his enquiry into New Spain (Mexico), *Essai Politique sur Le Royaume de la Nouvelle-Espagne* (1811), read differently by British, French, and German periodicals and public audiences. See Nicolaas A Rupke, "A Geography of Enlightenment: The Critical Reception of Alexander von Humboldt's Mexico Work," in *Geography and Enlightenment,* ed. Livingstone and Withers (n. 25 above, this chap.). See also Nicolaas A. Rupke, *Alexander von Humboldt: A Metabiography* (Chicago: University of Chicago Press, 2008).

68. MacLaren, "From Exploration to Publication," 51–52.

69. Marc H. Dawson, "The Many Minds of Sir Halford J. Mackinder: Dilemmas of Historical Editing," *History in Africa,* vol. 14 (1987).

70. David Henige, "In Quest of Error's Sly Imprimatur: The Concept of 'Authorial Intent' in Modern Textual Criticism," *History in Africa* 14 (1987): 92.

71. In a large literature on reception, see Lorraine Daston, "Taking Note(s)," *Isis,* vol. 95, no. 3 (2004); James A. Secord, *Victorian Sensation: The Extraordinary Publication, Reception, and Secret Authorship of "Vestiges of the Natural History of Creation"* (Chicago: University of Chicago Press, 2000), and "Knowledge in Transit" (n. 10 above, this chap.); Innes M. Keighren, *Bringing Geography to the Book: Ellen Semple and the Reception of Geographical Knowledge* (London: I. B. Tauris, 2010); Geoffrey Cantor and Sally Shuttleworth, eds., *Science Serialized: Representation of the Sciences in Nineteenth-Century Periodicals* (Cambridge MA: MIT Press, 2004).

72. Carpenter, *Seven Lives* (n 18 above, this chap.). On John Murray I, see Zachs, *First John Murray* (n. 18 above, this chap.).

73. James Raven, *The Business of Books: Booksellers and the English Book Trade, 1450–1850* (New Haven, CT: Yale University Press, 2007), 123.

74. On these issues, see Richard D. Altick, *The English Common Reader: A Social History of the Mass Reading Public, 1800–1900* (Chicago: University of Chicago Press, 1957); Alan Richardson, *Literature, Education, and Romanticism: Reading as Social Practice, 1780–1832* (Cambridge: Cambridge University Press, 1994); William St. Clair, *The Reading Nation in the Romantic Period* (Cambridge: Cambridge University Press, 2004); David Vincent, *Literacy and Popular Culture: England, 1750–1914* (Cambridge: Cambridge University Press, 1989); Michael Suarez, "Introduction," in *The Cambridge History of the Book in Britain,* vol. 5, *1695–1830,* ed. Michael F. Suarez and Michael L. Turner (Cambridge: Cambridge University Press, 2009); James Raven, "The Promotion and Constraints of Knowledge: The Changing Structure of Publishing in Victorian Britain," in *The Organisation of Knowledge in Victorian Britain,* ed. Martin Daunton (Oxford: Oxford University Press for the British Academy, 2005); Raven, *Business of Books;* Asa Briggs, *A History of Longmans and Their Books, 1724–1990: Longevity in Publishing* (London: British Library; New Castle, DE: Oak Knoll Press, 2008), 149–228.

75. Zachs, *First John Murray*; Carpenter, *Seven Lives*, 23. Carpenter estimates that of the first thousand books in the Murray imprint, 40 percent were copublished.

76. William B. C. Lister, *A Bibliography of Murray's Handbooks for Travellers and Biographies of Authors, Editors, Revisers and Principal Contributors* (Dereham, UK: Dereham Books, 1993); Carpenter, *Seven Lives*, 170–72; Gráinne Goodwin and Gordon Jackson, "Guidebook Publishing in the Nineteenth Century: John Murray's *Handbooks for Travellers*," *Studies in Travel Writing*, vol. 17, no. 1 (2013).

77. Leslie Howsam, *Kegan Paul: A Victorian Imprint* (London: Kegan Paul International; Toronto: University of Toronto Press, 1998), 7.

78. St. Clair, *Reading Nation*, 232–33.

79. David McKitterick, "Introduction," in *The Cambridge History of the Book in Britain*, vol. 6, *1830–1914*, ed. David McKitterick (Cambridge: Cambridge University Press, 2009), 4.

80. George Paston, *At John Murray's: Records of a Literary Circle, 1843–1892* (London: John Murray, 1932), 24.

81. John Murray [III] to Woodbine Parish, January 26, 1855, MS 41912, fol. 239, National Library of Scotland (hereafter NLS), Edinburgh.

82. Book formats are defined in the appendix.

83. Anonymous review of *Journal of an Expedition to Explore the Course and Termination of the Niger*, by Richard Lander and John Lander, *The Original*, no. 12 (May 19, 1832), 185.

84. Briggs, *History of Longmans*, 283.

85. Howsam, *Kegan Paul*, 106–7.

86. Zachs, *First John Murray* (n. 18 above, this chap.), 158–60.

87. Ina Ferris, "Antiquarian Authorship: D'Israeli's Miscellany of Literary Curiosity and the Question of Secondary Genres," *Studies in Romanticism*, vol. 45, no. 4 (2006); [Isaac D'Israeli], *Curiosities of Literature* (London: John Murray, 1791).

88. George Byron, *Childe Harold's Pilgrimage: A Romaunt* (London: John Murray; Edinburgh: William Blackwood; Dublin: John Cumming, 1812); Jane Austen, *Emma, a Novel*, 3 vols. (London: John Murray, 1816); Kathryn Sutherland, "Jane Austen's Dealings with John Murray and His Firm," *Review of English Studies*, vol. 64, no. 263 (2013).

89. Graham D. Caie, "Bringing the John Murray Archive to the National Library of Scotland," *Byron Journal* 34, no. 1 (2006), 49–55.

90. Victoria Cooper and Dave Russell, "Publishing for Leisure," in *The Cambridge History of the Book in Britain*, vol. 6, *1830–1914*, ed. David McKitterick (Cambridge: Cambridge University Press, 2009), 491.

91. *A Hand-Book for Travellers on the Continent: Being a Guide through Holland, Belgium, Prussia, and Northern Germany, and along the Rhine, from Holland to Switzerland* (London: John Murray and Son, 1836). On scientific tourism, including work published by Murray, see Aileen Fyfe, "Natural History and the Victorian Tourist: From Landscapes to Rock-Pools," in *Geographies of Nineteenth-Century Science*, ed. David N. Livingstone and Charles W. J. Withers (Chicago: University of Chicago Press, 2011).

92. Anonymous, "Handbooks for England," *Times* (September 22, 1859), 10.

93. Edmund W. Gilbert, "Richard Ford and His 'Hand-Book for Travellers in Spain,'" *Geographical Journal*, vol. 106, nos. 3–4 (1945).

94. William Zachs, Peter Isaac, Angus Fraser, and William Lister, "Murray Family," in *Oxford Dictionary of National Biography* (Oxford: Oxford University Press, 2004–13), unpaginated.

95. Mary Lutyens, "Murrays of Albemarle Street," *Times Literary Supplement* (October 24, 1968).

96. John Barrow to John Murray [II], August 16, 1816, MS 40555, fol. 64v, NLS. Barrow must be referring to the *Edinburgh Review:* that other "radical" Edinburgh-based periodical with which Murray's *Quarterly* was in competition, *Blackwood's Magazine*, was not published until 1817.

97. John Barrow, *An Auto-Biographical Memoir* (London: John Murray, 1847), v.] *for* Barrow *

98. Driver, *Geography Militant* (n. 55 above, this chap.), 31.

99. James Tuckey, *Narrative of an Expedition to Explore the River Zaire* (London: John Murray, 1818), ii–iii. See app.: 47.

100. Thomas Moore, *The Works of Lord Byron: With His Letters and Journals, and His Life*, 17 vols. (London: John Murray, 1834), 11:313.

101. Edward Chappell, *Narrative of a Voyage to Hudson's Bay in His Majesty's Ship* Rosamond (London: J. Mawman, 1817); [John Barrow], review of *Narrative of a Voyage to Hudson's Bay*, by Edward Chappell, *Quarterly Review* 18, no. 35 (October 1817): 199.

102. [Barrow], review of *Narrative of a Voyage*, 202.

103. Fleming, *Barrow's Boys* (n. 19 above, this chap.), 35.

104. Wheatley, "Arctic in the *Quarterly*" (n. 19 above, this chap.).

105. Francis Spufford, *I May Be Some Time: Ice and the English Imagination* (London: Faber, 1996), 49.

106. Sydney Parkinson, *A Journal of a Voyage to the South Seas, in His Majesty's Ship, the* Endeavour (London: Printed for Stanfield Parkinson and sold by Richardson & Urquhart; Evans; Hooper; Murray; Leacroft; Riley, 1773). See app.: 1.

107. Holmes, *Age of Wonder* (n. 2 above, this chap.), 44.

108. John Hawkesworth, *An Account of the Voyages Undertaken by the Order of His Present Majesty for Making Discoveries in the Southern Hemisphere*, 3 vols. (London: W. Strahan; T. Cadell, 1773).

109. Holmes, *Age of Wonder*, 44.

110. Denis J. Carr, ed., *Sydney Parkinson: Artist of Cook's* Endeavour *Voyage* (London: British Museum [Natural History]; Honolulu: University of Hawaii Press, 1983).

111. John Fothergill, *Explanatory Remarks on the Preface to Sydney Parkinson's* Journal of a Voyage to the South-Seas ([London, 1773]).

112. Holmes, *Age of Wonder*, 45; Sarah Johnston, "Missionary Positions: Romantic European Polynesians from Cook to Stevenson," in *Travel Writing in the Nineteenth Century: Filling the Blank Spaces*, ed. Tim Youngs (London: Anthem Press, 2006), 180.

113. Carole Fabricant, "Eighteenth-Century Travel Literature," in *The Cambridge History of English Literature, 1660–1780*, ed. John Richetti (Cambridge: Cambridge University Press, 2005), 707.] *general*

114. Anonymous review of *Travels Round the World*, by Pierre Marie François de Pagès, *Monthly Review* 5 (May 1791): 32.

115. Shef Rogers, "Enlarging the Prospects of Happiness: Travel Reading and Travel Writing," in *The Cambridge History of the Book in Britain*, vol. 5, *1695–1830*, ed. Michael F. Suarez and Michael L. Turner (Cambridge: Cambridge University Press 2009), 786.

116. Leask, *Curiosity* (n. 9 above, this chap.), 13.

117. Roger Chartier, *The Order of Books: Readers, Authors, and Libraries in Europe between the Fourteenth and Eighteenth Centuries*, trans. Lydia G. Cochrane (Cambridge: Polity Press, 1994).

CHAPTER TWO

1. William E. Parry to John Murray [II], August 24, 1826, MS 42689 (no folio), National Library of Scotland (hereafter NLS), Edinburgh.

2. William Henry Edwards, *A Voyage up the River Amazon, Including a Residence at Pará* (London: John Murray, 1847), iv; emphasis in original. See app.: 184.

3. George W. Featherstonhaugh, *Excursion through the Slave States: From Washington on the Potomac to the Frontier of Mexico*, 2 vols. (London: John Murray, 1844), 1:22, 23–24; emphasis in original. See app.: 166. On Featherstonhaugh's work in the United States, see Edmund Berkeley

and Dorothy Smith Berkely, *George William Featherstonhaugh: The First U.S. Government Geologist* (Tuscaloosa: University of Alabama Press, 1988).

4. Featherstonhaugh, *Excursion through the Slave States*, 1:xii; emphasis in original. Featherstonhaugh had earlier proposed a different title (though what that title was is not clear). Murray III praised the revised and final title when he acknowledged receipt of the first part of the book in manuscript for the press: John Murray [III] to George Featherstonhaugh, March 4, 1844, MS 41911, fol. 155, NLS.

5. Mansfield Parkyns, *Life in Abyssinia: Being Notes Collected during Three Years' Residence and Travels in That Country*, 2 vols. (London: John Murray, 1853), 1:1. See app.: 218.

6. Richard Wilbraham, *Travels in the Trans-Caucasian Provinces of Russia, and along the Southern Shore of the Lakes of Van and Urumiah, in the Autumn and Winter of 1837* (London: John Murray, 1839), vii. See app.: 145.

7. Anonymous review of *Rough Notes Taken during Some Rapid Journeys*, by Francis Bond Head, *New Monthly Magazine* 21, no. 73 (January 1, 1827), 5; emphasis in original. Francis Bond Head's *Rough Notes* was sufficiently popular to merit five further editions by 1861 (see app.: 95).

8. [Lawrence Sterne], *A Sentimental Journey through France and Italy*, 2 vols. (London: T. Becket; P. A. DeHondt), 1:27-28.

9. This distinction is from Percy G. Adams, *Travelers and Travel Liars, 1660-1800* (Berkeley: University of California Press, 1962), 8.

10. Nigel Leask, *Curiosity and the Aesthetics of Travel Writing, 1770-1840* (Oxford: Oxford University Press, 2002). See also Mary Louise Pratt, *Imperial Eyes: Travel Writing and Transculturation*, 2nd ed. (New York: Routledge, 2008).

11. The phrase "the practical illustration of geographical science" is from the opening lines of William Moorcroft and George Trebeck, *Travels in the Himalayan Provinces of Hindustan and the Panjab*, 2 vols. (London: John Murray, 1841), 1:i. "The practical illustration of Geographical Science has at no period been prosecuted in this country with more unremitting diligence than in the present day."

12. The works in question—for the full titles see the appendix—are Kinneir (1813), Elphinstone (1815), Malcolm (1815), Ellis (1817), Tuckey (1818), John Ross (1819), Bowdich (1819), Parry (1821), Graham (two separate works in 1824), Parry (1824), Lyon (1821), Lyon (1824), Laing (1825), Lyon (1825), Beechey (1826), Denham (1826), Graham (1826), King (1827), Parry (1826), Franklin (1827), Parry (two separate works in 1828), Franklin (1829), Clapperton (1829), Lander (1832), Burnes (1834), Parry (1835), Back (1836), Wellsted (1838), Wood (1841), Moorcroft (1841), Burnes (1842), and James Clark Ross (1847).

13. This is evident in the many papers in the British Library relating to Burnes's geographical and mapping activities; see, in particular, British Library (hereafter BL), London. MS Eur D1165/13, MS Eur D153, MS Eur B256, MS Eur F208/112 and the materials in the India Office Library papers: IOR/F/4/1265/50902; IOR/L/ML/9/144/204-07; IOR/L/PS/19/25; IOR/L/PS/19/41; IOR/F/4/1247/50208. In one of his dispatches, Burnes addresses part of his "Military and Geographic Memoir on the Indus and Punjab Rivers" as "Enclosure to the Secret Letter to the Honbl. The Secret Committee" of the British Government in India: February 14, 1832, IOR/L/PS/19/41, BL. For a fuller account of Burnes's work in relation to that of his predecessors and to Malcolm's coordination of geographical inquiry in late Enlightenment Central Asia, see Charles W. J. Withers, "On Enlightenment's Margins: Geography, Imperialism and Mapping in Central Asia, *c.*1798-*c.*1838," *Journal of Historical Geography*, vol. 39 (2013).

14. Wood wrote how "Burnes has been accused of *under*-estimating the size of the Indus; but with what degree of justice, the result of experimental steam-voyages will by this time have shown. It would, indeed, have been the safer side to err on, but he has done just the contrary, and drawn a too favorable picture of the capabilities of this river, both in his published work and practical notes." John Wood, *A Personal Narrative of a Journey to the Source of the River Oxus* (John Murray: London, 1841), 73; emphasis in original.

15. Phillip Parker King, *Narrative of a Survey of the Intertropical and Western Coasts of Australia*, 2 vols. (John Murray: London, 1827), 1:xxvii. See app.: 99.

16. Ibid., 1:xxxi.

17. Ibid., 1:vii.

18. James Kingston Tuckey, *Narrative of an Expedition to Explore the River Zaire* (John Murray: London, 1818), xxxvi. See app.: 47.

19. Ibid., xxxv.

20. Ibid., xli.

21. Ibid., xli–xlii.

22. Marie-Noëlle Bourguet, " A Portable World: The Notebooks of European Travellers (Eighteenth to Nineteenth Centuries)," *Intellectual History Review*, vol. 20, no. 3 (2010).

23. Francis Galton, *The Art of Travel*, 2nd ed. (London: John Murray, 1856), 152–56. See app.: 221.

24. John Barrow to John Murray [II], May 8, 1817, MS 40055, fol. 70, NLS. Barrow notes: "I have been desired to superintend their publication & I have agreed to do so, without any stipulation than this, that the profits of the work be they what they may, should be distributed to the widows & children of the unhappy sufferers." Although Barrow is silent as to who "desired" this work, it was almost certainly his superior, John Wilson Croker.

25. Anonymous review of *Narrative of an Expedition*, by James Tuckey, *British Critic* 10 (August 1818): 154–55.

26. Anonymous review of *Narrative of an Expedition*, by James Tuckey, *Monthly Review* 86 (June 1818): 113–29.

27. John Barrow to John Murray [II], August 25, 1817, MS 40055, fol. 73, NLS.

28. John Barrow to John Murray [II], September 1, 1817, MS 40055, fols. 77v–78, NLS. What Barrow actually wrote was: "But if she comes I will endeavour to frighten her so far as to prevent on her to endeavour to receive the M.S." But his intent is clear, if not his wording.

29. Copy Ledger B, fol. 62, MS 42725, NLS.

30. This total includes the works—for full titles see the appendix—of John Ross (1819), Parry (in 1821, 1824, 1826, 1828 [two separate works], and 1835), Franklin (three separate works of 1823, 1828, and 1829), Lyon (1824 and 1825), Back (1836, 1838), Sir James Clark Ross (1847), and McClintock (1859). Barrow's 1818 work is not here included in being, essentially, a compilation and chronology of polar works for an earlier period (although it ends with George Back's accounts of the 1830s).

31. David McClay, "John Murray and the Publishing of Scottish Polar Explorers," *Geographer: The Newsletter of the Royal Scottish Geographical Society* (Winter 2011-12); Adriana Craciun, "Writing the Disaster: Franklin and Frankenstein," *Nineteenth-Century Literature*, vol. 65, no. 4 (2011).

32. Janice Cavell, *Tracing the Connected Narrative: Arctic Exploration in British Print Culture, 1818-1860* (Toronto: University of Toronto Press, 2008).

33. Adriana Craciun, "Writing the Disaster"; Kim Wheatley, "The Arctic in the *Quarterly Review*," *European Romantic Review*, vol. 20, no. 4 (2009).

34. On which topics, see Francis Spufford, *I May Be Some Time: Ice and the English Imagination* (London: Faber, 1996); Eric G. Wilson, *The Spiritual History of Ice: Romanticism, Science, and the Imagination* (New York: Palgrave Macmillan, 2003). On Arctic science in this period, see Trevor H. Levere, *Science and the Canadian Arctic: A Century of Exploration, 1818-1918* (Cambridge: Cambridge University Press, 1993).

35. John Franklin, *Narrative of a Second Expedition to the Shores of the Polar Sea* (London: John Murray, 1828), 319. See app.: 103.

36. Maria Graham to John Murray [II], May 31, 1821, MS 40185 (no folio but item 4), NLS.

37. In writing to Sir James Clark Ross over what was in time to be his 1847 work, Murray III was in no doubt that Ross should understand his concern in this respect: "I am sure you cannot be aware how serious a matter to me is the long continued delay of your wk When I agreed with

you for its publication I expected it would have been out within *12 months[.]* *The Public Interest* has greatly cooled with the lapse of time[.] The work has almost ceased to be asked for (except by) our Friend Sir John Barrow at whose instigation I now write." John Murray [III] to Sir James Clark Ross, October 5, 1846, MS 41912, fols. 11-12, NLS.

38. John Ross, *A Voyage of Discovery, Made under the Orders of the Admiralty, in His Majesty's Ships* Isabella *and* Alexander (London: John Murray, 1819), 9. See app.: 55.

39. Ibid., 9, 11-12, 13.

40. William E. Parry, *Journal of a Voyage for the Discovery of a North-West Passage from the Atlantic to the Pacific* (London; John Murray, 1821), xxiv-xix. See app.: 66.

41. See Richard C. Davis, ed., *Sir John Franklin's Journals and Correspondence: The First Arctic Land Expedition, 1819-1822* (Toronto: Champlain Society, 1995), 285-88, 297-301.

42. Ross, *Voyage of Discovery*, i-ii.

43. Ibid., ii-iii.

44. This notebook of 1821 survives as MS 42237, NLS.

45. For a fuller account of Lyon's exploratory narratives, see Charles W. J. Withers, "Travel, *En Route* Writing, and the Problem of Correspondence," in *Routes, Roads and Landscape*, ed. Mari Hvattum, Brita Brenna, Beate Elvebakk, and Janike Kampevold Larsen (Farnham, UK: Ashgate, 2011).

46. George F. Lyon, *A Narrative of Travels in Northern Africa* (London: John Murray, 1821). See app.: 65.

47. George F. Lyon, *The Private Journal of Captain G. F. Lyon* (London: John Murray, 1824), v. See app.: 83.

48. Ibid., vi.

49. William E. Parry, *Journal of a Second Voyage for the Discovery of a North-West Passage*, 2 vols. (London: John Murray, 1824), 1:xiii. See app.: 84.

50. Copy Ledger B, fol. 93, MS 42725, NLS. From 2,250 copies (1,500 first edition, 750 second edition), Murray made a profit by January 1823 of £2,353 16s. 10d.

51. Anonymous review of *Narrative of a Journey to the Shores of the Polar Sea*, by John Franklin, *London Literary Gazette*, no. 625 (January 10, 1829), 24.

52. Hugh S. Tremenheere, *Notes on Public Subjects, Made during a Tour in the United States and Canada* (London: John Murray, 1852), 4-5, 69. See app.: 212.

53. James Riley to John Murray [II], January 10, 1817, MS 41020 (no folio), NLS.

54. Alexander Burnes, *Travels into Bokhara*, 3 vols. (London: John Murray, 1834), 1:213, 223. See app.: 126.

55. This is not to say China was unknown before this period. Jesuit scholars and merchants both provided information about the country. But Western geographical knowledge was nonetheless limited, based on partial insight or even wholly fictitious: see John J. Clarke, *Oriental Enlightenment: The Encounter between Asian and Western Thought* (London: Routledge, 1997); Frank Dikotter, "China," in *The Cambridge History of Science*, vol. 4, *Eighteenth-Century Science*, ed. Roy Porter (Cambridge: Cambridge University Press, 2003).

56. Henry Ellis, *Journal of the Proceedings of the Late Embassy to China*, 2 vols. (London: John Murray, 1817), 1:iii. See app.: 38. Ellis, who later served under Sir John Malcolm in Persia, also assisted Dixon Denham in preparation of his African travels with Clapperton and Oudney (Dixon Dehnam, Hugh Clapperton, and Walter Oudney, *Narrative of Travels and Discoveries in Northern and Central Africa, in the Years 1822, 1823, and 1824*, 2 vols. [London: John Murray, 1826]): Henry Ellis to John Murray [II], December 30, 1826, MS 40377 (no folio), NLS.

57. J. L. Cranmer-Byng, "Lord Macartney's Embassy to Peking in 1793," *Journal of Oriental Studies*, vol. 4, nos. 1-2 (1957-58); Robert A. Bickers, ed., *Ritual and Diplomacy: The Macartney Mission to China, 1792-1794* (London: British Association for Chinese Studies, 1993).

58. Ellis, *Journal of the Proceedings*, 1:197.

59. Ibid., 2:440.

60. John Francis Davis to John Murray [II], August 5, 1834, MS 40319 (no folio), NLS. Davis's *The Chinese: A General Description of the Empire of China and Its Inhabitants* was first published in two volumes by Charles Knight in 1836 and revised by Knight in 1844. Murray published a further revised edition in 1857 (see app.: 233). Interestingly, Davis drew extensively on other's China narratives rather than Ellis's: "At the head of *travels*, both as to date and excellence, stand the authentic account of Lord Macartney's Mission by Staunton, and Barrow's China, to both of which works it will be seen that reference has been more than once made in the following pages" (John Francis Davis, *China*, 2 vols. [London: John Murray, 1857], 1:v–vi [see app.: 233]). The works referred to are: George Staunton, *An Authentic Account of an Embassy from the King of Great Britain to the Emperor of China*, 3 vols. (London: G. Nicol, 1797); and John Barrow, *Travels in China* (London: T. Cadell & W. Davies, 1804).

61. John Francis Davis to John Murray [II], October 28, 1822, MS 40319 (no folio), NLS. In his 1857 *China*, Davis outlined what he saw as a hierarchy among Western authors on China: "At the head of *travels*, both as to date and excellence, stand the authentic account of Lord Macartney's Mission by Staunton, and Barrow's on China" (for the Staunton volume, see app.: 72). Davis, *China*, 1:xvi. Barrow was much taken with John Davis's China work published by Knight: "I am reading a very clever work [he wrote to Murray II]—the best I have seen—on China by Mr. J. Davis who from his long residence in that country, his official situation, his thorough knowledge of the almost unknown language—his extensive reading and his translations of their dramatic and poetic writings—possesses advantages which no other European can pretend to." John Barrow to John Murray [II], April 30, 1837, MS. 40058, fols. 20–20v, NLS.

62. George Staunton to John Murray [II], January 20, 1834, MS. 41152 (no folio but item 20), NLS.

63. William Martin Leake, *Journal of a Tour in Asia Minor* (London: John Murray, 1824), iii, vi. See app.: 82.

64. James Rennell, *The Geographical System of Herodotus, Examined; and Explained, by a Comparison with Those of Other Ancient Authors, and with Modern Geography* (London: Printed for the author and sold by G. & W. Nichol, 1800), and *Illustrations, (Chiefly Geographical,) of the History of the Expedition of Cyrus, from Sardis to Babylonia; and the Retreat of the Ten Thousand Greeks, from Thence to Trebisonde, and Lydia* (London: G. & W. Nicol, 1816). These works, notably his *Geographical System of Herodotus*, were the only published parts of Rennell's never-completed project for a comparative geography (ancient and modern) of western Asia.

65. Leake, *Journal of a Tour in Asia Minor*, xvi; Richard Pococke, *A Description of the East, and Some Other Countries*, 2 vols. (London: Printed for the author and sold by J. & P. Knapton; W. Innys; W. Meadows; G. Hawkins; S. Brit; T. Longman; C. Hitch; R. Dodsley; J. Nourse; J. Rivington, 1743).

66. Francis Beaufort, *Karamania; or, A Brief Description of the South Coast of Asia Minor and of the Remains of Antiquity* (London: R. Hunter, 1817).

67. The quote is from Leake, *Journal of a Tour in Asia Minor*, 2.

68. Ibid., 78. On Leake's methods as part of classical geography, see William A. Koelsch, *Geography and the Classical World: Unearthing Historical Geography's Forgotten Past* (London: I. B. Tauris, 2012), 36–40.

69. James R. Wellsted, *Travels in Arabia*, 2 vols. (London: John Murray, 1838), 1:v. See app.: 139.

70. James Raymond Wellsted to John Murray [II?], February 28, 1837, MS 41258 (no folio), NLS.

71. Anonymous review of *Travels in Arabia*, by James Wellsted, *Athenaeum*, no. 533 (January 13, 1838), 29–30. The references here to "Albemarle Street or the Row" are, respectively, to the location in London of Murray's business offices (50 Albemarle Street) and to Savile Row, then the address of the Royal Geographical Society.

72. Josiah Leslie Porter to John Murray [III], August 6, 1856, MS. 40969 (no folio), NLS.

73. Brian Yothers, *The Romance of the Holy Land in American Travel Writing, 1790–1876* (Aldershot: Ashgate, 2007), 21.

74. Edward Robinson and Eli Smith, *Biblical Researches in Palestine, Mount Sinai and Arabia Patraea*, 3 vols. (London: John Murray, 1841), 1:377–78. See app.: 153.

75. Ibid., 1:47.

76. Robinson's Holy Land travels are more fully discussed in Innes M. Keighren and Charles W. J. Withers, "The Spectacular and the Sacred: Narrating Landscape in Works of Travel," *cultural geographies* 19, no. 1 (2012).

77. Thomas Bowdich, *Mission from Cape Coast Castle to Ashantee* (London: John Murray, 1819), 161. See app.: 48.

78. Charles Fellows, *A Journal Written during an Excursion in Asia Minor* (London: John Murray, 1839), 158. See app.: 140.

79. Charles Fellows to John Murray [II?], June 5, 1840, MS. 40395 (no folio), NLS.

80. John Macdonald Kinneir, *Journey through Asia Minor, Armenia, and Koordistan* (London: John Murray, 1818), viii–ix. See app.: 45.

81. James Brooke, *Narrative of Events in Borneo and Celebes*, 2 vols. (London: John Murray, 1848), 1:3. See app.: 189.

82. Henry L. Maw, *Journal of a Passage from the Pacific to the Atlantic* (London: John Murray, 1829), 4. See app.: 114.

83. Wilbraham, *Travels in the Trans-Caucasian Provinces of Russia* (n. 6 above, this chap.), 283.

84. Francis Bond Head, *Rough Notes Taken during Some Rapid Journeys across the Pampas and among the Andes* (London: John Murray, 1826), x.

85. George A. F. FitzClarence, *Journal of a Route across India, through Egypt, to England* (London: John Murray, 1819), 355. See app.: 51.

86. Basil Hall, *Account of a Voyage of Discovery to the West Coast of Corea, and the Great Loo-Choo Island* (London: John Murray, 1818), 42. See app.: 44.

87. We take this point from Barbara Korte, *English Travel Writing from Pilgrimages to Postcolonial Explorations*, trans. Catherine Matthias (Basingstoke: Macmillan, 2000), 11. See also Jonathan Raban, *For Love and Money: Writing, Reading, Travelling, 1969–1987* (London: Collins Harvill, 1987), 246–48.

88. Roy C. Bridges, "Nineteenth-Century East African Travel Records, with an Appendix on 'Armchair Geographers' and Cartography," *Paideuma* 33 (1987): 179.

89. For arguments in support, see Marie-Noelle Bourguet, "The Explorer," in *Enlightenment Portraits*, ed. Michel Vovelle (Chicago: University of Chicago Press, 1997); Adriana Craciun, "What Is an Explorer?" *Eighteenth-Century Studies*, vol. 45, no. 1 (2011).

90. Craciun, "What Is an Explorer?" 35.

91. Drummond-Hay's motive for his travels in northern Africa was royal imperative: in search of "a barb of the purest blood" to improve Queen Victoria's stock of racehorses: John H. Drummond-Hay, *Western Barbary: Its Wild Tribes and Savage Animals* (London: John Murray, 1844), iii. See app.: 169.

92. John Davidson, "Letter from Mr Davidson to the Secretary of the Geographical Society, Dated Wednoon [*sic*], 22nd May, 1836," *Journal of the Royal Geographical Society of London*, vol. 6 (1836); John Davidson and William Willshire, "Extracts from the Correspondence of the Late Mr. Davidson, during His Residence in Morocco; with an Account of His Further Progress in the Desert," *Journal of the Royal Geographical Society of London*, vol. 7 (1837); John Davidson, *Notes Taken during Travels in Africa* (London: J. L. Cox, 1839).

93. Drummond-Hay, *Western Barbary*, 102.

94. Alexander G. Laing, *Travels in the Timannee, Kooranko, and Soolima Countries, in Western Africa* (London: John Murray, 1825). See app.: 88.

95. John L. Burckhardt, *Travels in Nubia* (London: John Murray, 1819), v. See app.: 49.

96. On this, see the essays in Christopher Lawrence and Steven Shapin, eds., *Science Incarnate: Historical Embodiments of Natural Knowledge* (Chicago: University of Chicago Press, 1998); and

Johannes Fabian, *Out of Our Minds: Reason and Madness in the Exploration of Central Africa* (Berkeley: University of California Press, 2000).

97. Julian R. Jackson, *What to Observe; or, The Traveller's Remembrancer* (London: James Madden, 1841).

98. Andrew S. Goudie, "Colonel Julian Jackson and His Contribution to Geography," *Geographical Journal*, vol. 144, no. 2 (1978).

99. "Royal Premium," *Journal of the Royal Geographical Society of London* 2 (1832): vii.

100. "Regarding the Labours of the Roy. Geographl. Society by A Member [Julian R. Jackson]," fol. 2, MS AP 8, Royal Geographical Society (with the Institute of British Geographers) archives (hereafter RGS-IBG), London.

101. John F. W. Herschel, *A Manual of Scientific Enquiry* (London: John Murray, 1849), 138. See app.: 194. The *Manual's* chapters and their authors are as follows: "Astronomy" by George B. Airy; "Magnetism" by Edward Sabine; "Hydrography" by Capt. Beechey; "Tides" by William Whewell; "Geography" by William Hamilton; "Geology" by Charles Darwin; "Earthquakes" by R. Mallet; "Mineralogy" by Henry de la Beche; "Meteorology" by Sir John Herschel; "Atmospheric Waves" by W. R. Birt; "Zoology" by Richard Owen; "Botany" by Sir William Hooker; "Ethnology" by James Pritchard; "Medicine and Medical Statistics" by Dr. Bryson; and "Statistics" by G. R. Porter.

102. Herschel, *Manual of Scientific Enquiry*, 138.

103. Ibid., iii.

104. We take this point about Naval men and the development of science from Randolph Cock, "Scientific Servicemen in the Royal Navy and the Professionalisation of Science, 1816–55," in *Science and Beliefs: From Natural Philosophy to Natural Science, 1700–1900*, ed. David M. Knight and Matthew D. Eddy (Aldershot: Ashgate, 2005).

105. William N. Glascock, *The Naval Service or Officers' Manual for Every Grade in His Majesty's Ships*, 2 vols. (London: Saunders & Otley, 1836). At one point, Glascock observed that "no public register is so faulty in its notation of time, so careless of construction, or so truly unimportant in 'remark,' as those [logbooks] which are cautiously treasured in the archives of Somerset-House" (Glascock, *Naval Service*, 1:95). In correspondence of 1826, Lyon refers more than once to "my friend Glascock," the context making clear that this is W. N. Glascock (George Francis Lyon to John Murray [II], July 11, 1826; September 18, 1826, MS 40731 [no folio], NLS).

106. William John Hamilton, "Geography," in *A Manual of Scientific Enquiry*, ed. John F. W. Herschel (London: John Murray, 1849), 127.

107. Ibid., 128, 129.

108. William John Hamilton, *Researches in Asia Minor, Pontus and Armenia*, 2 vols. (London: John Murray, 1842), 1:vi–vii. See app.: 157.

109. J. F. W. Herschel to John Murray [III], June 28, 1848, MS 40553 (no folio), NLS. In this letter, Herschel likens the intended layout of the book "to the form of [Humboldt's] 'Kosmos.'"

110. In the period in question, the members of this committee were Sir Roderick Impey Murchison (then President of the RGS), Col. Yorke, Lt. Henry Raper, Captain Robert Fitzroy (of *Beagle* fame), Mr. Arrowsmith (the mapmaker), Col. Sykes, and Col. Edward Sabine (who had accompanied John Ross and William Parry on their Arctic voyages).

111. RGS Committee Minute Book 1841–1865, November 15, 1852, fol. 56, RGS-IBG archives.

112. RGS Committee Minute Book 1841–1865, December 20, 1852, fol. 57, RGS-IBG archives.

113. Francis Galton, ed., *Hints to Travellers*, 4th ed. (London: Edward Stanford for the Royal Geographical Society, 1878), 9. On *Hints to Travellers*, see Felix Driver, "Scientific Exploration and the Construction of Geographical Knowledge: *Hints to Travellers*," *Finisterra: Revista Portuguesa de Geografia*, vol. 33, no. 65 (1998), and *Geography Militant: Cultures of Exploration and Empire* (Oxford: Blackwell, 2001); Charles W. J. Withers, "Science, Scientific Instruments and Questions of Method in Nineteenth-Century British Geography," *Transactions of the Institute of British Geographers*, vol. 38, no. 1 (2013), and *Geography and Science in Britain, 1831–1939: A*

Study of the British Association for the Advancement of Science (Manchester: Manchester University Press, 2010).

CHAPTER THREE

1. Regina Akel, *Maria Graham: A Literary Biography* (Amherst, NY: Cambria Press, 2009); Innes M. Keighren and Charles W. J. Withers, "Questions of Inscription and Epistemology in British Travelers' Accounts of Early Nineteenth-Century South America," *Annals of the Association of American Geographers* 101, no. 6 (2011): 1331–46.

2. Ángela Pérez-Majía, *A Geography of Hard Times: Narratives about Travel to South America, 1780–1849*, trans. Dick Cluster (Albany: State University of New York Press, 2004).

3. Mary Louise Pratt, *Imperial Eyes: Travel Writing and Transculturation*, 2nd ed. (New York: Routledge, 2008), 141; emphasis in original.

4. Nigel Leask, *Curiosity and the Aesthetics of Travel Writing, 1770–1840* (Oxford: Oxford University Press, 2002), 54.

5. Maria Graham, *Journal of a Residence in India* (Edinburgh: Archibald Constable; London: Longman, Hurst, Rees, Orme, & Brown, 1813), vi.

6. Leask, *Curiosity*, 207.

7. Anonymous review of *Journal of a Residence in India*, by Maria Graham, *Eclectic Review* 10 (December 1813): 579.

8. Ibid. On women's contribution to nineteenth-century botany, see, e.g., Anne Secord, "Botany on a Plate: Pleasure and the Power of Pictures in Promoting Early Nineteenth-Century Scientific Knowledge," *Isis*, vol. 93, no. 1 (2002); Ann B. Shteir, *Cultivating Women, Cultivating Science: Flora's Daughters and Botany in England, 1760–1860* (Baltimore: Johns Hopkins University Press, 1996).

9. Anonymous review of *Journal of a Residence in India*, by Maria Graham, *Eclectic Review* 10 (December 1813): 579.

10. Maria Graham, *Journal of a Voyage to Brazil, and a Residence There, during Part of the Years 1821, 1822, 1823* (London: Longman, Hurst, Rees, Orme, Brown, & Green; John Murray, 1824), 89. See app.: 79. For a discussion of the characteristics of philosophical travel, see, e.g., Peter Burke, "The Philosopher as Traveller: Bernier's Orient," in *Voyages and Visions: Towards a Cultural History of Travel*, ed. Jaś Elsner and Joan-Pau Rubiés (London: Reaktion, 1999); Reinhold Schiffer, *Oriental Panorama: British Travellers in 19th Century Turkey* (Amsterdam: Editions Rodopi, 1999).

11. Alexander von Humboldt and Aimé Bonpland, *Personal Narrative of Travels to the Equinoctial Regions of the New Continent, during the Years 1799–1804*, 7 vols., trans. Helen Maria Williams (London: Longman, Hurst, Rees, Orme, & Brown; J. Murray; H. Colburn, 1814–29) (app.: 30); Robert Southey, *History of Brazil*, 3 vols. (London: Longman, Hurst, Rees, & Orme, 1810–19).

12. Maria Graham to John Murray [II], September 22, 1821, MS 40185 (no folio), National Library of Scotland (hereafter NLS), Edinburgh.

13. Ibid.

14. John Hamilton, *Travels through the Interior Provinces of Columbia*, 2 vols. (London: John Murray, 1827), 1:157. See app.: 98.

15. Akel, *Maria Graham*, 97.

16. Martina Kölbl-Ebert, "Observing Orogeny—Maria Graham's Account of the Earthquake in Chile in 1822," *Episodes*, vol. 22, no. 1 (1999); Michelle Medeiros, "Crossing Boundaries into a World of Scientific Discoveries: Maria Graham in Nineteenth-Century Brazil," *Studies in Travel Writing*, vol. 16, no. 3 (2012); Carl Thompson, "Earthquakes and Petticoats: Maria Graham, Geology, and Early Nineteenth-Century 'Polite' Science," *Journal of Victorian Culture*, vol. 17, no. 3 (2012).

17. Maria Graham, *Journal of a Residence in India*, 2nd ed. (Edinburgh: Archibald Constable; London: Longman, Hurst, Rees, Orme, & Brown, 1813); Maria Graham, *Letters on India* (London:

Longman, Hurst, Rees, Orme, & Brown; Edinburgh: A. Constable, 1814); Maria Graham, *Three Months Passed in the Mountains East of Rome, during the Year 1819* (London: Longman, Hurst, Rees, Orme, & Brown; Edinburgh: A. Constable, 1820).

18. Longman & Co. to Constable & Co., March 30, 1813, MS 1393 1/98/35, University of Reading, Special Collections Service.

19. Maria Graham to John Murray [II], January 2, 1815, MS.40186 (no folio), NLS.

20. Maria Graham to John Murray [II], March 31, 1821, MS.40185 (no folio), NLS.

21. Maria Graham to John Murray [II], January 1824, MS.40185 [no folio], NLS.

22. Anonymous review of *Journal of a Voyage to Brazil*, by Maria Graham, *London Literary Gazette*, no. 377 (April 10, 1824); Maria Graham to John Murray [II], April 27, 1824, MS.40185 (no folio), NLS.

23. Maria Graham to John Murray [II], April 28, 1824, MS.40185 (no folio), NLS.

24. Johann Baptist von Spix and Carl Friedrich Philipp von Martius, *Travels in Brazil, in the Years 1817–1820*, 2 vols (London: Longman, Hurst, Rees, Orme, Brown, & Green, 1824).

25. Percy G. Adams, *Travelers and Travel Liars, 1660–1800* (Berkeley: University of California Press, 1962), 6; Steven Shapin, *Never Pure: Historical Studies of Science as If It was Produced by People with Bodies, Situated in Time, Space, Culture, and Society, and Struggling for Credibility and Authority* (Baltimore: Johns Hopkins University Press, 2010), 61.

26. Maria Graham to John Murray [II], [December 1826?], MS 40186 (no folio but item 35), NLS.

27. Bruno Latour, *Science in Action: How to Follow Scientists and Engineers through Society* (Cambridge, MA: Harvard University Press, 1987), 227.

28. Ibid., 215.

29. Daniela Bleichmar, "The Geography of Observation: Distance and Visibility in Eighteenth-Century Botanical Travel," in *Histories of Scientific Observation*, ed. Lorraine Daston and Elizabeth Lunbeck (Chicago: University of Chicago Press, 2011), 376.

30. Barbara J. Shapiro, "Testimony in Seventeenth-Century English Natural Philosophy: Legal Origins and Early Development," *Studies In History and Philosophy of Science* 33, no. 2 (2002): 249.

31. Lorraine Daston, "Marvelous Facts and Miraculous Evidence in Early Modern Europe," *Critical Inquiry* 18, no. 1 (1991): 93–124; Daniel Carey, "Travel, Geography, and the Problem of Belief: Locke as a Reader of Travel Literature," in *History and Nation*, ed. Julia Rudolph (Lewisburg, PA: Bucknell University Press, 2006).

32. Adams, *Travelers and Travel Liars*; Mary B. Campbell, *The Witness and the Other World: Exotic European Travel Writing, 400–1600* (Ithaca, NY: Cornell University Press, 1988); Zweder R. W. M. von Martels, ed., *Travel Fact and Travel Fiction: Studies on Fiction, Literary Tradition, Scholarly Discovery, and Observation in Travel Writing* (Leiden: Brill, 1994).

33. Steven Shapin, "The House of Experiment in Seventeenth-Century England," *Isis* 79, no. 3 (1988): 376, and *A Social History of Truth: Civility and Science in Seventeenth-Century England* (Chicago: University of Chicago Press, 1994); John Dunn, "The Concept of 'Trust' in the Politics of John Locke," in *Philosophy in History: Essays on the Historiography of Philosophy*, ed. Richard Rorty, Jerome B. Schneewind, and Quentin Skinner (Cambridge: Cambridge University Press, 1984).

34. Palmira Fontes da Costa, "The Making of Extraordinary Facts: Authentication of Singularities of Nature at the Royal Society of London in the First Half of the Eighteenth Century," *Studies in History and Philosophy of Science* 33, no. 2 (2002): 267.

35. Fontes da Costa, "Making of Extraordinary Facts," 269.

36. Charles W. J. Withers, *Geography, Science and National Identity: Scotland Since 1520* (Cambridge: Cambridge University Press, 2001), 70; emphasis in original.

37. Charles L. Batten, *Pleasurable Instruction: Form and Convention in Eighteenth-Century Travel Literature* (Berkeley: University of California Press, 1978); Percy G. Adams, *Travel Literature and the Evolution of the Novel* (Lexington: University Press of Kentucky, 1983); Jonathan P. A. Sell, *Rhetoric and Wonder in English Travel Writing, 1560–1613* (Aldershot: Ashgate, 2006).

38. Donna Brown, "Travel Books," in *An Extensive Republic: Print, Culture, and Society in the New Nation, 1790–1840*, ed. Robert A. Gross and Mary Kelley (Chapel Hill: University of North Carolina Press, 2010), 451.

39. Lorraine Daston and Peter Galison, *Objectivity*, 2nd ed. (New York: Zone Books, 2010).

40. Henry Ellis, *Journal of the Proceedings of the Late Embassy to China* (London: John Murray, 1817), 479–80. See app.: 38.

41. Robert J. Mayhew, "Mapping Science's Imagined Community: Geography as a Republic of Letters, 1600–1800," *British Journal for the History of Science* 38, no. 1 (2005): 77.

42. Richard Lander and John Lander, *Journal of an Expedition to Explore the Course and Termination of the Niger*, 3 vols. (London: John Murray, 1832), 3:xvii. See app.: 124.

43. For a discussion of this question for later nineteenth-century African exploration, see Lawrence Dritsas, "Expeditionary Science: Conflicts of Method in Mid-Nineteenth-Century Geographical Discovery," in *Geographies of Nineteenth-Century Science*, ed. David N. Livingstone and Charles W. J. Withers (Chicago: University of Chicago Press, 2011); and David Lambert, *Mastering the Niger: James MacQueen's African Geography and the Struggle over African Slavery* (Chicago: University of Chicago Press, 2013).

44. Hugh Murray, *Historical Account of Discoveries and Travels in Africa*, 2 vols. (Edinburgh: Archibald Constable; London: Longman, Hurst, Rees, Orme, & Brown, 1817); James Rennell, *The Geographical System of Herodotus, Examined; and Explained* (London: Printed for the author and sold by G. & W. Nichol, 1800).

45. Thomas Maurice, *Observations on the Ruins of Babylon* (London: John Murray, 1816), 5, iii–iv. See app.: 36.

46. For discussion of Rennell's work being used in this way in Central Asia by several of Murray's authors, see Charles W. J. Withers, "On Enlightenment's Margins: Geography, Imperialism and Mapping in Central Asia, c.1798–c.1838," *Journal of Historical Geography*, vol. 39 (2013).

47. Archibald Edmonstone, *A Journey to Two of the Oases of Upper Egypt* (London: John Murray, 1822), 122. See app.: 70.

48. Ibid., 112, 113.

49. Henry N. Stevens, *Ptolemy's Geography: A Brief Account of All Printed Editions Down to 1730*, 2nd ed. (London: Henry Stevens, Son & Stiles, 1908).

50. Anthony J. Mills, "Sir Archibald Edmonstone: a Scottish Traveller Asserts a Claim," in *Egyptian Encounters*, ed. Jason Thompson (Cairo: American University in Cairo Press, 2002).

51. Edmonstone, *Journey*, xiv.

52. William Martin Leake, *Journal of a Tour in Asia Minor* (London: John Murray, 1824), ix. See app.: 82.

53. Ibid., vi.

54. Ibid., xvi, xiii, xv.

55. Ibid., xi.

56. Ibid., 151.

57. Leask, *Curiosity*, 2.

58. Charles MacFarlane, *Constantinople in 1828* (London: Saunders & Otley, 1829), and *Reminiscences of a Literary Life* (London: John Murray, 1917).

59. Charles MacFarlane to John Murray [II], [1830], MS 40741 (no folio), NLS.

60. William Martin Leake, *Travels in the Morea*, 3 vols. (London: John Murray, 1830). See app.: 121.

61. Charles MacFarlane to John Murray [II], [1830], MS 40741 (no folio), NLS.

62. Ibid.

63. William Bullock, *Six Months' Residence and Travels in Mexico* (London: John Murray, 1824), 97. See app.: 77.

64. Edward Robinson, *Biblical Researches in Palestine, Mount Sinai and Arabia Petræa*, 3 vols. (London: John Murray, 1841), 1:48. See app.: 153.

65. Innes M. Keighren, and Charles W. J. Withers, "The Spectacular and the Sacred: Narrating Landscape in Works of Travel," *cultural geographies*, vol. 19, no. 1 (2012).

66. Robinson, *Biblical Researches*, 1:167n1.

67. Ibid.

68. Walter Hamilton, *A Geographical, Statistical, and Historical Description of Hindostan, and the Adjacent Countries*, 2 vols. (London: John Murray, 1820), 1:xi. See app.: 60.

69. Ibid.

70. Ibid.

71. Walter Hamilton to John Murray [II], October 24, 1817, MS 40516 (no folio), NLS.

72. Ibid.

73. Walter Hamilton to John Murray [II], June 30, 1820, MS 40516 (no folio), NLS.

74. Walter Hamilton to John Murray [II], October 24, 1817, MS 40516 (no folio), NLS.

75. Samuel Smiles, *A Publisher and His Friends: Memoir and Correspondence of the Late John Murray, with an Account of the Origin and Progress of the House, 1768–1843*, 2 vols. (London: John Murray, 1891), 1:278.

76. Walter Hamilton to John Murray [II], undated, MS 40516 (no folio), NLS.

77. Walter Hamilton to John Murray [II], 22 March 1822, MS 40516 (no folio), NLS.

78. Ibid.

79. Walter Hamilton to John Murray [II], February 16, 1824, and May 30, 1827—both in MS 40516 (no folio), NLS.

80. Simon Schaffer, Lissa Roberts, Kapil Raj, and James Delbourgo, eds., *The Brokered World: Go-Betweens and Global Intelligence, 1770–1820* (Sagamore Beach, MA: Science History Publications, 2009).

81. Felix Driver and Lowri Jones, *Hidden Histories of Exploration: Researching the RGS-IBG Collections* (Egham: Royal Holloway, University of London, 2009), 5.

82. John Taylor, *Travels from England to India, in the year 1789*, 2 vols. (London: John Murray, 1799), 1:316–17. See app.: 16.

83. Edmonstone, *Journey*, 43. On disguise among women travelers and questions of gendered identity in later periods, see Dúnlaith Bird, *Travelling in Different Skins: Gender Identity in European Women's Oriental Travelogues, 1850–1950* (Oxford: Oxford University Press, 2012).

84. Leake, *Journal*, 7.

85. Ibid., 97, 49, 343.

86. Ibid., 161, 227, 349.

87. Anonymous review of *Journal of a Tour in Asia Minor*, by William Martin Leake, *Oriental Herald* 2, no. 7 (July 1824): 429.

88. Ibid.

89. Ibid.

90. Frederick W. Beechey and Henry W. Beechey, *Proceedings of the Expedition to Explore the Northern Coast of Africa* (London: John Murray, 1828), 6. See app.: 102.

91. George Francis Lyon, *A Narrative of Travels in Northern Africa, in the Years 1818, 19, and 20* (London: John Murray, 1821), 8. See app.: 65.

92. Beechey and Beechey, *Proceedings*, 7.

93. Ibid., 7.

94. Ibid., 7.

95. Denham Dixon, Hugh Clapperton, and Walter Oudney, *Narrative of Travels and Discoveries in Northern and Central Africa*, 2 vols. (London: John Murray, 1826). See app.: 93.

96. Ibid., 1:14.

97. Frank T. Kryza, *The Race for Timbuktu: In Search of Africa's City of Gold* (New York: Ecco, 2006).

98. Beechey and Beechey, *Proceedings*, 33.

99. Ibid., 97.

100. Ibid., 120, 127.

101. Ibid., 127.

102. Ibid., 328, 329.

103. Taylor, *Travels from England to India*, 1:200.

104. Robert Fortune, *Three Years' Wanderings in the Northern Provinces of China* (London: John Murray, 1847), ix. See app.: 185.

105. Taylor, *Travels from England to India*, 1:38; Richard Wilbraham, *Travels in the Trans-Caucasian Provinces of Russia* (London: John Murray, 1839), 39. See app.: 145.

106. Robinson, *Biblical Researches*, 1:92.

107. Ibid., 1:61.

108. Ibid., 1:222, 239, 212.

109. Godfrey Mundy, *Pen and Pencil Sketches*, 2 vols. (London: John Murray, 1832), 1:viii. See app.: 125.

110. Ibid., 2:32.

111. Ibid., 2:32-33.

112. Ibid., 2:33.

113. Anonymous review of *Pen and Pencil Sketches*, by Godfrey Mundy, *Edinburgh Review* 57, no. 116 (July 1832), 358, 359.

114. Ibid., 359.

115. Anonymous review of *Pen and Pencil Sketches*, by Godfrey Mundy, *Examiner* (May 27, 1832), 340.

116. Leake, *Journal*, 151.

117. Mundy, *Pen and Pencil Sketches*, 1:10.

118. Anonymous review of *Pen and Pencil Sketches*, by Godfrey Mundy, *Frazer's Magazine for Town and Country* 6, no. 32 (September 1832): 148.

119. Mundy, *Pen and Pencil Sketches*, 1:258.

120. Ibid., 1:297-98.

121. Jack Meadows, *The Victorian Scientist: The Growth of a Profession* (London: British Library, 2004).

122. Harold M. Otness, "Passenger Ship Libraries," *Journal of Library History* 14, no. 4 (1979): 486.

123. Bill Bell, "Bound for Australia: Shipboard Reading in the Nineteenth Century," *Journal of Australian Studies* 25, no. 68 (2001): 5-18; David M. Hovde, "Sea Colportage: The Loan Library System of the American Seamen's Friend Society, 1859-1967," *Libraries and Culture*, vol. 29, no. 4 (1994); Harry R. Skallerup, *Books Afloat and Ashore: A History of Books, Libraries, and Reading among Seamen during the Age of Sail* (Hamden, CT: Archon Books, 1974).

124. John Ross, *A Voyage of Discovery, Made under the Orders of the Admiralty, in His Majesty's Ships* Isabella *and* Alexander (London: John Murray, 1819), xx. See app.: 55.

125. Ibid., xxii.

126. Alexander Mackenzie, *Voyages from Montreal, on the River St. Laurence, through the Continent of North America, to the Frozen Pacific Oceans; in the Years 1789 and 1793* (London: T. Cadell, Jun. & W. Davies; Corbert & Morgan; Edinburgh: W. Creech, 1801); Samuel Hearne, *A Journey from Prince of Wales's Fort in Hudson's Bay* (London: A. Strahan; T. Cadell, 1795); Thomas Falkner, *A Description of Patagonia, and the Adjoining Parts of South America* (Hereford: T. Lewis, 1774); William Dampier, *Voyages and Descriptions*, 3 vols. (London: James Knapton, 1699). On Dampier's "piratical Enlightenment" and his work's publishing history, see Adriana Craciun, "Oceanic Voyages, Maritime Books, and Eccentric Inscriptions," *Atlantic Studies* 10, no. 2 (2013): 174-79.

127. A. D. Morrison-Low, *Making Scientific Instruments in the Industrial Revolution* (Aldershot: Ashgate, 2007).

128. Ross, *Voyage of Discovery*, 245.

129. Janice Cavell, *Tracing the Connected Narrative: Arctic Exploration in British Print Culture, 1818-1860* (Toronto: University of Toronto Press, 2008), 68.

130. Maurice J. Ross, *Polar Pioneers: A Biography of John and James Clark Ross* (Montreal: McGill-Queen's University Press, 1994), 52.

131. Sir John Ross Memoirs, vol. 1 (covering the period 1818-19, written [from internal evidence] in 1848), unpaginated, MS 655/1, University of Cambridge, Scott Polar Research Institute.

132. Ross, *Voyage of discovery*, ii.

133. Ross Memoirs, vol. 1, unpaginated.

134. Ibid.

135. On these points, see Michael S. Reidy, *Tides of History: Ocean Science and Her Majesty's Navy* (Chicago: University of Chicago Press, 2008).

136. Copy Ledger B, fol. 71, MS 42725, NLS.

137. Ross Memoirs, vol. 1, unpaginated.

138. Ibid.

139. Cavell, *Tracing the Connected Narrative* (n. 129 above, this chap.), 70, 71. For the critical reception of Ross's narrative, see Cavell, *Tracing the Connected Narrative*; Fergus Fleming, *Barrow's Boys: A Stirring Story of Daring, Fortitude and Outright Lunacy* (London: Granta; 1998); Ross, *Polar Pioneers.*

140. Edward Sabine, *Remarks on the Account of the Late Voyage of Discovery to Baffin's Bay, Published by Captain J. Ross, R. N.* (London: John Booth, 1819).

141. Ibid., 4.

142. John Ross, *An Explanation of Captain Sabine's Remarks on the Late Voyage of Discovery to Baffin's Bay* (London: John Murray, 1819). See app.: 56.

143. Ross Memoirs, vol. 1, unpaginated.

144. Alexander Caldcleugh, *Travels in South America, during the Years 1819-20-21*, 2 vols. (London: John Murray, 1825), 1:304.

145. Ibid., 1:304-5.

146. Alexander Gordon Laing, *Travels in the Timannee, Kooranko, and Soolima Countries, in Western Africa* (London: John Murray, 1825), 260. See app.: 88.

147. Ibid., 464.

148. Thomas Nelson, *A Biographical Memoir of the Late Dr W. Oudney, and Captain Hugh Clapperton* (Edinburgh: Waugh & Innes; London: Whittaker, Treacher, 1830), 141.

149. Simon Schaffer, "Easily Cracked: Scientific Instruments in States of Disrepair," *Isis*, vol. 102, no. 4 (2011).

150. Johannes Fabian, *Out of Our Minds: Reason and Madness in the Exploration of Central Africa* (Berkeley: University of California Press, 2000).

151. Bruce Hevly, "The Heroic Science of Glacier Motion," *Osiris*, vol. 11 (1996).

152. Joseph Beete Jukes, *Excursions in and about Newfoundland, during the Years 1839 and 1840*, 2 vols. (London: John Murray, 1842), 1:15. See app.: 158.

153. Ibid., 1:46.

154. Ibid., 2:15.

155. Ibid., 2:14.

156. Murray's ledger books record a total loss, by mid-June 1862, of £35 12s. 3d. on a print run of 750 copies (Copy Ledger D, fols. 204, 279, MS 42729, NLS).

CHAPTER FOUR

1. George Francis Lyon, *The Private Journal of Captain G. F. Lyon, of H.M.S. Hecla, during the Recent Voyage of Discovery under Captain Parry* (London: John Murray, 1824), v. See app.: 83.

2. Anonymous review of *Travels in North America during the Years 1834, 1835, & 1836*, by Charles Augustus Murray and *A Diary in America*, by Frederick Marryat, *Dublin Review* 7, no. 14 (November 1839): 399.

3. Fergus Fleming, *Barrow's Boys: A Stirring Story of Daring, Fortitude and Outright Lunacy* (London: Granta, 1998), 95. Lyons's self-description is to his friend and patron, Lord (Lyon to Bayntun, December 23, 1829, MS DD/H1/D/553, Bayntun Somerset Archives and Record Office).

4. Adriana Craciun, "What Is an Explorer?" *Eighteenth-Century Studies* 45, no. 1 (2011): 33.

5. Walter Hamilton to John Murray [II], February 9, 1820, November 2, 1820, November 10, 1820, November 16, 1820, January 21, 1821, MS 40516, fols. 32, 39, 41, 42, 44, respectively, National Library of Scotland (hereafter NLS), Edinburgh.

6. Henry Milton to John Murray [III], June 111, 846, MS 40828 (no folio), NLS.

7. John Murray [III] to Charles MacFarlane, October 12, 1847, MS 41912, fol. 96, NLS.

8. The complex publication history of Livingstone's *Missionary Travels* has been subject to sustained scrutiny elsewhere and so is not detailed here. See Louise Henderson, "Geography, Travel and Publishing in Mid-Victorian Britain" (PhD diss., Royal Holloway, University of London, 2012), and "An Authentic, Invented Africa: David Livingstone's *Missionary Travels* and Nineteenth-Century Practices of Illustration," *Scottish Geographical Journal*, vol. 129, nos. 3–4 (2013); Justin Livingstone, "The Meaning and Making of *Missionary Travels*: The Sedentary and Itinerant Discourses of a Victorian Bestseller," *Studies in Travel Writing*, vol. 15, no. 3 (2011), and "*Missionary Travels*, Missionary Travails: David Livingstone and the Victorian Publishing Industry," in *David Livingstone: Man, Myth and Legacy*, ed. Sarah Worden (Edinburgh: National Museums of Scotland, 2012).

9. David Livingstone to John Murray [III], May 31, 1857, MS 42420 (no folio), NLS.

10. Stuart Sherman, *Telling Time: Clocks, Diaries, and English Diurnal Form, 1660–1785* (Chicago: University of Chicago Press, 1996), 180.

11. Kevin Dunn, *Pretexts of Authority: The Rhetoric of Authorship in the Renaissance Preface* (Stanford, CA: Stanford University Press, 1994); Jonathan P. A. Sell, *Rhetoric and Wonder in English Travel Writing, 1560–1613* (Aldershot: Ashgate, 2006).

12. Dunn, *Pretexts of Authority*, 4.

13. Lena Wahlgren-Smith, "On the Composition of Herbert Losinga's Letter Collection," *Classica et Mediaevalia* 55 (2004): 236.

14. Dunn, *Pretexts of Authority*, 5–6.

15. Mary B. Campbell, *The Witness and the Other World: Exotic European Travel Writing, 400–1600* (Ithaca, NY: Cornell University Press, 1988), 230–31.

16. Sell, *Rhetoric and Wonder*, 66.

17. See Roger Chartier, *The Order of Books: Readers, Authors, and Libraries in Europe between the Fourteenth and Eighteenth Centuries*, trans. Lydia G. Cochrane (Cambridge: Polity Press, 1994); Michel Foucault, "What Is an Author?" in *The Foucault Reader*, ed. Paul Rabinow (London: Penguin, 1991); Adrian Johns, *The Nature of the Book: Print and Knowledge in the Making* (Chicago: University of Chicago Press, 1998).

18. Foucault, "What Is an Author?" 102.

19. Ibid.,108.

20. Roger Chartier, "Foucault's Chiasmus: Authorship between Science and Literature in the Seventeenth and Eighteenth Centuries," in *Scientific Authorship: Credit and Intellectual Property in Science*, ed. Mario Biagioli and Peter Galison (New York: Routledge, 2003); Martha Woodmansee, "On the Author Effect: Recovering Collectivity," in *The Construction of Authorship: Textual Appropriation in Law and Literature*, ed. Martha Woodmansee and Peter Jaszi (Durham, NC: Duke University Press, 1994).

21. Steven Shapin, *A Social History of Truth: Civility and Science in Seventeenth-Century Europe* (Chicago: University of Chicago Press, 1994).

22. Chartier, "Foucault's Chiasmus," 22.

23. Adrian Johns, "The Ambivalence of Authorship in Early Modern Natural Philosophy," in *Scientific Authorship*, ed. Biagioli and Galison, 79.

24. Shapin, *Social History of Truth*, 179; emphasis in original.

25. Foucault, "What Is an Author?" 109.

26. Steven Shapin, "Pump and Circumstance: Robert Boyle's Literary Technology," *Social Studies of Science* 14, no. 4 (1984): 497.

27. Kenneth J. E. Graham, *The Performance of Conviction: Plainness and Rhetoric in the Early English Renaissance* (Ithaca, NY: Cornell University Press, 1994).

28. Lisa Jardine, *Francis Bacon: Discovery and the Art of Discourse* (London: Cambridge University Press, 1974), 152.

29. Donna J. Haraway, *Modest_Witness@Second_Millennium.FemaleMan©_Meets_OncoMouse™* (New York: Routledge, 1997), 26.

30. Joe Snader, *Caught between Worlds: British Captivity Narratives in Fact and Fiction* (Lexington: University Press of Kentucky, 2000), 50.

31. Peter C. Mancall, "Introduction: Observing More Things and More Curiously," *Huntington Library Quarterly* 70, no. 1 (2007), 2.

32. Sell, *Rhetoric and Wonder* (n. 11 above, this chap.), 15.

33. On the emergence of the professional author, see, e.g., Dustin Griffin, "The Rise of the Professional Author?" in *The Cambridge History of the Book in Britain*, vol. 5, *1695–1830*, ed. Michael F. Suarez and Michael L. Turner (Cambridge: Cambridge University Press, 2009); Patrick Leary and Andrew Nash, "Authorship," in *The Cambridge History of the Book in Britain*, vol. 6, *1830–1914*, ed. David McKitterick (Cambridge: Cambridge University Press, 2009).

34. John Kinnear to John Murray [III], August 3, 1840, MS 10654 (no folio), NLS. The book was published as *Cairo, Petra, and Damascus in 1839* (1841). See app.: 151.

35. Letters dated from April 14, 1837, to May 21, 1874, MS 10908, NLS.

36. The house of Murray did, however, recommend anonymous publication of certain books of European travel, including two by women: Elizabeth Rigby's *A Residence on the Shores of the Baltic* (1841) and Rebecca McCoy's *The Englishwoman in Russia* (1855). See Anthony Cross, "Two English 'Lady Travellers' in Russia and the House of Murray," *Slavonica*, vol. 17, no. 1 (2011).

37. Brian Maidment, "Periodicals and Serial Publications," in *The Cambridge History of the Book in Britain*, vol. 5, ed. Suarez and Turner; James Raven, "The Anonymous Novel in Britain and Ireland, 1750–1830," in *The Faces of Anonymity: Anonymous and Pseudonymous Publication from the Sixteenth to the Twentieth Century*, ed. Robert J. Griffin (Basingstoke: Palgrave Macmillan, 2003).

38. Robert J. Griffin, "Anonymity and Authorship," *New Literary History*, vol. 30, no. 4 (1999); James A. Secord, *Victorian Sensation: The Extraordinary Publication, Reception, and Secret Authorship of "Vestiges of the Natural History of Creation"* (Chicago: University of Chicago Press, 2000).

39. The role and significance of the shipwreck narrative has been described by, among others, Josiah Blackmore, *Manifest Perdition: Shipwreck Narrative and the Disruption of Empire* (Minneapolis: University of Minnesota Press, 2002); Margarette Lincoln, "Shipwreck Narratives of the Eighteenth and Early Nineteenth Century: Indicators of Culture and Identity," *British Journal for Eighteenth-Century Studies*, vol. 20, no. 2 (1997); Carl Thompson, *The Suffering Traveller and the Romantic Imagination* (Oxford: Oxford University Press, 2007).

40. [Robert Young], *An Account of the Loss of His Majesty's Ship* Deal Castle (London: J. Murray, 1787), 3. See app.: 7.

41. Ibid.

42. [George Robertson], *A Short Account of a Passage from China* (London: J. Murray, 1788). See app.: 8.

43. On the function and changing nature of title page anonymity see, e.g., Marcy L. North, *The Anonymous Renaissance: Cultures of Discretion in Tudor-Stuart England* (Chicago: University

of Chicago Press, 2003); Richard B. Sher, *The Enlightenment and the Book: Scottish Authors and Their Publishers in Eighteenth-Century Britain, Ireland, and America* (Chicago: University of Chicago Press, 2006).

44. [Robertson], *Short Account*, iii.

45. George Robertson, *Charts of the China Navigation, Principally Laid Down upon the Spot, and Corrected from the Latest Observations, with Particular Views of the Land* (London: J. Murray, 1788); Andrew Cook, "Alexander Dalrymple and the Hydrographic Office," in *Pacific Empires: Essays in Honour of Glyndwr Williams*, ed. Alan Frost and Jane Samson (Melbourne: Melbourne University Press, 1999).

46. [Robertson], *Short Account*, 1.

47. Ibid., 3.

48. George Robertson, *Memoir of a Chart of the China Sea* (London: Printed for the author and sold by J. Murray, 1791), i.

49. Ibid., ii.

50. "Presents Made to the Royal Society from November 1790 to June 1791," *Philosophical Transactions of the Royal Society of London*, vol. 81, no. 1 (1791); "Donations Presented to the Royal Society of Edinburgh," *Transactions of the Royal Society of Edinburgh*, vol. 4 (1798); Charles D. Waterston and Angus Macmillan Shearer, *Biographical Index of Former Fellows of the Royal Society of Edinburgh, 1783–2002* (Edinburgh: Royal Society of Edinburgh, 2006).

51. Michael Ashcroft, "Robert Charles Dallas," *Jamaica Journal*, vol. 44 (1980).

52. [Robert C. Dallas], *A Short Journey in the West Indies*, 2 vols. (London: Printed for the author and sold by J. Murray and J. Forbes, 1790), 1:3. See app.: 10.

53. Ibid., 1:92.

54. Ibid., 1:109.

55. Ibid., 1:109.

56. Ibid., 1:109–10.

57. Anonymous review of *A Short Journey in the West Indies*, by Robert Dallas, *Monthly Review* 4 (March 1791): 336.

58. Ibid., 336.

59. Ibid., 337; emphasis in original.

60. Peter W. Graham, "Byron and the Publishing Business," in *The Cambridge Companion to Byron*, ed. Drummond Bone (Cambridge: Cambridge University Press, 2004).

61. George Carter, *A Narrative of the Loss of the* Grosvenor *East Indiaman* (London: J. Murray / William Lane, 1791). See app.: 12.

62. Ian E. Glenn, "The Wreck of the *Grosvenor* and the Making of South Africa Literature," *English in Africa*, vol. 22, no. 2 (1995).

63. Stephen Taylor, *The Caliban Shore: The Fate of the* Grosvenor *Castaways* (London: Faber, 2004).

64. Michael Titlestad, "'The Unhappy Fate of Master Law': George Carter's *A Narrative of the Loss of the* Grosvenor, *East Indiaman* (1791)," *English Studies in Africa* 51, no. 2 (2008): 21.

65. Ibid., 21.

66. Anonymous review of *A Narrative of the Loss of the* Grosvenor *East Indiaman*, by George Carter, *Monthly Review* 7 (April 1792): 470; emphasis in original.

67. Craciun, "What Is an Explorer?" (n. 4 above, this chap.), 33.

68. Mancall, "Introduction" (n. 31 above, this chap.), 6.

69. On captivity narratives as a genre, see, e.g., Linda Colley, *Captives: Britain, Empire and the World, 1600–1850* (London: Jonathan Cape, 2002); Snader, *Caught between Worlds* (n. 30 above, this chap.).

70. Charles W. J. Withers, "Geography, Enlightenment and the Book: Authorship and Audience in Mungo Park's African Texts," in *Geographies of the Book*, ed. Miles Ogborn and Charles W. J. Withers (Farnham: Ashgate, 2010).

71. Charles Hansford Adams, ed., *The Narrative of Robert Adams, a Barbary Captive: A Critical Edition* (New York: Cambridge University Press, 2005), ix; [Robert Adams], *The Narrative of Robert Adams* (London: John Murray, 1816), xii. See app.: 34.

72. Ann Fabian, *The Unvarnished Truth: Personal Narratives in Nineteenth-Century America* (Berkeley: University of California Press, 2000).

73. [Adams], *Narrative of Robert Adams*, xi.

74. Ibid., xi.

75. Ibid., xiii, xii.

76. Ibid., xiii.

77. Anonymous review of *Journal d'un Voyage a Temboctou et a Jenné*, by René Caillié, *Quarterly Review* 42, no. 84 (March 1830): 453.

78. [Adams], *Narrative of Robert Adams*, xiii.

79. Ibid., xvi.

80. Ibid., xvii, xix.

81. Fabian, *Unvarnished Truth*, 29.

82. Ibid., 35.

83. Anonymous review of *The Narrative of Robert Adams*, by [Robert Adams], *Quarterly Review* 14, no. 14 (January 1816): 473.

84. Anonymous review of *The Narrative of Robert Adams*, by [Robert Adams], *Edinburgh Review* 26, no. 52 (June 1816): 385.

85. Anonymous review of *The Narrative of Robert Adams*, by [Robert Adams], *Augustan Review* 3, no. 15 (July 1816): 59; emphasis in original.

86. Anonymous review of *The Narrative of Robert Adams*, by [Robert Adams], *Monthly Review* 82 (January 1817): 26.

87. Anonymous review of *The Narrative of Robert Adams*, by [Robert Adams], *North American Review* 5, no. 2 (July 1817): 204.

88. Michael Heffernan, "'A Dream as Frail as Those of Ancient Time': The In-credible Geographies of Timbuctoo," *Environment and Planning D: Society and Space* 19, no. 2 (2001): 209.

89. Anonymous review of *The Narrative of Robert Adams*, by [Robert Adams], *British Lady's Magazine*, vol. 4, no. 19 (July 1816).

90. Heffernan, "In-credible Geographies of Timbuctoo," 209.

91. James Riley, *Loss of the American Brig* Commerce (London: John Murray, 1817), x. See app.: 42.

92. R. Gerald McMurtry, "The Influence of Riley's *Narrative* upon Abraham Lincoln," *Indiana Magazine of History*, vol. 30, no. 2 (1934).

93. Ghislaine Lydon, "Writing Trans-Saharan History: Methods, Sources and Interpretations across the African Divide," *Journal of North African Studies*, vol. 10, nos. 3–4 (2005).

94. Riley, *Loss*, xiv. For a biographical account of Eddy, see Silvio A. Bedini, *With Compass and Chain: Early American Surveyors and Their Instruments* (Frederick, MD: Professional Surveyors Publishing, 2001).

95. Heffernan, "In-credible Geographies of Timbuctoo"; Charles W. J. Withers, "Mapping the Niger, 1798–1832: Trust, Testimony and 'Ocular Demonstration' in the Late Enlightenment," *Imago Mundi*, vol. 56, no. 2 (2004).

96. James Riley, *An Authentic Narrative of the Loss of the American Brig* Commerce (New York: T. & W. Mercein, 1817), iv.

97. John W. Francis, *Old New York; or, Reminiscences of the Past Sixty Years* (New York: Charles Roe, 1858), 69.

98. Riley, *Authentic Narrative*, vii.

99. Ibid., vii.

100. Mancall, "Introduction" (n. 31 above, this chap.), 7.

101. McMurtry, "Influence of Riley's *Narrative*," 134.

102. Ibid. Riley's text has been subject to significant scholarly attention. See, e.g., Robert J. Allison, *The Crescent Obscured: The United States and the Muslim World, 1776–1815* (New York: Oxford University Press, 1995); Paul Baepler, "The Barbary Captivity Narrative in American Culture," *Early American Literature*, vol. 39, no. 2 (2004); Hester Blum, "Pirated Tars, Piratical Texts: Barbary Captivity and American Sea Narratives," *Early American Studies*, vol. 1, no. 2 (2003); Dean King, *Skeletons on the Zahara* (London: Arrow, 2004); David Lambert, "'Taken Captive by the Mystery of the Great River': Towards an Historical Geography of British Geography and Atlantic Slavery," *Journal of Historical Geography*, vol. 35, no. 1 (2009).

103. James Riley to John Murray [II], January 10, 1817, MS 41020 (no folio), NLS.

104. Anonymous review of *An Authentic Narrative of the Loss of the American Brig* Commerce, by James Riley, *Quarterly Review* 16, no. 32 (January 1817): 287.

105. James Riley to John Murray [II], January 10, 1817, MS 41020 (no folio), NLS.

106. Ibid.

107. Anonymous review of *Loss of the American Brig* Commerce, by James Riley, *British Critic* 7 (June 1817): 616.

108. Ibid., 633.

109. Anonymous review of *Loss of the American Brig* Commerce, by James Riley, *Monthly Review* 84 (October 1817): 128–29.

110. Anonymous review of *Loss of the American Brig* Commerce, by James Riley, *Eclectic Review*, n.s., 11 (January 1819): 66.

111. Samuel Smiles, *A Publisher and His Friends: Memoir and Correspondence of the Late John Murray, with an Account of the Origin and Progress of the House, 1768–1843*, 2 vols. (London: John Murray, 1891), 2:28.

112. James Riley to John Murray [II], March 23, 1819, MS 41020 (no folio), NLS.

113. Craciun, "What Is an Explorer?" (n. 4 above, this chap.), 33.

114. Jan Borm, "Defining Travel: On the Travel Book, Travel Writing and Terminology," in *Perspectives on Travel Writing*, ed. Glen Hooper and Tim Youngs (Aldershot: Ashgate, 2004), 17.

115. Anonymous review of *Narrative of Travels and Discoveries in Northern and Central Africa*, by Denham Dixon, Hugh Clapperton, and Walter Oudney, *Edinburgh Review* 44, no. 87 (June 1826): 174.

116. Anonymous review of *A Narrative of Travels in Northern Africa*, by George Francis Lyon, *Literary Chronicle and Weekly Review*, no. 97 (March 24, 1821), 177.

117. George Francis Lyon, *A Narrative of Travels in Northern Africa, in the Years 1818, 19, and 20* (London: John Murray, 1821), vi. See app.: 65.

118. Ibid., v.

119. Charles W. J. Withers, "Memory and the History of Geographical Knowledge: The Commemoration of Mungo Park, African Explorer," *Journal of Historical Geography*, vol. 30, no. 2 (2004); Withers, "Geography, Enlightenment, and the Book," 212–20.

120. Lyon, *Narrative of Travels*, vi–vii.

121. George Francis Lyon to John Murray [II], April 18, 1827, MS 40731 (no folio), NLS. There is no clue in this correspondence as to the reasons for the change in title, or why the variant titles were proposed and held and by whom.

122. Anonymous review of *A Narrative of Travels in Northern Africa*, by George Francis Lyon, *Quarterly Review* 25, no. 49 (April 1821): 25.

123. Anonymous review of *A Narrative of Travels in Northern Africa*, by George Francis Lyon," *London Literary Gazette*, no. 218 (24 March 1821), 179.

124. Ibid.

125. Anonymous review of *A Narrative of Travels in Northern Africa*, by George Francis Lyon, *Monthly Review* 96 (October 1821): 113.

126. Anonymous review of *A Narrative of Travels in Northern Africa*, by George Francis Lyon, *British Review* 18, no. 35 (September 1821): 178.

127. Fleming, *Barrow's Boys* (n. 3 above, this chap.), 106.

128. Ibid., 178.

129. See, e.g., Frank T. Kryza, *The Race for Timbuktu: In Search of Africa's City of Gold* (New York: Ecco, 2006); Jamie Bruce Lockhart, *A Sailor in the Sahara: The Life and Travels in Africa of Hugh Clapperton, Commander RN* (London: I. B. Tauris, 2008); Jamie Bruce Lockhart, "Hugh Clapperton, 1788–1827," in *Geographers Biobibliographical Studies*, ed. Hayden Lorimer and Charles W. J. Withers, vol. 28 (London: Continuum, 2009).

130. Withers, "Mapping the Niger" (n. 95 above, this chap.).

131. Dixon Denham to John Murray [II], July 28, 1825, MS 40324 (no folio), NLS.

132. Dixon Denham, Hugh Clapperton, and Walter Oudney, *Narrative of Travels and Discoveries in Northern and Central Africa*. 2 vols. (London: John Murray, 1826), 1:v. See app.: 93.

133. Christopher Fyfe, "Denham, Dixon (1786–1828)," in *Oxford Dictionary of National Biography* (Oxford University Press, 2004–13), doi:10.1093/ref:odnb/7476.

134. Dixon et al., *Narrative of Travels*, 1:vi; emphasis in original.

135. Ibid., 1:vii–viii.

136. Lockhart, "Hugh Clapperton."

137. Dixon et al., *Narrative of Travels*, vol. 2, n.p.

138. Christopher Lloyd, *Mr Barrow of the Admiralty: A Life of Sir John Barrow, 1764–1848* (London: Collins, 1970), 122.

139. Dixon et al., *Narrative of Travels*, vol. 2, n.p.

140. Lockhart, *Sailor in the Sahara*, 188.

141. [John Barrow], review of *Narrative of Travels and Discoveries in Northern and Central Africa*, by Denham Dixon, Hugh Clapperton, and Walter Oudney, *Quarterly Review* 33, no. 66 (March 1826): 545.

142. Hugh Clapperton and Richard Lander, *Journal of a Second Expedition into the Interior of Africa* (London: John Murray, 1829), xi, xi–xii. See app.: 109.

143. Fleming, *Barrow's Boys* (n. 3 above, this chap.), 199.

144. Clapperton and Lander, *Journal of a Second Expedition*, 178. Clapperton's inscriptive practices are detailed in Jamie Bruce Lockhart, "In the Raw: Some Reflections on Transcribing and Editing Lieutenant Hugh Clapperton's Writings on the Borno Mission of 1822–25," *History in Africa*, vol. 26 (1999).

145. Richard Lander, *Records of Captain Clapperton's Last Expedition to Africa*, 2 vols. (London: Henry Colburn and Richard Bently, 1830), 1:x–xi.

146. Fleming, *Barrow's Boys*, 225.

147. Clapperton and Lander, *Journal of a Second Expedition*, 178.

148. Fleming, *Barrow's Boys*, 229.

149. Clapperton and Lander, *Journal of a Second Expedition*, xviii.

150. Ibid., xviii.

151. Lander, *Records*, 1:x.

152. Jamie Bruce Lockhart and Paul E. Lovejoy, eds., *Hugh Clapperton into the Interior of Africa: Records of the Second Expedition, 1825–1827* (Leiden: Brill, 2005), 69.

153. Ibid., 69–70.

154. Ibid., 69.

155. Ibid.

156. Anonymous review of *Journal of a Second Expedition into the Interior of Africa*, by Hugh Clapperton and Richard Lander, *Oriental Herald* 20, no. 63 (March 1829): 466.

157. Anonymous review of *Journal of a Second Expedition into the Interior of Africa*, by Hugh Clapperton and Richard Lander, *Eclectic Review*, n.s., 1 (February 1829): 162.

158. Anonymous review of *Journal of a Second Expedition into the Interior of Africa*, by Hugh Clapperton and Richard Lander, *British Critic* 6, no. 11 (July 1829): 48.

159. Lander, *Records* (n. 145 above, this chap.), 1:xxi.

160. Anonymous review of *the Journal of the Royal Geographical Society of London for the Year 1830-31, Quarterly Review* 46, no. 91 (November 1831): 74.

161. Ibid.

162. The editing and production of this volume is described by Jamie Bruce Lockhart, "Documents Relating to the Lander Brothers' Niger Expedition of 1830," accessed December 16, 2011, http://www.tubmaninstitute.ca/documents_relating_to_the_lander_brothers_niger_expedition_of_1830.

163. Richard Lander and John Lander, *Journal of an Expedition to Explore the Course and Termination of the Niger*, 3 vols. (London: John Murray, 1832), 3:147, 153. See app.: 124.

164. Mercedes Mackay, *The Indomitable Servant* (London: Rex Collins, 1978), 315.

165. Lander and Lander, *Journal of an Expedition*, 1:ix.

166. Robin Hallett, ed., *The Niger Journal of Richard and John Lander* (London: Routledge & Kegan Paul, 1965), 33.

167. Lander and Lander, *Journal of an Expedition*, 1:lxiii-lxiv.

168. John Lander to Murray [II], [1831], MS 40668 (no folio), NLS.

169. [Sarah Lushington], *Narrative of a Journey from Calcutta to Europe* (London: John Murray, 1829), x-xi. See app.: 113.

CHAPTER FIVE

1. Gérard Genette, *Paratexts: Thresholds of Interpretation*, trans. Jane Lewin. (Cambridge: Cambridge University Press, 1997), 1.

2. William Zachs, *The First John Murray and the Late Eighteenth-Century London Book Trade: With a Checklist of His Publications* (Oxford: Oxford University Press for the British Academy, 1998), 373.

3. [Thomas Beddoes], *Alexander's Expedition down the Hydaspes & the Indus to the Indian Ocean* (London: John Murray / James Phillips, 1792), vi. See app.: 14. The "Rennel" referred to is the geographer and mapmaker James Rennell, author in 1780 of *A Bengal Atlas* (which Beddoes cites as an authoritative source in his 1792 work). The "Mr. Bewick" referred to is the wood-engraver and illustrator Thomas Bewick (1753-1828).

4. William Alexander, *Picturesque Representations of the Dress and Manners of the Turks* (London: John Murray, 1814), vi. See app.: 28.

5. Ibid., iii.

6. John Francis Davis, *Chinese Moral Maxims, with a Free and Verbal Translation; Affording Examples of the Grammatical Structure of the Language* (London: John Murray; Macao: Honorable Company's Press, 1823), v.

7. George Catlin, *Caitlin's Notes of Eight Years' Travels and Residence in Europe, with His North American Indian Collection*, 2 vols. (London: Published by the author, 1848), 1:51, and *Letters and Notes on the Manners, Customs, and Condition of the North American Indians*, 2 vols. (London: Published by the author, 1841).

8. John Murray [III] to Robert Fortune, November 14, 1846, MS 41912 (no folio), National Library of Scotland (hereafter NLS).

9. Janice Cavell, "Making Books for Mr. Murray: The Case of Edward Parry's Third Arctic Narrative," *Library*, vol. 14, no. 1 (2013).

10. John Barrow, *Voyages of Discovery and Research within the Arctic Regions* (London: John Murray, 1846), v-vi. See app.: 178.

11. George Blakiston Wilkinson, *South Australia; Its Advantages and Its Resources* (London: John Murray, 1848), vi. See app.: 192.

12. George Blakiston Wilkinson, *The Working Man's Handbook to South Australia* (London: John Murray, 1849), v. See app.: 201.

13. Léon de Laborde, *Voyage de l'Arabie Pétrée* (Paris: Giard, 1830).

14. Léon de Laborde, *Journey through Arabia Petræa to Mount Sinai* (London: John Murray, 1836), xviii. See app.: 133.

15. Ibid., xix.

16. Léon de Laborde to John Murray [II], January 12, 1838, MS 4066 (no folio), NLS.

17. John Gardiner Kinnear, *Cairo, Petra, and Damascus* (London: John Murray, 1841), vi. See app.: 151.

18. John Kinnear to John Murray [II?], n.d., MS 10654 (no folio), NLS.

19. Mansfield Parkyns to John Murray [III], May 27, 1853, MS 12604/1917 (no folio), NLS.

20. There were other sound reasons for including several accounts within a single volume: together they corroborate an expedition's findings in ways that a single-authored account would not.

21. Francis Bond Head to John Murray [II], [1831], MS 42278-79 (no folio), NLS.

22. Francis Galton to John Murray [III], March 3, 1853, MS 12604/1435 (no folio), NLS. Galton first proposed the title "Narrative of an Explorer in Tropical Africa," noting that "tropical *South* Africa would be more explicit but perhaps too lengthy." Murray, it seems, adopted Galton's suggestion without demur.

23. Francis Galton, *The Art of Travel; or, Shifts and Contrivances Available in Wild Countries*, 2nd ed. (London: John Murray, 1856), title page. See app.: 221.

24. Genette, *Paratexts* (n. 1 above, this chap.), 74.

25. Janine Barchas, *Graphic Design, Print Culture, and the Eighteenth Century Novel* (Cambridge: Cambridge University Press, 2003), 22.

26. Abraham Salamé to John Murray [II], July 11, 1818, MS 41056 (no folio), NLS.

27. Abraham Salamé to John Murray [II], July 12, 1823, MS 41056 (no folio), NLS.

28. Francis Bond Head to John Murray [II], September 28, 1830, MS 42278-79 (no folio), NLS. Judging by the finally published version, Murray seems to have taken this advice.

29. On Burnes's map work and the cartographic and diplomatic context in which he worked, see Charles W. J. Withers, "On Enlightenment's Margins: Geography, Imperialism and Mapping in Central Asia, *c.*1798–*c.*1838," *Journal of Historical Geography*, vol. 39 (2013).

30. Alexander Burnes to John Murray [II], January 25, 1834, MS 42048 (no folio), NLS.

31. Alexander Burnes to John Murray [II], June 13, 1834, MS 42048, NLS.

32. Alexander Burnes to John Murray [II], December 28, 1834, MS 42048, NLS.

33. Henry T. Prinsep, *Origin of the Sikh Power in the Punjab, and Political Life of Muha-Raja Runjeet Singh* (Calcutta: Military Orphan Press, 1834).

34. Copy Ledger D, fol. 137, MS 42729, NLS.

35. Alexander Burnes, *Travels into Bokhara*, 3 vols., 2nd ed. (London: John Murray, 1835), 1:vi. See app.: 126.

36. Genette, *Paratexts* (n. 1 above, this chap.), 41.

37. James Clark Ross to John Murray [III], June 7, 1847, MS 41038 (no folio), NLS.

38. Frederick W. Beechey to John Murray [II], February 18, 1825, MS 12604/1076 (no folio), NLS.

39. Frederick W. Beechey to John Murray [II], October 20, 1829, MS 12604/1076, NLS.

40. Richard Lander and John Lander, *Journal of an Expedition to Explore the Course and Termination of the Niger*, 3 vols. (John Murray, 1832), 1:lxiii. See app.: 124.

41. [Charlotte Maitland], *Letters from Madras, during the Years 1836-1839* (London: John Murray, 1843), iii, iv. See app.: 161.

42. Julia Charlotte Maitland to John Murray [III], July 9, 1846, MS 40760 (no folio), NLS.

43. Regina Akel, *Maria Graham: A Literary Biography* (Amherst, NY: Cambria Press, 2009), 84; Kathleen L. Skinner, "Ships, Logs, and Voyages: Maria Graham Navigates the Journey of H.M.S. *Blonde*" (honor's thesis, University of Texas at Austin, 2010).

44. Unsigned and undated letter presenting Murray II with "an abstract" of "a letter [dated September 9, 1831] . . . from Captain Munday," September 1831, MS 40858 (no folio), NLS.

316 NOTES TO PAGES 151–158

45. [John Malcolm], *Sketches of Persia*, 2 vols. (London: John Murray, 1827), 1:12. See app.: 100.

46. Withers, "On Enlightenment's Margins" (n. 29 above, this chap.).

47. Genette, *Paratexts*, 160.

48. Lander and Lander, *Journal of an*, 1:xiv. The original work is Richard Millikin, *The River-Side: A Poem, in Three Books* (Cork: Printed by J. Conner, 1807).

49. Millikin, *River-Side*, 104.

50. Charles W. J. Withers, "Voyages et Credibilité: Vers une Géographie de la Confiance," *Géographie et Cultures*, vol. 33 (2000); Miles Bredin, *The Pale Abyssinian: A Life of James Bruce, African Explorer and Adventurer* (London: Harper Collins, 2000).

51. Charles Dickens, *The Speeches of Charles Dickens*, ed. K. J. Fielding (Oxford: Clarendon Press, 1960), 157.

52. Anthony Glinoer, "Collaboration and Solidarity: The Collective Strategies of the Romantic Cenacle," in *Models of Collaboration in Nineteenth-Century French Literature*, ed. Seth Whitten (Farnham: Ashgate, 2009), 40.

53. Bruno Latour, *Science in Action: How to Follow Scientists and Engineers through Society* (Cambridge, MA: Harvard University Press, 1987), 33.

54. John Martin, *An Account of the Natives of the Tonga Islands, in the South Pacific Ocean*, 2 vols. (London: John Murray, 1817), 1:iii. See app.: 39.

55. James Raymond Wellsted, *Travels in Arabia*, 2 vols. (London: John Murray, 1838), 1:vii. See app.: 139.

56. Michael Twyman, "The Illustration Revolution," in *The Cambridge History of the Book in Britain*, vol. 6, *1830–1914*, ed. David McKitterick (Cambridge: Cambridge University Press, 2009), 120.

57. Ibid., 127.

58. Bernard Smith, *Imagining the Pacific: In the Wake of the Cook Voyages* (New Haven, CT: Yale University Press, 1992), 59.

59. Robert G. David, *The Arctic in the British Imagination, 1818–1914* (Manchester: Manchester University Press, 2000), 11.

60. William Gilpin, *Three Essays: On Picturesque Beauty; On Picturesque Travel; and On Sketching Landscape* (London: R. Blamire, 1792), ii. On this point, see also John Bonehill, "'New Scenes Drawn by the Pencil of Truth': Joseph Banks' Northern Voyage," *Journal of Historical Geography*, vol. 43 (2014).

61. Claudio Greppi, "'On the Spot': Traveling Artists and the Iconographic Inventory of the World, 1769–1859," in *Tropical Visions in an Age of Empire*, ed. Felix Driver and Luciana Martens (Chicago: University of Chicago Press, 2005). On representation of the natural and human worlds in this period, see Charlotte Klonk, "Science, Art, and the Representation of the Natural World," in *The Cambridge History of Science*, vol. 4, *Eighteenth-Century Science*, ed. Roy Porter (Cambridge: Cambridge University Press, 2003); Bernard Smith, *European Vision and the South Pacific*, 2nd ed. (New Haven, CT: Yale University Press, 1985); Barbara Maria Stafford, *Voyage into Substance: Science, Nature, and the Illustrated Travel Account, 1760–1840* (Cambridge, MA: MIT Press, 1984).

62. Joseph Hooker to John Murray [III], [1854], MS 40573 (no folio), NLS.

63. Joseph Hooker, *Himalayan Journals*, 2 vols. (London: John Murray, 1854), 1 xvii–xviii. See app.: 220.

64. David Arnold, "Envisioning the Tropics: Joseph Hooker in India and the Himalayas, 1848–1850," in *Tropical Visions in an Age of Empire*, ed. Felix Driver and Luciana Martens (Chicago: University of Chicago Press, 2005), 154.

65. Copy Ledger E, fol. 229, MS 42730, NLS. Whether this was because of the illustrative material is not known. For Arnold, the book failed "to combine travelogue and scientific research in a single piece of writing," but there is no sense that the illustrations were at fault in this respect, and it is not clear why the 1855 edition should not sell when its predecessor had (David Arnold,

The Tropics and the Travelling Gaze: India, Landscape, and Science, 1800-1856 [Seattle: University of Washington Press, 2006], 41).

66. David Livingstone to John Murray [III], May 22, 1857, MS 42420 (no folio), NLS.

67. Louise Henderson, "John Murray and the Publication of David Livingstone's Missionary Travels," *Livingstone Online*, accessed January 6, 2012, http://www.livingstoneonline.ucl.ac.uk /companion.php?id=HIST2.

68. Archibald Edmonstone, *A Journey to Two of the Oases of Upper Egypt* (London: John Murray, 1822), xiii. See app.: 70.

69. Dixon Denham, Hugh Clapperton, and Walter Oudney, *Narrative of Travels and Discoveries in Northern and Central Africa*, 2 vols. (London: John Murray, 1826), 1:viii. See app.: 93.

70. John Barrow to John Murray [II], December 25, 1822, MS 40056, fol. 13, NLS.

71. John Barrow to John Murray [II], May 2, 1823, MS 40056, fol. 28v, NLS.

72. Denham, Clapperton, and Oudney, *Narrative of Travels*, lxiii-lxiv.

73. Woodbine Parish to John Murray [II?], July 10, 1838, MS 40908 (no folio), NLS.

74. Copy Ledger D, fol. 251, MS 42729, NLS.

75. George A. Barrow to John Murray [III], n.d., MS 40054 (no folio), NLS.

76. George Barrow, *Ceylon: Past and Present* (London: John Murray, 1857), iii. See app.: 232. Notwithstanding the quality of the map, Barrow's *Ceylon* incurred a loss for Murray (but not for Barrow who had agreed on the sale of his copyright rather than a share of profits and so benefited to the sum of over £50; Copy Ledger E, fol. 336, MS. 42730, NLS).

77. John Barrow to John Murray [II], June 26, 1834, MS 40057, fol. 108, NLS.

78. Thomas Stamford Raffles, *The History of Java*, 2 vols. (London: John Murray, 1817), 1:7. See app.: 41.

79. Basil Hall, *Account of a Voyage of Discovery to the West Cost of Corea* (London: John Murray, 1818), viii. See app.: 44. From sales of the print run of 1,750 copies, Hall, Murray, and Henry Clifford, who compiled the linguistic appendix in Hall's *Corea* ("Vocabulary of the Loo-Choo Language"), had each earned £135 7s. 10d. by the end of 1819 (Copy Ledger B, fol. 21, MS 42725, NLS).

80. On this point, see the essays in Mario Biagioli and Peter Galison, eds., *Scientific Authorship: Credit and Intellectual Property in Science* (New York: Routledge, 2003).

81. In 1815 Maurice had written to Murray II: "The attempt to sell my own books set the trade against me" (May 11, 1815, MS 40797 [no folio], NLS). In 1818, Maurice returned to the question of finances, asking the publisher to forgive his previous behavior and "make all due allowances for an author compelled by the successive bankruptcy of three booksellers . . . to become occasionally the vendor of his own wares" (Maurice to John Murray [II], January 15, 1818, MS 40797, NLS). Two other exceptions are to be found in Frederick Henniker's *Notes, during a Visit to Egypt* (1823; app.: 76), which offered readers a handsome illustrated title page and three other substantial illustrations, and Cooper's *Views in Egypt and Nubia* (app.: 81), a series that appeared in eight parts between 1824 and 1827, including lithographs taken from illustrations by Bossi.

82. For further the details surrounding Layard's dealings with Murray, see Frederick N. Bohrer, "The Printed Orient: The Production of A. H. Layard's Earliest Works," *Culture and History*, vol. 11 (1992).

83. John Murray [III] to Austen Henry Layard, March 21, 1848, MS 41912, fol. 62, NLS. The reference to the "work like that of Wilkinson" is to John Gardner Wilkinson's *Modern Egypt and Thebes*, which Murray had published in 1843 (app.: 165).

84. John Murray [III] to Austen Henry Layard, n.d., MS 42343 (no folio), NLS.

85. Shawn Malley, *From Archaeology to Spectacle in Victorian Britain: The Case of Assyria, 1845-1854* (Farnham: Ashgate, 2012), 49.

86. Austen Henry Layard, *Nineveh and Its Remains*, 2 vols. (London: John Murray, 1849), 1:x. See app.: 196.

87. John Gardiner Kinnear to John Murray [III], September 14, 1840, MS 10654 (no folio),

NLS: "As to the title of the book, I believe an attractive title is sometimes of use—How would this do? 'Egypt, Arabia Petrea and Syria, with some remarks, on the Government of Ahmet Ali, and the present prospects of these Countries.' Or 'Cairo, Petra, and Damascus in 1839, with some remarks on the Govt of Mehmet Ali, and the present prospects of Syria.'"

88. John Murray [III] to Austen Henry Layard, n.d., MS 42343 (no folio), NLS.

89. Imperial-sized paper measured 30 by 22 inches; colombier-sized paper measured 34.5 by 23.5 inches, see Silvie Turner, *The Book of Fine Paper* (New York: Thames and Hudson, 1998).

90. Copy Ledger E, fols. 86, 91, 151, 183, 213, 227, 246, 296, 360, MS 42730, NLS.

91. Austen Henry Layard, *A Popular Account of Discoveries at Nineveh* (London: John Murray, 1851), iii. See app.: 196.

92. Anonymous review of *A Popular Account of Discoveries at Nineveh*, by Austen Henry Layard, *Eclectic Review*, n.s., 3 (May 1852): 634.

93. Austen Henry Layard, *Autobiography and Letters from His Childhood until His Appointment as H.M. Ambassador at Madrid*, 2 vols. (London: John Murray, 1903), 2:202.

94. "Mr. Layard's Second Expedition to Assyria," *Athenaeum*, no. 1325 (19 March 1853), 339.

95. Layard, *Nineveh and Its Remains*, v. See app.: 196.

96. Genette, *Paratexts* (n. 1 above, this chap.), 407.

CHAPTER SIX

1. Alan C. Dooley, *Author and Printer in Victorian England* (Charlottesville: University Press of Virginia, 1992), 8.

2. Karl Marx, *Grundrisse: Foundations of the Critique of Political Economy*, trans. Martin Nicolaus (Harmondsworth: Penguin, 1973), 110.

3. Leslie Howsam, *Cheap Bibles: Nineteenth-Century Publishing and the British and Foreign Bible Society* (Cambridge: Cambridge University Press, 1991); Aileen Fyfe, *Steam-Powered Knowledge: William Chambers and the Business of Publishing, 1820–1860* (Chicago: University of Chicago Press, 2012), and "Steam and the Landscape of Knowledge: W. & R. Chambers in the 1830s–1850s," in *Geographies of the Book*, ed. Miles Ogborn and Charles W. J. Withers (Farnham: Ashgate, 2010).

4. Robert Darnton, "What Is the History of Books?" *Daedalus* 111, no. 3 (1982): 67.

5. John Murray [III] to Robert Fortune, November 14, 1846, MS 41912 (no folio), National Library of Scotland (hereafter NLS).

6. Angus Fraser, "A Publishing House and Its Readers, 1841–1889: The Murrays and the Miltons," *Papers of the Bibliographical Society of America* 90, no. 1 (1996): 11.

7. Ibid., 17.

8. William Napier, *The Life and Opinions of General Sir Charles James Napier, G.C.B.*, 4 vols. (London: John Murray, 1857).

9. Fraser, "A Publishing House and Its Readers," 29.

10. K. D. Reynolds, "FitzClarence, George Augustus Frederick, First Earl of Munster (1794–1842)," in *Oxford Dictionary of National Biography* (Oxford: Oxford University Press, 2004–13), doi:10.1093/ref:odnb/9542.

11. George FitzClarence, *Journal of a Route Across India, to England* (London: John Murray 1819), v. See app.: 51.

12. George FitzClarence to John Murray [II], November 8, 1818, MS 40404 (no folio), NLS.

13. FitzClarence, *Journal of a Route Across India*, vii.

14. Copy Ledger B, fol. 156, MS 42725, NLS.

15. Walter Hamilton to John Murray [II], November 25, 1816, MS 40516 (no folio), NLS.

16. Ibid.

17. Walter Hamilton, *The East India Gazetteer* (London: John Murray, 1815), 142, and *A Geo-*

graphical, Statistical, and Historical Description of Hindostan, 2 vols. (London: John Murray, 1820), 1:222. See app.: 32 and 60, respectively.

18. David Livingstone to John Murray [III], May 31, 1857, MS 42420 (no folio), NLS.

19. David Livingstone to John Murray [III], May 30, 1857, MS 42420 (no folio), NLS.

20. For a fuller account of the relationships between David Livingstone, *Missionary Travels*, and the publishing industry, not just his connections with the house of Murray, see Justin Livingstone, *"Missionary Travels*, Missionary Travails: David Livingstone and the Victorian Publishing Industry," in *David Livingstone: Man, Myth and Legacy*, ed. Sarah Worden (Edinburgh: National Museums Scotland, 2012).

21. John Richardson, *Arctic Searching Expedition: A Journal of a Boat-Voyage through Rupert's Land and the Arctic Sea, in Search of the Discovery Ships under Command of Sir John Franklin*, 2 vols. (London: Longman, Brown, Green, & Longmans, 1851).

22. John Richardson to John Murray [III], December 13, 1850, MS 41017 (no folio), NLS.

23. There is, unfortunately, nothing in either the John Murray Archive materials or in the archives of the Longman publishing house to suggest why Richardson published with Longman rather than with Murray, although the sense that Murray may have overplayed his hand in suggesting these substantial changes is probably at the heart of his decision.

24. Joseph John Gurney to John Murray [III?], January 17, 1841, MS 40499 (no folio), NLS.

25. Mary Margaret Busk to John Murray [III?], April 1838, MS 40017 (no folio), NLS.

26. By number of texts and percentage of the total travel imprint with which we are concerned (app.), the printing firms used by Murray were: Clowes, ninety-two texts (39 percent); Spottiswoode, twenty-five (11 percent); Thomas Davison, eighteen (8 percent); C. Roworth, sixteen (7 percent); Woodfall and Kinder, sixteen (7 percent); Bradbury and Evans, twelve (5 percent); Bulmer, seven (3 percent); John and Richard Taylor, four (2 percent); Stewart and Murray, three (1 percent); Cox and Baylis, two (1 percent); Maurice, two (1 percent); William Nicol, two (1 percent); W. Pople, two (1 percent); and A. Strahan, two (1 percent). A further twenty different printers each printed one text each.

27. [Frances Bond Head], "The Printer's Devil," *Quarterly Review* 65, no. 129 (December 1839): 8.

28. Ibid., 9-10.

29. Ibid., 15.

30. For further details on these stages of proof production, see Allan C. Dooley, *Author and Printer in Victorian England* (Charlottesville: University Press of Virginia, 1992).

31. James Hamilton to John Murray [III], October 25, 1857, MS 40435 (no folio), NLS.

32. Howard Robinson, *The British Post Office: A History* (Princeton, NJ: Princeton University Press, 1948).

33. Barron Field to John Murray [II], December 13, 1821, MS 40401 (no folio), NLS.

34. John Murray [III] to Edward Robinson, October 12, 1852, MS 41912 (no folio), NLS.

35. Edward Strutt Abdy to John Murray [II], [1836], MS 40002 (no folio), NLS.

36. Phillip King to John Murray [II], [February 1825], MS 40650 (no folio), NLS.

37. Phillip King to John Murray [II], April 1, 1826, MS 40650 (no folio), NLS.

38. Phillip King to John Murray [II], May 26, 1826, MS 40650 (no folio), NLS.

39. William Parry to John Murray [II], 9 February 1826, MS 40921 (no folio), NLS.

40. William Parry to John Murray [II], 7 March 1826, MS 40921 (no folio), NLS.

41. William Parry to John Murray [II], 14 August 1826, MS 40921 (no folio), NLS.

42. "Stereotyping is a process whereby, from a mould taken of the original typset page, one or more duplicate plates is cast"; see D. C. Greetham, *Textual Scholarship: An Introduction* (New York: Garland Publishing, 1994).

43. Edward Robinson to John Murray [III?], 20 October 1840, MS 41029 (no folio), NLS.

44. Advertisement in *Quarterly Literary Advertiser* (June 1841), 2.

45. James Ross to John Murray [III], June 29, 1847, MS 41038 (no folio), NLS.

46. Frederick Henniker to John Murray [II], [July 1822], MS 40546 (no folio), NLS.

47. Alexander Burnes to John Murray [II], May 28, 1834, MS 42048 (no folio), NLS.

48. Copy Ledger E, fols. 186, 273, MS 42730, NLS.

49. John Murray [III] to Austen Henry Layard, May 3, 1849, MS 42343 (no folio), NLS.

50. Gordon Waterfield, *Layard of Nineveh* (London: John Murray, 1963).

51. [Sara Austen], review of *Nineveh and Its Remains*, by Austen Henry Layard, *Times*, February 9, 1849, 5.

52. Shawn Malley, *From Archaeology to Spectacle in Victorian Britain: The Case of Assyria, 1845-1854* (Farnham: Ashgate, 2012), 47.

53. John Gardiner Kinnear, *Cairo, Petra, and Damascus, in 1839* (London: John Murray, 1841), vii. See app.: 151.

54. William Hamilton to John Murray [III], November 21, 1839, MS 40517 (no folio), NLS.

55. William Hamilton, *Researches in Asia Minor, Pontus, and Armenia*, 2 vols. (London: John Murray, 1842), 1:viii. See app.: 157.

56. Richard Lander and John Lander, *Journal of an Expedition to Explore the Course and Termination of the Niger*, 3 vols. (London: John Murray, 1832), 1:viii. See app.: 124.

57. Ibid., 1:viii, ix.

58. [Sarah Lushington], *Narrative of a Journey from Calcutta to Europe* (London: John Murray, 1829), x-xi. See app.: 113.

59. Charles Lushington, *The History, Design, and Present State of the Religious, Benevolent and Charitable Institutions, Founded by the British in Calcutta and Its Vicinity* (Calcutta: Hindostanee Press, 1824).

60. Frederick Henniker, *Notes, during a Visit to Egypt, Nubia, the Oasis, Mount Sinai, and Jerusalem* (London: John Murray, 1823), v. See app.: 76.

61. [Richard H. Horne], *Exposition of the False Medium and Barriers Excluding Men of Genius from the Public* (London: Effingham Wilson, 1833), 244-45.

62. William St. Clair, *The Reading Nation in the Romantic Period* (Cambridge: Cambridge University Press, 2004), 32.

63. Ibid.

64. Ibid.

65. Richard Altick, *The English Common Reader: A Social History of the Mass Reading Public, 1800-1900* (Chicago: University of Chicago Press, 1957), 274.

66. Fyfe, *Steam-Powered Knowledge* (n. 3 above, this chap.), 1.

67. Pierre Bourdieu, *The Field of Cultural Production: Essays on Art and Literature*, ed. Randal Johnson (Cambridge: Polity Press 1993).

68. Patrick Brantlinger, *The Reading Lesson: The Threat of Mass Literacy in Nineteenth-Century British Fiction* (Bloomington: Indiana University Press, 1998), 12.

69. Andrew King and John Plunkett, eds., *Victorian Print Media: A Reader* (Oxford: Oxford University Press, 2005).

70. James J. Barnes and Patience P. Barnes, "Reassessing the Reputation of Thomas Tegg, London Publisher, 1776-1846," *Book History*, vol. 3 (2000).

71. Walter Scott, *The Life of Napoleon Buonaparte, Emperor of the French*, 9 vols. (Edinburgh: Cadell; London: Longman, Rees, Orme, Brown, & Green, 1827).

72. John G. Lockhart, *Memoirs of the Life of Sir Walter Scott, Bart.*, 7 vols. (Edinburgh: Robert Cadell; London: John Murray / Whittaker, 1838), 7:154.

73. Advertisement in *Quarterly Literary Advertiser* (April 1829), 1.

74. Anonymous review of *Memoirs of the Life of Napoleon Buonaparte*, by John G. Lockhart, *London Literary Gazette*, no. 636 (28 March 1829), 203.

75. "Noctes Ambrosianae," *Blackwood's Edinburgh Magazine*, vol. 25, no. 153 (June 1829): 798.

76. "The Family Library, Vols. I. to IX.," *Monthly Review*, vol. 13, no. 54 (February 1830): 252.

77. Advertisement in *Quarterly Literary Advertiser* (July 1829), 13.

78. Anonymous review of *Memoirs of the Life of Napoleon Buonaparte*, by John G. Lockhart," *Quarterly Review* 39, no. 78 (April 1829): 495.

79. John Feather, *A History of British Publishing* (London: Croom Helm, 1988), 143.

80. Scott Bennett, "John Murray's Family Library and the Cheapening of Books in Early Nineteenth Century Britain," *Studies in Bibliography*, vol. 29 (1976).

81. Gelien Matthews, *Caribbean Slave Revolts and the British Abolitionist Movement* (Baton Rouge: Louisiana State University Press, 2006).

82. Feather, *History of British Publishing*, 143.

83. Fyfe, "Steam and the Landscape of Knowledge" (n. 3 above, this chap.), 71–72.

84. Bennett, "John Murray's Family Library."

85. *Mr. Murray's List of New Works* (London: John Murray, November 1861), 11.

86. John Murray [III] to Charles Darwin, April 4, 1845, MS 41911 fols. 207–8, NLS.

87. Frederick Burkhardt and Sydney Smith, eds., *The Correspondence of Charles Darwin*, vol. 3, *1844–1846* (Cambridge: Cambridge University Press, 1987), 191.

88. John Tallmadge, "From Chronicle to Quest: The Shaping of Darwin's 'Voyage of the Beagle,'" *Victorian Studies*, vol. 23, no. 3 (1980). In the preface to the 1845 Murray-published edition of his *Journal of Researches*, Darwin drew his readers' attention to the abridgment in content in order to find a place in the Home and Colonial series: "This volume contains, in the form of a Journal, a history of our voyage, and a sketch of those observations in Natural History and Geology, which I think will possess some interest for the general reader. I have in this edition largely condensed and corrected some parts, and have added a little to others, in order to render the volume more fitted for popular reading; but I trust that naturalists will remember, that they must refer for details to the larger publications, which comprise the scientific details of the Expedition" (Charles Darwin, *Journal of Researches into the Natural History and Geology of the Countries Visited during the Voyage of H. M. S. Beagle* [London: John Murray, 1845], v; see app.: 174).

89. Richard Ruland, "Melville and the Fortunate Fall: Typee as Eden," *Nineteenth-Century Fiction*, vol. 23, no. 3 (1968).

90. "Prospectus and Specimen of Mr. Murray's Colonial and Home Library," *Quarterly Literary Advertiser* (August 1843), unpaginated.

91. Fraser, "John Murray's Colonial and Home Library," 357.

92. [Head], "Printer's Devil," 30.

93. "Murray's Colonial and Home Library," *Examiner* (28 October 1843), 677.

94. "Prospectus and Specimen of Mr. Murray's Colonial and Home Library," *Quarterly Literary Advertiser* (August 1843), unpaginated.

95. Priya Joshi, "Trading Places: The Novel, the Colonial Library, and India," in *Print Areas: Book History in India*, ed. Abhijit Gupta and Swapan Chakravarty (Delhi: Permanent Black, 2004), 28.

96. John Barrow, *The Life, Voyages, and Exploits of Admiral Sir Francis Drake, Knt.* (London: John Murray, 1843).

97. Altick, *English Common Reader* (n. 64 above, this chap.), 297.

98. These figures are taken from Fraser, "John Murray's Colonial and Home Library."

99. Joshi, "Trading Places," 27.

100. John Barnes, Bill Bell, Rimi B. Chatterjee, Wallace Kirsop, and Michael Winship, "A Place in the World," in *The Cambridge History of the Book in Britain*, vol. 6, *1830–1914*, ed. David McKitterick (Cambridge: Cambridge University Press, 2009), 600.

101. *Mr. Murray's List of New Works Now Ready* (London: John Murray, January 1852), 24.

102. Anonymous, "Summary," *Literary Gazette*, no. 1878 (January 15, 1853), 64.

103. William Zachs, Peter Isaac, Angus Fraser, and William Lister, "Murray Family," in *Ox-*

ford Dictionary of National Biography (Oxford: Oxford University Press, 2004–13), doi:10.1093/ref:odnb/64922.

104. Charles W. J. Withers, *Geography and Science in Britain, 1831–1939: A Study of the British Association for the Advancement of Science* (Manchester: Manchester University Press, 2010), 90, 147–48, 152, 158.

105. Martyn Lyons, *A History of Reading and Writing in the Western World* (Basingstoke: Palgrave Macmillan, 2010), 137–52.

CHAPTER SEVEN

1. William Allen, *Picturesque Views on the River Niger, Sketched during Lander's Last Visit in 1832–33* (London: John Murray / Hodgson & Graves / Ackerman, 1840), 12. See app.: 146.

2. Richard B. Sher, *The Enlightenment and the Book: Scottish Authors and Their Publishers in Eighteenth-Century Britain, Ireland, and America* (Chicago: University of Chicago Press, 2006), xv.

3. Richard D. Altick, *The English Common Reader: A Social History of the Mass Reading Public, 1800–1900*, 2nd ed. (Columbus: Ohio State University Press, 1998); Christopher A. Bayly, *Imperial Meridian: The British Empire and the World, 1780–1830* (London: Longman, 1989).

4. Anonymous review of *A Pilgrimage in Europe and America*, by J. C. Beltrami, *Quarterly Review* 37, no. 74 (March 1828): 448.

5. William H. Smyth and Frederick Lowe, *Narrative of a Journey from Lima to Para, across the Andes and down the Amazon* (London: John Murray), 94–95. See app.: 134.

6. Archibald Edmonstone, *A Journey to Two of the Oases of Upper Egypt* (London: John Murray, 1822), xiii. See app.: 70.

7. Because use of the camel in Arabian and African travel was so common (see plate 11), checking the camel's speed and ensuring a measure of correspondence between different types of camel in different parts of the world was a problem for assessing travel time in terms of "camel days travel," e.g, and not linear distances. On this, see James Rennell, "On the Rate of Travelling, as Performed by Camels; and Its Applications, as a Scale, to the Purposes of Geography," *Philosophical Transactions of the Royal Society of London*, vol. 81 (1791).

8. John Murray [I] to Mr Kinglake, 22 November 1805, MS 41098 (no folio), National Library of Scotland (hereafter NLS), Edinburgh.

9. George Francis Lyon to John Murray [II], [April 10, 1827], MS 40731 (no folio), NLS.

10. F. S. Newbold to John Murray, July 17, 1839, MS 40872 (no folio), NLS.

11. Francis Bond Head to John Murray [III], [December 1826], MSS 42278–79 fols. 466A–C, NLS.

12. John Murray [II] to J. W. Croker, [1821], MS 41908, fol. 207v, NLS. The work in question to which Murray was objecting was [Alexander Fisher], *Journal of a Voyage of Discovery, to the Arctic Regions, Performed betweem the 4th of April and the 18th of November, 1818, in His Majesty's Ship Alexander* (London: Richard Philips, 1819).

13. James Eli Adams, *A History of Victorian Literature* (Chichester: Wiley-Blackwell, 2012).

14. William St. Clair, *The Reading Nation in the Romantic Period* (Cambridge: Cambridge University Press, 2004), 42.

15. M. Dumont, *A Treatise on Juridical Evidence, Extracted from the Manuscripts of Jeremy Bentham, Esq.* (London: Baldwin, Cradock, & Joy, 1825); Harriet Martineau, *How to Observe: Morals and Manners* (London: Charles Knight, 1838).

16. Anonymous review of *Narrative of a Year's Journey through Central and Eastern Arabia*, by William G. Palgrave, *Westminster Review* 84, no. 165 (October 1865): 179.

17. Anonymous review of *Narrative of Travels and Discoveries in Northern and Central Africa*, by Denham Dixon, Hugh Clapperton, and Walter Oudney, *Edinburgh Review* 44, no. 87 (June 1826): 174.

18. Humphry Sandwith to John Murray [III], December 29, 1859, MS 41062 (no folio), NLS. The "last" book to which Sandwith makes reference is his 1856 *A Narrative of the Siege of Kars*, which went into several editions (app.: 229). The book to which Sandwith is here referring was published in two volumes as *The Hekim Bashi; or, The Adventures of Giuseppe Antonelli, a Doctor in the Turkish Service* (London: Smith, Elder, 1864) and is usually seen as a work of medical fiction.

19. Peter Hulme and Tim Youngs, "Introduction," in *The Cambridge Companion to Travel Writing*, ed. Peter Hulme and Tim Youngs (Cambridge: Cambridge University Press, 2002), 10.

20. Mary Baine Campbell, "Travel Writing and Its Theory," in *Cambridge Companion to Travel Writing*, ed. Hulme and Youngs, 269.

21. Robert Darnton, "What Is the History of Books?" *Daedalus*, vol. 111, no. 3 (1982); Leslie Howsam, *Old Books and New Histories: An Orientation to Studies in Book and Print Culture* (Toronto: University of Toronto Press, 2006).

22. We take these words from Ian S. MacLaren, "In Consideration of the Evolution of Explorers and Travellers into Authors: A Model," *Studies in Travel Writing* 15, no. 3 (2011): 225, who is citing William H. Sherman, "Stirrings and Searchings (1500-1720)," in *Cambridge Companion to Travel Writing*, ed. Hulme and Youngs, 36.

23. Leah Price, *How to Do Things with Books in Victorian Britain* (Princeton, NJ: Princeton University Press, 2012), 37.

24. Campbell, "Travel Writing and Its Theory," 263.

25. MacLaren, "In Consideration of the Evolution of Explorers and Travellers into Authors," 221 and 223, respectively.

26. David Livingstone, *Missionary Travels and Researches in South Africa* (London: John Murray, 1857), 8. See app.: 235.

27. A point made by Campbell, "Travel Writing and Its Theory," 269; see also Thomas Richards, *The Imperial Archive: Knowledge and the Fantasy of Empire* (London: Verso, 1993); Peter Burke, *A Social History of Knowledge: From Gutenberg to Diderot* (Cambridge: Polity Press, 2000).

28. Carolyn Steedman, *Dust* (Manchester: Manchester University Press, 2001); Joan M. Schwartz and Terry Cook, "Archives, Records, and Power: The Making of Modern Memory," *Archival Science*, vol. 2, nos. 1-2 (2002).

29. The term "communicative action" is from James A. Secord, "Knowledge in Transit," *Isis*, vol. 95, no. 4 (2004).

30. Adrian Johns, *The Nature of the Book: Print and Knowledge in the Making* (Chicago: University of Chicago Press, 1998), 623.

Bibliography

NOTE ON THE BIBLIOGRAPHY

References made to the manuscript items examined in the John Murray Archive of the National Library of Scotland have been given as endnotes with respect to each chapter and so the class-marks for individual manuscripts are not here listed again. The full bibliographical details of the John Murray published books that have been the focus of our study appear in the appendix and so these individual works do not appear here, as might more usually be the case. The bibliographical reference value of the appendix for our primary Murray works is enhanced by the inclusion there of cost, format, and other publication information, including the edition history, of the house of Murray's non-European books of travel and exploration, 1773–1859.

PRIMARY SOURCES

British Library

MS Eur D153
MS Eur D1165/13
MS Eur B256
MS Eur F208/12

Royal Geographical Society (with the Institute of British Geographers)

MS AP 8
CB Correspondence Block Papers
RGS Committee Minute Books, 1841–1865

Senate House Library, University of London

AL308 Letter from John Taylor to Alexander Dalrymple, 17 February 1795

Scott Polar Research Institute

MS 655/1 Memoirs of Sir John Ross

Somerset Archives and Record Office

MS DD/H1/D/553 Bayntun MSS

University of Reading Special Collections

MS 1391 1/98/35 Longman to Constable and Co.

SECONDARY SOURCES

Adams, Charles Hansford, ed. *The Narrative of Robert Adams, a Barbary Captive: A Critical Edition*. New York: Cambridge University Press, 2005.

Adams, James Eli. *A History of Victorian Literature*. Chichester: Wiley-Blackwell, 2012.

Adams, Percy G. *Travelers and Travel Liars, 1660–1800*. Berkeley: University of California Press, 1962.

——. *Travel Literature and the Evolution of the Novel*. Lexington: University Press of Kentucky, 1983.

Adams, Thomas R., and Nicholas Barker. "A New Model for the History of the Book." In *A Potencie of Life: Books in Society*, edited by Nicholas Barker, 5–43. London: British Library, 1993.

Advertisement for Biblical Researches in Palestine, by Edward Robinson. *Quarterly Literary Advertiser* (June 1841), 1–2.

Advertisement for The Family Library. Quarterly Literary Advertiser (April 1829), 1–3.

——. *Quarterly Literary Advertiser* (July 1829), 12–16.

Akel, Regina. *Maria Graham: A Literary Biography*. Amherst, NY: Cambria Press, 2009.

Allison, Robert J. *The Crescent Obscured: The United States and the Muslim World, 1776–1815*. New York: Oxford University Press, 1995.

Altick, Richard D. *The English Common Reader: A Social History of the Mass Reading Public, 1800–1900*. Chicago: University of Chicago Press, 1957.

——. *The English Common Reader: A Social History of the Mass Reading Public, 1800–1900*. 2nd ed. Columbus: Ohio State University Press, 1998.

Anonymous. Review of *An Authentic Narrative of the Loss of the American Brig* Commerce, by James Riley. *Quarterly Review* 16, no. 32 (January 1817): 287–321.

——. Review of *Journal d'un Voyage a Temboctou et a Jenné*, by René Caillié. *Quarterly Review* 42, no. 84 (March 1830): 450–75.

——. Review of *Journal of an Expedition to Explore the Course and Termination of the Niger*, by Richard Lander and John Lander. *Original*, no. 12 (May 19, 1832), 185–86.

——. Review of *Journal of a Residence in India*, by Maria Graham. *Eclectic Review* 10 (December 1813): 569–80.

——. Review of *Journal of a Second Expedition into the Interior of Africa*, by Hugh Clapperton and Richard Lander. *British Critic* 6, no. 11 (July 1829): 46–81.

——. Review of *Journal of a Second Expedition into the Interior of Africa*, by Hugh Clapperton and Richard Lander. *Eclectic Review*, n.s., 1 (February 1829): 161–78.

——. Review of *Journal of a Second Expedition into the Interior of Africa*, by Hugh Clapperton and Richard Lander. *Oriental Herald* 20, no. 63 (March 1829): 465–72.

———. Review of *Journal of a Tour in Asia Minor,* by William Martin Leake. *Oriental Herald* 2, no. 7 (July 1824), 429–30.

———. Review of *Journal of a Voyage to Brazil,* by Maria Graham. *London Literary Gazette,* no. 377 (April 10, 1824), 227–28.

———. Review of *The Journal of the Royal Geographical Society of London for the Year 1830–31. Quarterly Review* 46, no. 91 (November 1831): 55–80.

———. Review of *Loss of the American Brig* Commerce, by James Riley. *British Critic* 7 (June 1817): 616–33.

———. Review of *Loss of the American Brig* Commerce, by James Riley. *Eclectic Review,* n.s., 11 (January 1819): 64–84.

———. Review of *Loss of the American Brig* Commerce, by James Riley. *Monthly Review* 84 (October 1817): 127–39.

———. Review of *Memoirs of the Life of Napoleon Buonaparte,* by John G. Lockhart. *London Literary Gazette,* no. 636 (March 28, 1829), 203.

———. Review of *Memoirs of the Life of Napoleon Buonaparte,* by John G. Lockhart. *Quarterly Review* 39, no. 78 (April 1829): 475–521.

———. Review of *Narrative of a Journey to the Shores of the Polar Sea,* by John Franklin. *London Literary Gazette,* no. 625 (January 10, 1829), 24.

———. Review of *Narrative of an Expedition,* by James Tuckey. *British Critic* 10 (August 1818): 137–56.

———. Review of *Narrative of an Expedition,* by James Tuckey. *Monthly Review* 86 (June 1818): 113.

———. Review of *Narrative of a Year's Journey through Central and Eastern Arabia,* by William G. Palgrave. *Westminster Review* 84, no. 165 (October 1865): 178–88.

———. Review of *The Narrative of Robert Adams,* by [Robert Adams]. *Augustan Review* 3, no. 15 (July 1816): 58–69.

———. Review of *The Narrative of Robert Adams,* by [Robert Adams]. *British Lady's Magazine* 4, no. 19 (July 1816): 25–29.

———. Review of *The Narrative of Robert Adams,* by [Robert Adams]. *Edinburgh Review* 26, no. 52 (June 1816): 385–402.

———. Review of *The Narrative of Robert Adams,* by [Robert Adams]. *Monthly Review* 82 (January 1817): 26–38.

———. Review of *The Narrative of Robert Adams,* by [Robert Adams]. *North American Review* 5, no. 2 (July 1817): 204–24.

———. Review of *The Narrative of Robert Adams,* by [Robert Adams]. *Quarterly Review* 14, no. 14 (January 1816): 453–73.

———. Review of *A Narrative of the Loss of the* Grosvenor *East Indiaman,* by George Carter. *Monthly Review* 7 (April 1792): 470.

———. Review of *Narrative of Travels and Discoveries in Northern and Central Africa,* by Denham Dixon, Hugh Clapperton, and Walter Oudney. *Edinburgh Review* 44, no. 87 (June 1826): 173–219.

———. Review of *A Narrative of Travels in Northern Africa,* by George Francis Lyon. *British Review* 18, no. 35 (September 1821): 178–98.

———. Review of *A Narrative of Travels in Northern Africa,* by George Francis Lyon. *Literary Chronicle and Weekly Review,* no. 97 (March 24, 1821), 177–80.

———. Review of *A Narrative of Travels in Northern Africa,* by George Francis Lyon. *London Literary Gazette,* no. 218 (March 24, 1821), 179–81.

———. Review of *A Narrative of Travels in Northern Africa,* by George Francis Lyon. *Monthly Review* 96 (October 1821): 113–27.

———. Review of *A Narrative of Travels in Northern Africa,* by George Francis Lyon. *Quarterly Review* 25, no. 49 (April 1821): 25–50.

———. Review of *Pen and Pencil Sketches*, by Godfrey Mundy. *Edinburgh Review* 57, no. 116 (July 1832): 358–70.

———. Review of *Pen and Pencil Sketches*, by Godfrey Mundy. *Examiner* (May 27, 1832): 340.

———. Review of *Pen and Pencil Sketches*, by Godfrey Mundy. *Frazer's Magazine for Town and Country* 6, no. 32 (September 1832): 148–60.

———. Review of *A Pilgrimage in Europe and America*, by J. C. Beltrami. *Quarterly Review* 37, no. 74 (March 1828): 448–58.

———. Review of *A Popular Account of Discoveries at Nineveh*, by Austen Henry Layard. *Eclectic Review*, n.s., 3 (May 1852): 633–34.

———. Review of *Rough Notes Taken during Some Rapid Journeys*, by Francis Bond Head. *New Monthly Magazine* 21, no. 73 (January 1, 1827): 5–6.

———. Review of *Travels in Arabia*, by James Wellsted. *Athenaeum*, no. 533 (January 13, 1838): 29–32.

———. Review of *Travels in North America during the Years 1834, 1835, & 1836*, by Charles Augustus Murray, and *A Diary in America*, by Frederick Marryat. *Dublin Review* 7, no. 14 (November 1839): 399–429.

———. Review of *Travels Round the World*, by Pierre Marie François de Pagès. *Monthly Review* 5 (May 1791): 32–38.

Arnold, David. "Envisioning the Tropics: Joseph Hooker in India and the Himalayas, 1848–1850." In *Tropical Visions in an Age of Empire*, edited by Felix Driver and Luciana Martens, 137–55. Chicago: University of Chicago Press, 2005.

———. *The Tropics and the Travelling Gaze: India, Landscape, and Science, 1800–1856*. Seattle: University of Washington Press, 2006.

[Apperley, Charles]. *The Chase, the Turf, and the Road*. London: John Murray, 1837.

Ashcroft, Michael. "Robert Charles Dallas." *Jamaica Journal* 44 (1980): 94–101.

Austen, Jane. *Emma, a Novel*. 3 vols. London: John Murray, 1816.

[Austen, Sara.] Review of *Nineveh and Its Remains*, by Austen Henry Layard. *Times*, February 9, 1849, 5.

Bacon, Francis. *Of the Proficience and Advancement of Learning*. London: Printed for Henry Tomes, 1605.

Baepler, Paul. "The Barbary Captivity Narrative in American Culture." *Early American Literature* 39, no. 2 (2004): 217–46.

Barchas, Janine. *Graphic Design, Print Culture, and the Eighteenth-Century Novel*. Cambridge: Cambridge University Press, 2003.

Barnes, James J., and Patience P. Barnes. "Reassessing the Reputation of Thomas Tegg, London Publisher, 1776–1846." *Book History* 3 (2000): 45–60.

Barnes, John, Bill Bell, Rimi B. Chatterjee, Wallace Kirsop, and Michael Winship. "A Place in the World." In *The Cambridge History of the Book in Britain*. Vol. 6: *1830–1914*, edited by David McKitterick, 595–634. Cambridge: Cambridge University Press, 2009.

Barnett, Clive. "Impure and Worldly Geography: The Africanist Discourse of the Royal Geographical Society, 1831–73." *Transactions of the Institute of British Geographers* 23, no. 19 (1998): 239–51.

Barrow, John. *An Auto-Biographical Memoir*. London: John Murray, 1847.

———. *The Life, Voyages, and Exploits of Admiral Sir Francis Drake, Knt*. London: John Murray, 1843.

[———]. Review of *Narrative of a Voyage to Hudson's Bay*, by Edward Chappell. *Quarterly Review* 18, no. 35 (October 1817): 199–223.

[———]. Review of *Narrative of Travels and Discoveries in Northern and Central Africa*, by Denham Dixon, Hugh Clapperton, and Walter Oudney. *Quarterly Review* 33, no. 66 (March 1826): 518–49.

———. *Travels in China, Containing Descriptions, Observations, and Comparisons, Made and Collected in the Course of a Short Residence at the Imperial Palace of Yuen-Min-Yuen, and on a Subsequent Journey through the Country from Pekin to Canton.* London: T. Cadell & W. Davies, 1804.

Batten, Charles L. *Pleasurable Instruction: Form and Convention in Eighteenth-Century Travel Literature.* Berkeley: University of California Press, 1978.

Bayly, Christopher A. *Imperial Meridian: The British Empire and the World, 1780–1830.* London: Longman, 1989.

Beaufort, Francis. *Karamania; or, A Brief Description of the South Coast of Asia Minor and of the Remains of Antiquity.* London: R. Hunter, 1817.

Bedini, Silvio A. *With Compass and Chain: Early American Surveyors and Their Instruments.* Frederick, MD: Professional Surveyors Publishing, 2001.

Bell, Bill. "Bound for Australia: Shipboard Reading in the Nineteenth Century." *Journal of Australian Studies* 25, no. 68 (2001): 5–18.

———. "The Secret History of Smith and Elder: *The Publisher's Circular* as a Source for Publishing History." In *A Genius for Letters: Booksellers and Bookselling from the 16th to the 20th Century,* edited by Robin Myers and Michael Harris, 67–81. Winchester: St Paul's Bibliographies.

Bennett, Scott. "John Murray's Family Library and the Cheapening of Books in Early Nineteenth Century Britain." *Studies in Bibliography* 29: 140–67.

Berkeley, Edmund, and Dorothy Smith Berkeley. *George William Featherstonhaugh: The First U.S. Government Geologist.* Tuscaloosa: University of Alabama Press, 1988.

Biagioli, Mario, and Peter Galison, eds. *Scientific Authorship: Credit and Intellectual Property in Science.* New York: Routledge, 2003.

Bickers, Robert A., ed. *Ritual and Diplomacy: The Macartney Mission to China, 1792–1794.* London: British Association for Chinese Studies, 1993.

Bird, Dúnlaith. *Travelling in Different Skins: Gender Identity in European Women's Oriental Travelogues, 1850–1950.* Oxford: Oxford University Press, 2012.

Blackmore, Josiah. *Manifest Perdition: Shipwreck Narrative and the Disruption of Empire.* Minneapolis: University of Minnesota Press, 2002.

Bleichmar, Daniela. "The Geography of Observation: Distance and Visibility in Eighteenth-Century Botanical Travel." In *Histories of Scientific Observation,* edited by Lorraine Daston and Elizabeth Lunbeck, 373–95. Chicago: University of Chicago Press, 2011.

Blum, Hester. "Pirated Tars, Piratical Texts: Barbary Captivity and American Sea Narratives." *Early American Studies* 1, no. 2 (2003): 133–58.

Blunt, Alison. *Travel, Gender, and Imperialism: Mary Kingsley and West Africa.* New York: Guilford Press, 1994.

Bohrer, Frederick N. "The Printed Orient: The Production of A. H. Layard's Earliest Works." *Culture and History* 11 (1992): 85–105.

Bonehill, John. "'New Scenes Drawn by the Pencil of Truth': Joseph Banks' Northern Voyage." *Journal of Historical Geography* 45 (2014): 9–27.

Borm, Jan. "Defining Travel: On the Travel Book, Travel Writing and Terminology." In Hooper and Youngs, *Perspectives on Travel Writing,* 13–26.

Borrow, George. *The Bible in Spain; or, The Journeys, Adventures, and Imprisonments of an Englishman, in an Attempt to Circulate the Scriptures in the Peninsula.* London: John Murray, 1843.

Bourdieu, Pierre. *The Field of Cultural Production: Essays on Art and Literature.* Edited by Randal Johnson. Cambridge: Polity Press.

Bourguet, Marie-Noëlle. "The Explorer." In *Enlightenment Portraits,* edited by Michel Vovelle, 257–315. Chicago: University of Chicago Press, 1997.

———. "A Portable World: The Notebooks of European Travellers (Eighteenth to Nineteenth Centuries)." *Intellectual History Review* 20, no. 3 (2010): 377–400.

Brantlinger, Patrick. *The Reading Lesson: The Threat of Mass Literacy in Nineteenth-Century British Fiction.* Bloomington: Indiana University Press, 1998.

Bravo, Michael. "Ethnographic Navigation and the Geographical Gift." In *Geography and Enlightenment*, edited by David N. Livingstone and Charles W. J. Withers, 199–235. Chicago: University of Chicago Press, 1999.

Bredin, Miles. *The Pale Abyssinian: A Life of James Bruce, African Explorer and Adventurer*. London: Harper Collins, 2000.

Breen, Benjamin. "No Man Is an Island: Early Modern Globalization, Knowledge Networks, and George Psalmanazara's Formosa." *Journal of Early Modern History* 17, no. 4 (2013): 391–417.

Bridges, Roy C. "Exploration and Travel Outside Europe (1720–1914)." In Hulme and Youngs, *The Cambridge Companion to Travel Writing*, 53–69.

———. "Nineteenth-Century East African Travel Records, with an Appendix on 'Armchair Geographers' and Cartography." *Paideuma* 33 (1987): 117–35.

Briggs, Asa. *A History of Longmans and Their Books 1724–1990: Longevity in Publishing*. London: British Library; New Castle, DE: Oak Knoll Press, 2008.

Brown, Dona. "Travel Books." In *An Extensive Republic: Print, Culture, and Society in the New Nation, 1790–1840*, edited by Robert A. Gross and Mary Kelley, 449–58. Chapel Hill: University of North Carolina Press, 2010.

Burke, Peter. "The Philosopher as Traveller: Bernier's Orient." In *Voyages and Visions: Towards a Cultural History of Travel*, edited by Jaś Elsner and Joan-Pau Rubiés, 124–37. London: Reaktion, 1999.

———. *A Social History of Knowledge: From Gutenberg to Diderot*. Cambridge: Polity Press, 2000.

Burkhardt, Frederick, and Sydney Smith, eds. *The Correspondence of Charles Darwin*. Vol. 3: *1844–1856*. Cambridge: Cambridge University Press, 1987.

Burnett, D. Graham. *Masters of All They Surveyed: Exploration, Geography, and a British El Dorado*. Chicago: University of Chicago Press, 2000.

Byron, George. *Childe Harold's Pilgrimage: A Romaunt*. London: London: John Murray; Edinburgh: William Blackwood; Dublin: John Cumming, 1812.

Caie, Graham D. "Bringing the John Murray Archive to the National Library of Scotland." *Byron Journal* 34, no. 1 (2006): 49–55.

Campbell, Mary B. "Travel Writing and Its Theory." In Hulme and Youngs, *The Cambridge Companion to Travel Writing*, 261–78.

———. *The Witness and the Other World: Exotic European Travel Writing, 400–1600*. Ithaca, NY: Cornell University Press, 1988.

Cantor, Geoffrey, and Sally Shuttleworth, eds. *Science Serialized: Representation of the Sciences in Nineteenth-Century Periodicals*. Cambridge MA: MIT Press, 2004.

Carey, Daniel. "Travel, Geography, and the Problem of Belief: Locke as a Reader of Travel Literature." In *History and Nation*, edited by Julia Rudolph, 97–136. Lewisburg, PA: Bucknell University Press, 2006.

Carpenter, Humphrey. *The Seven Lives of John Murray: The Story of a Publishing Dynasty 1768–2002*. London: John Murray, 2008.

Carr, Denis J., ed. *Sydney Parkinson: Artist of Cook's Endeavour Voyage*. London: British Museum (Natural History); Honolulu: University of Hawaii Press, 1983.

Catlin, George. *Caitlin's Notes of Eight Years' Travels and Residence in Europe, with His North American Indian Collection*. 2 vols. London: Published by the author, 1848.

———. *Letters and Notes on the Manners, Customs, and Condition of the North American Indians*. 2 vols. London: Published by the author, 1841.

Cavell, Janice. "The Hidden Crime of Dr Richardson." *Polar Record* 43, no. 2 (2007): 155–64.

———. "Making Books for Mr. Murray: The Case of Edward Parry's Third Arctic Narrative." *Library* 14, no. 1 (2013): 45–69.

———. "Representing Akaitcho: European Vision and Revision in the Writing of John Franklin's *Narrative of a Journey to the Shores of the Polar Sea*. . . ." *Polar Record* 44, no. 1 (2003): 25–34.

———. *Tracing the Connected Narrative: Arctic Exploration in British Print Culture, 1818–1860*. Toronto: University of Toronto Press, 2008.

Chappell, Edward. *Narrative of a Voyage to Hudson's Bay in His Majesty's Ship* Rosamond. London: J. Mawman, 1817

Chartier, Roger. "Foucault's Chiasmus: Authorship between Science and Literature in the Seventeenth and Eighteenth Centuries." In *Scientific Authorship: Credit and Intellectual Property in Science*, edited by Mario Biagioli and Peter Galison, 13-32. New York: Routledge, 2003.

——. *The Order of Books: Readers, Authors, and Libraries in Europe between the Fourteenth and Eighteenth Centuries*. Translated by Lydia G. Cochrane. Cambridge: Polity Press, 1994.

Clarke, John J. *Oriental Enlightenment: The Encounter between Asian and Western Thought*. London: Routledge, 1997.

Clegg, Cyndia Susan. "History of the Book: An Undisciplined Discipline." *Renaissance Quarterly* 54, no. 1 (2001): 221-45.

Clifford, James, and George E. Marcus, ed. *Writing Culture: The Poetics and Politics of Ethnography*. Berkeley: University of California Press, 1986.

Cock, Randolph. "Scientific Servicemen in the Royal Navy and the Professionalisation of Science, 1816-55." In *Science and Beliefs: From Natural Philosophy to Natural Science, 1700-1900*, edited by David M. Knight and Matthew D. Eddy, 95-111. Aldershot: Ashgate, 2005.

Colbert, Benjamin. "Bibliography of British Travel Writing." *Cardiff Corvey* 13 (2004). http://www.cardiff.ac.uk/encap/journals/corvey/articles/cc13_n01.html.

Colley, Linda. *Captives: Britain, Empire and the World, 1600-1850*. London: Jonathan Cape, 2002.

Cook, Andrew. "Alexander Dalrymple and the Hydrographic Office." In *Pacific Empires: Essays in Honour of Glyndwr Williams*, edited by Alan Frost and Jane Samson, 53-68. Melbourne: Melbourne University Press, 1999.

Cooper, Victoria, and Dave Russell. "Publishing for Leisure." In *The Cambridge History of the Book in Britain*. Vol. 6: *1830-1914*, edited by David McKitterick, 475-99. Cambridge: Cambridge University Press, 2009.

Craciun, Adriana. "Oceanic Voyages, Maritime Books, and Eccentric Inscriptions." *Atlantic Studies* 10, no. 2 (2013): 170-96.

——. "What Is an Explorer?" *Eighteenth-Century Studies* 45, no. 1 (2011): 29-51.

——. "Writing the Disaster: Franklin and Frankenstein." *Nineteenth-Century Literature* 65, no. 4 (2011): 433-80.

Cranmer-Byng, J. L. "Lord Macartney's Embassy to Peking in 1793." *Journal of Oriental Studies* 4, nos. 1-2 (1857-58): 117-87.

Crary, Jonathan. *Techniques of the Observer: On Vision and Modernity in the Nineteenth Century*. Cambridge, MA: MIT Press, 1990.

Cross, Anthony. "Two English 'Lady Travellers' in Russia and the House of Murray." *Slavonica* 17, no 1. (2011): 1-14.

Cunningham, Andrew, and Nicholas Jardine, ed. *Romanticism and the Sciences*. Cambridge: Cambridge University Press, 1990.

Dampier, William. *Voyages and Descriptions*. 3 vols. (London: James Knapton, 1699).

Darnton, Robert. "What Is the History of Books?" *Daedalus* 111, no. 3 (1982): 65-83.

Daston, Lorraine. 1991. "The Ideal and the Reality of the Republic of Letters in the Enlightenment." *Science in Context* 4, no. 2 (1992): 367-86.

——. "Marvelous Facts and Miraculous Evidence in Early Modern Europe." *Critical Inquiry* 18, no. 1 (1991): 93-124.

——. "Taking Note(s)." *Isis* 95, no. 3 (2004): 443-48.

Daston, Lorraine, and Peter Galison. *Objectivity*. 2nd ed. New York: Zone Books, 2010.

David, Robert G. *The Arctic in the British Imagination 1818-1914*. Manchester: Manchester University Press, 2000.

Davidson, John. "Letter from Mr Davidson to the Secretary of the Geographical Society, Dated Wednoon, 22nd May, 1836." *Journal of the Royal Geographical Society of London* 6 (1836): 429-33.

——. *Notes Taken during Travels in Africa*. London: J. L. Cox, 1839.

Davidson, John, and William Willshire. "Extracts from the Correspondence of the Late Mr. Davidson, during His Residence in Morocco; with an Account of His Further Progress in the
Desert." *Journal of the Royal Geographical Society of London* 7 (1837): 144–72.

Davis, John Francis. *The Chinese: A General Description of the Empire of China and its Inhabitants.*
2 vols. London: Charles Knight, 1836.

———. *Chinese Moral Maxims, with a Free and Verbal Translation; Affording Examples of the Grammatical Structure of the Language.* London: John Murray; Macao: Honorable Company's Press, 1823.

Davis, Richard C., ed. *Sir John Franklin's Journals and Correspondence: The First Arctic Land
Expedition, 1819–1822.* Toronto: Champlain Society, 1995.

Dawson, Marc H. "The Many Minds of Sir Halford J. Mackinder: Dilemmas of Historical Editing."
History in Africa 14 (1987): 27–42.

Dickens, Charles. *The Speeches of Charles Dickens.* Edited by K. J. Fielding. Oxford: Clarendon
Press, 1960.

Dikotter, Frank. "China." In *The Cambridge History of Science.* Vol. 4: *Eighteenth-Century Science,*
edited by Roy Porter, 688–97. Cambridge: Cambridge University Press, 2003.

[D'Israeli, Isaac.] *Curiosities of Literature.* London: John Murray, 1791.

"Donations Presented to the Royal Society of Edinburgh." *Transactions of the Royal Society of Edinburgh* 4, no. 1 (1798): 38–39.

Dooley, Allan C. *Author and Printer in Victorian England.* Charlottesville: University Press of Virginia, 1992.

Dritsas, Lawrence. "Expeditionary Science: Conflicts of Method in Mid-Nineteenth-Century Geographical Discovery." In *Geographies of Nineteenth-Century Science,* edited by David N. Livingstone and Charles W. J. Withers, 255–75. Chicago: University of Chicago Press, 2011.

———. *Zambesi: David Livingstone and Expeditionary Science in Africa.* London: I. B. Tauris, 2010.

Driver, Felix. *Geography Militant: Cultures of Exploration and Empire.* Oxford: Blackwell, 2001.

———. "Scientific Exploration and the Construction of Geographical Knowledge: *Hints to Travellers.*" *Finisterra: Revista Portuguesa de Geografia* 33, no. 65 (1998): 21–30.

Driver, Felix, and Lowri Jones. *Hidden Histories of Exploration: Researching the RGS-IBG Collections.*
Egham: Royal Holloway, University of London, 2009.

Dumont, M. *A Treatise on Juridical Evidence, Extracted from the Manuscripts of Jeremy Bentham,
Esq.* London: Baldwin, Cradock, & Joy, 1825.

Duncan, James, and Derek Gregory, eds. *Writes of Passage: Reading Travel Writing.* London: Routledge, 1999.

Dunn, John. "The Concept of 'Trust' in the Politics of John Locke." In *Philosophy in History: Essays on the Historiography of Philosophy,* edited by Richard Rorty, Jerome B. Schneewind, and
Quentin Skinner, 279–302. Cambridge: Cambridge University Press, 1984.

Dunn, Kevin. *Pretexts of Authority: The Rhetoric of Authorship in the Renaissance Preface.* Stanford,
CA: Stanford University Press, 1994.

Edney, Matthew. "Cartography: Disciplinary History." In *Sciences of the Earth: An Encyclopedia of
Events, People, and Phenomena,* edited by Gregory A. Good, 81–85. New York: Garland, 1998.

———. *Mapping an Empire: The Geographical Construction of British India, 1765–1843.* Chicago: University of Chicago Press, 1998.

Eisenstein, Elizabeth. *The Printing Press as an Agent of Change: Communications and Cultural Transformations in Early-Modern Europe.* 2 vols. Cambridge: Cambridge University Press, 1979.

Eliot, Simon. *Some Patterns and Trends in British Publishing, 1800–1819.* London: Bibliographical
Society, 1994.

———. "Some Trends in British Book Production, 1800–1919." In *Literature in the Marketplace:
Nineteenth-Century British Publishing and Reading Practices,* edited by John O. Jordan and Robert L. Patten, 19–43. Cambridge: Cambridge University Press, 1995.

Elsner, Jaś, and Joan-Pau Rubiés, eds. *Voyages and Visions: Towards A Cultural History of Travel.*
London: Reaktion, 1999.

Fabian, Ann. *The Unvarnished Truth: Personal Narratives in Nineteenth-Century America*. Berkeley: University of California Press, 2000.

Fabian, Johannes. *Out of Our Minds: Reason and Madness in the Exploration of Central Africa*. Berkeley: University of California Press, 2000.

Fabricant, Carole. "Eighteenth-Century Travel Literature." In *The Cambridge History of English Literature, 1660–1780*, edited by John Richetti, 707–44. Cambridge: Cambridge University Press, 2005.

Falkner, Thomas. *A Description of Patagonia, and the Adjoining Parts of South America*. Hereford: T. Lewis, 1774.

"The Family Library, Vols. I. to IX." *Monthly Review* 13, no. 54 (February 1830): 252–58.

Feather, John. *A History of British Publishing*. London: Croom Helm, 1988.

Ferris, Ina. "Antiquarian Authorship: D'Israeli's Miscellany of Literary Curiosity and the Question of Secondary Genres." *Studies in Romanticism* 45, no. 4 (2006): 523–42.

Finkelstein, David. *The House of Blackwood: Author-Publisher Relations in the Victorian Era*. University Park: Pennsylvania State University Press, 2002.

———. "Unraveling Speke: The Unknown Revision of an African Exploration Classic." *History in Africa* 30 (2003): 117–32.

[Fisher, Alexander.] *Journal of a Voyage of Discovery, to the Arctic Regions, Performed between the 4th of April and the 18th of November, 1818, in His Majesty's Ship* Alexander. London: Richard Philips, 1819.

Fleming, Fergus. *Barrow's Boys: A Stirring Story of Daring, Fortitude and Outright Lunacy*. London: Granta, 1998.

Fontes da Costa, Palmira. "The Making of Extraordinary Facts: Authentication of Singularities of Nature at the Royal Society of London in the First Half of the Eighteenth Century." *Studies in History and Philosophy of Science* 33, no. 2 (2002): 265–88.

Ford, Brian J. *Images of Science: A History of Scientific Illustration*. New York: Oxford University Press, 1982.

Fothergill, John. *Explanatory Remarks on the Preface to Sydney Parkinson's "Journal of a Voyage to the South-Seas."* [London, 1773?].

Foucault, Michel. "What Is an Author?" In *The Foucault Reader*, edited by Paul Rabinow, 101–20. London: Penguin, 1991.

Francis, John W. *Old New York; or, Reminiscences of the Past Sixty Years*. New York: Charles Roe, 1858.

Frank, Michael C. "Imaginative Geography as a Travelling Concept: Foucault, Said and the Spatial Turn." *European Journal of English Studies* 13, no. 1 (2009): 61–77.

Fraser, Angus. "A Publishing House and Its Readers, 1841–1889: The Murrays and the Miltons." *Papers of the Bibliographical Society of America* 90, no. 1 (1996): 4–47.

Fraser, Elisabeth A. "Books, Prints, and Travel: Reading in the Gaps of the Orientalist Archive." *Art History* 31, no. 3 (2008): 342–67.

Fulford, Tim, Debbie Lee, and Peter J. Kitson. *Literature, Science and Exploration in the Romantic Era: Bodies of Knowledge*. Cambridge: Cambridge University Press, 2004.

Fyfe, Aileen. "Natural History and the Victorian Tourist: From Landscapes to Rock-Pools." In *Geographies of Nineteenth-Century Science*, edited by David N. Livingstone and Charles W. J. Withers, 371–98. Chicago: University of Chicago Press, 2011.

———. "Steam and the Landscape of Knowledge: W. & R. Chambers in the 1830s–1850s." In *Geographies of the Book*, edited by Miles Ogborn and Charles W. J. Withers, 51–78. Farnham: Ashgate, 2010.

———. *Steam-Powered Knowledge: William Chambers and the Business of Publishing, 1820–1860*. Chicago: University of Chicago Press, 2012.

Fyfe, Christopher. "Clapperton, Hugh (1788–1827)." In *Oxford Dictionary of National Biography*. Oxford University Press, 2004–13. doi:10.1093/ref:odnb/7476.

Galton, Francis, ed. *Hints to Travellers*. 4th ed. London: Edward Stanford for the Royal Geographical Society, 1878.

Gascoigne, John. *The Enlightenment and the Origins of European Australia*. Cambridge: Cambridge University Press, 2002.

———. *Science in the Service of Empire: Joseph Banks, the British State and the Uses of Science in the Age of Revolution*. Cambridge: Cambridge University Press, 1998.

Genette, Gérard. *Paratexts: Thresholds of Interpretation*. Translated by Jane Lewin. Cambridge: Cambridge University Press, 1997.

Gilbert, Edmund W. "Richard Ford and His 'Hand-Book for Travellers in Spain.'" *Geographical Journal* 106, nos. 3–4 (1945): 144–51.

Gilpin, William. *Three Essays: On Picturesque Beauty; On Picturesque Travel; and On Sketching Landscape*. London: R. Blamire, 1792.

Glascock, William N. *The Naval Service or Officers' Manual for Every Grade in His Majesty's Ships*. 2 vols. London: Saunders & Otley, 1836.

Glenn, Ian E. "The Wreck of the *Grosvenor* and the Making of South Africa Literature." *English in Africa* 22, no. 2 (1995): 1–18.

Glinoer, Anthony. "Collaboration and Solidarity: The Collective Strategies of the Romantic Cenacle." In *Models of Collaboration in Nineteenth-Century French Literature*, edited by Seth Whitten, 37–54. Farnham: Ashgate, 2009.

Godlewska, Anne Marie Claire. *Geography Unbound: French Geographic Science from Cassini to Humboldt*. Chicago: University of Chicago Press, 1999.

Goldgar, Anne. *Impolite Learning: Conduct and Community in the Republic of Letters, 1680–1850*. New Haven, CT: Yale University Press, 1995.

Goodwin, Gráinne, and Gordon Johnston. "Guidebook Publishing in the Nineteenth Century: John Murray's *Handbooks for Travellers*." *Studies in Travel Writing* 17, no. 1 (2013): 43–61.

Goren, Haim. *Dead Sea Level: Science, Exploration and Imperial Interests in the Near East*. London: I. B. Tauris, 2011.

Goudie, Andrew S. "Colonel Julian Jackson and His Contribution to Geography." *Geographical Journal* 144, no. 2 (1978): 264–70.

Graham, Kenneth J. E. *The Performance of Conviction: Plainness and Rhetoric in the Early English Renaissance*. Ithaca, NY: Cornell University Press, 1994.

Graham, Maria. *Journal of a Residence in India*. 2nd ed. Edinburgh: Archibald Constable; London: Longman, Hurst, Rees, Orme, & Brown, 1813.

———. *Letters on India*. London: Longman, Hurst, Rees, Orme, & Brown; Edinburgh: A. Constable, 1814.

———. *Three Months Passed in the Mountains East of Rome, during the Year 1819*. London: Longman, Hurst, Rees, Orme, & Brown; Edinburgh: A. Constable, 1820.

Graham, Peter W. "Byron and the Publishing Business." In *The Cambridge Companion to Byron*, edited by Drummond Bone, 27–43. Cambridge: Cambridge University Press, 2004.

Greppi, Claudio. "'On the Spot': Traveling Artists and the Iconographic Inventory of the World, 1769–1859." In *Tropical Visions in an Age of Empire*, edited by Felix Driver and Luciana Martens, 23–42. Chicago: University of Chicago Press, 2005.

Griffin, Dustin. "The Rise of the Professional Author?" In *The Cambridge History of the Book in Britain*. Vol. 5: *1695–1830*, edited by Michael F. Suarez and Michael L. Turner, 132–45. Cambridge: Cambridge University Press, 2009.

Griffin, Robert J. "Anonymity and Authorship." *New Literary History* 30, no. 4 (1999): 877–95.

Guelke, Leonard, and Jeanne Kay Guelke. "Imperial Eyes on South Africa: Reassessing Travel Narratives." *Journal of Historical Geography* 30, no. 1 (2004): 11–31.

Hallett, Robin, ed. *The Niger Journal of Richard and John Lander*. London: Routledge & Kegan Paul, 1965.

Hamilton, William John. "Geography." In *A Manual of Scientific Enquiry; Prepared for the Use of Her*

Majesty's Navy: And Adapted for Travellers in General, edited by John F. W. Herschel, 127–55. London: John Murray. 1849.

A Hand-Book for Travellers on the Continent: Being a Guide through Holland, Belgium, Prussia, and Northern Germany, and along the Rhine, from Holland to Switzerland. London: John Murray and Son, 1836.

"Handbooks for England." *Times,* September 22, 1859, 10.

Haraway, Donna J. *Modest_Witness@Second_Millennium.FemaleMan©_Meets_OncoMouse™.* New York: Routledge, 1997.

Harley, J. B. *The New Nature of Maps: Essays in the History of Cartography,* edited by Paul Laxton. Baltimore: Johns Hopkins University Press, 2001.

Hawkesworth, John. *An Account of the Voyages Undertaken by the Order of His Present Majesty for Making Discoveries in the Southern Hemisphere.* 3 vols. London: W. Strahan / T. Cadell, 1773.

Hayden, Judy, ed. *Travel Narratives, the New Science, and Literary Discourse, 1569–1750.* Farnham: Ashgate, 2012.

[Head, Frances Bond.] "The Printer's Devil." *Quarterly Review* 65, no. 129 (December 1839): 1–30.

Hearne, Samuel. *A Journey from Prince of Wales's Fort in Hudson's Bay, to the Northern Ocean.* London: A. Strahan / T. Cadell, 1795.

Heffernan, Michael. "'A Dream as Frail as Those of Ancient Time': The In-credible Geographies of Timbuctoo." *Environment and Planning D: Society and Space* 19, no. 2 (2001): 203–25.

Henderson, Louise. "An Authentic, Invented Africa: David Livingstone's *Missionary Travels* and Nineteenth-Century Practices of Illustration." *Scottish Geographical Journal* 129, nos. 3–4 (2013): 179–93.

———. "Geography, Travel and Publishing in Mid-Victorian Britain." PhD diss., Royal Holloway, University of London, 2012.

———. "John Murray and the Publication of David Livingstone's Missionary Travels." In *Livingstone Online.* Accessed January 6, 2012. http://www.livingstoneonline.ucl.ac.uk/companion .php?id=HIST2.

Henige, David. "In Quest of Error's Sly Imprimatur: The Concept of 'Authorial Intent' in Modern Textual Criticism." *History in Africa* 14 (1987): 87–112.

———. "'Twixt the Cup and the Lip': Field Notes on the Way to Print." *History in Africa* 25 (1998): 119–31.

Henry, John. *The Scientific Revolution and the Origins of Modern Science.* Basingstoke: Macmillan, 1997.

Hevly, Bruce. "The Heroic Science of Glacier Motion." *Osiris* 11 (1996): 66–86.

Holloway, John. *A Month in Norway.* London: John Murray, 1853.

Holmes, Richard. *The Age of Wonder: How the Romantic Generation Discovered the Beauty and Terror of Science.* London: Harper Press, 2008.

Hooper, Glenn, and Tim Youngs, ed. *Perspectives on Travel Writing.* Aldershot: Ashgate, 2004.

[Horne, Richard H.] *Exposition of the False Medium and Barriers Excluding Men of Genius from the Public.* London: Effingham Wilson, 1833.

Hovde, David M. "Sea Colportage: The Loan Library System of the American Seamen's Friend Society, 1859–1967." *Libraries and Culture* 29, no. 4 (1994): 389–414.

Howsam, Leslie. *Cheap Bibles: Nineteenth-Century Publishing and the British and Foreign Bible Society.* Cambridge: Cambridge University Press, 1991.

———. "The History of the Book in Britain, 1801–1914." In *The Oxford Companion to the Book,* edited by Michael Suarez and H. R. Woudhuysen, 1:180–87. Oxford: Oxford University Press, 2010.

———. *Kegan Paul: A Victorian Imprint.* London: Kegan Paul International; Toronto: University of Toronto, 1998.

———. *Old Books and New Histories: An Orientation to Studies in Book and Print Culture.* Toronto: University of Toronto Press, 2006.

Hulme, Peter, and Tim Youngs, eds. *The Cambridge Companion to Travel Writing*. Cambridge: Cambridge University Press, 2002.

———. "Introduction." In Hulme and Youngs, *The Cambridge Companion to Travel Writing*, 1–13.

Iliffe, Rob. "Author-Mongering: The 'Editor' between Producer and Consumer." In *The Consumption of Culture, 1600–1800: Image, Object, Text*, edited by Ann Bermingham and John Brewer, 166–92. Abingdon: Routledge, 1995.

———. "Science and Voyages of Discovery." In *The Cambridge History of Science*. Vol. 4: *Eighteenth-Century Science*, edited by Roy Porter, 618–45. Cambridge: Cambridge University Press, 2003.

Jackson, Julian R. "On Picturesque Description in Books of Travels." *Journal of the Royal Geographical Society of London* 5 (1835): 381–87.

———. *What to Observe; or, The Traveller's Remembrancer*. London: John Madden, 1841.

Jacob, Christian. *The Sovereign Map: Theoretical Approaches in Cartography throughout History*. Translated by Tom Conley. Edited by Edward H. Dahl. Chicago: University of Chicago Press, 2006.

Jardine, Lisa. *Francis Bacon: Discovery and the Art of Discourse*. London: Cambridge University Press, 1974.

Johns, Adrian. "The Ambivalence of Authorship in Early Modern Natural Philosophy." In *Scientific Authorship: Credit and Intellectual Property in Science*, edited by Mario Biagioli and Peter Galison, 67–90. New York: Routledge, 2003.

———. *The Nature of the Book: Print and Knowledge in the Making*. Chicago: University of Chicago Press, 1998.

———. "Science and the Book in Modern Cultural Historiography." *Studies in the History and Philosophy of Science* 29, no. 2 (1998): 167–94.

Johnston, Sarah. "Missionary Positions: Romantic European Polynesians from Cook to Stevenson." In *Travel Writing in the Nineteenth Century: Filling the Blank Spaces*, edited by Tim Youngs, 179–200. London: Anthem Press, 2006.

Joshi, Priya. "Trading Places: The Novel, the Colonial Library, and India." In *Print Areas: Book History in India*, edited by Abhijit Gupta and Swapan Chakravarty, 17–64. Delhi: Permanent Black, 2004.

Kearns, Gerry. "The Imperial Subject: Geography and Travel in the Work of Mary Kingsley and Halford Mackinder." *Transactions of the Institute of British Geographers* 22, no. 4 (1997): 450–72.

Keighren, Innes M. *Bringing Geography to the Book: Ellen Semple and the Reception of Geographical Knowledge*. London: I. B. Tauris, 2010.

Keighren, Innes M., and Charles W. J. Withers. "Questions of Inscription and Epistemology in British Travelers' Accounts of Early Nineteenth-Century South America." *Annals of the Association of American Geographers* 101, no. 6 (2011): 1131–46.

———. "The Spectacular and the Sacred: Narrating Landscape in Works of Travel." *cultural geographies* 19, no. 1 (2012): 10–29.

Kemp, Martin. "Taking It on Trust: Form and Meaning in Naturalistic Representation." *Archives of Natural History* 17 (1990): 127–88.

Kennedy, Dane. *The Last Blank Spaces: Exploring Africa and Australia*. Cambridge, MA: Harvard University Press, 2013.

———. *Reinterpreting Exploration: The West in the World*. New York: Oxford University Press, 2014.

King, Andrew, and John Plunkett, eds. *Victorian Print Media: A Reader*. Oxford: Oxford University Press, 2005.

King, Dean. *Skeletons on the Zahara*. London: Arrow, 2004.

Klonk, Charlotte. "Science, Art, and the Representation of the Natural World." In *The Cambridge History of Science*. Vol. 4: *Eighteenth-Century Science*, edited by Roy Porter, 584–617. Cambridge: Cambridge University Press, 2003.

Koelsch, William A. *Geography and the Classical World: Unearthing Historical Geography's Forgotten Past*. London: I. B. Tauris, 2012.

Koivunen, Leila. *Visualizing Africa in Nineteenth-Century British Travel Accounts*. New York: Routledge, 2009.

Kölbl-Ebert, Martina. "Observing Orogeny—Maria Graham's Account of the Earthquake in Chile in 1822." *Episodes* 22, no. 1 (1999): 36–40.

Korte, Barbara. *English Travel Writing from Pilgrimages to Postcolonial Explorations*. Translated by Catherine Matthias. Basingstoke: Macmillan, 2000.

Kryza, Frank T. *The Race for Timbuktu: In Search of Africa's City of Gold*. New York: Ecco, 2006.

Kuehn, Julia, and Paul Smethurst, eds. 2009. *Travel Writing, Form, and Empire: The Poetics and Politics of Mobility*. New York: Routledge, 2009.

Laborde, Léon de. *Voyage de l'Arabie Pétrée*. Paris: Giard, 1830.

Lambert, David. *Mastering the Niger: James MacQueen's African Geography and the Struggle over Atlantic Slavery*. Chicago: University of Chicago Press, 2013.

———. "'Taken Captive by the Mystery of the Great River': Towards an Historical Geography of British Geography and Atlantic Slavery." *Journal of Historical Geography* 35, no. 1 (2009): 44–65.

Lander, Richard. *Records of Captain Clapperton's Last Expedition to Africa*. 2 vols. London: Henry Colburn and Richard Bentley, 1830.

Latour, Bruno. *Science in Action: How to Follow Scientists and Engineers through Society*. Cambridge, MA: Harvard University Press, 1987.

Lawrence, Christopher, and Steven Shapin, eds. *Science Incarnate: Historical Embodiments of Natural Knowledge*. Chicago: University of Chicago Press, 1998.

Layard, Austen Henry. *Autobiography and Letters from His Childhood until His Appointment as H.M. Ambassador at Madrid*. 2 vols. London: John Murray, 1903.

Leary, Patrick, and Andrew Nash. "Authorship." In *The Cambridge History of the Book in Britain*. Vol. 6: *1830–1914*, edited by David McKitterick, 172–213. Cambridge: Cambridge University Press, 2009.

Leask, Nigel. *Curiosity and the Aesthetics of Travel Writing, 1770–1840*. Oxford: Oxford University Press, 2002.

Levere, Trevor H. *Science and the Canadian Arctic: A Century of Exploration, 1818–1918*. Cambridge: Cambridge University Press, 1993.

Liebersohn, Harry. "Scientific Ethnography and Travel." In *The Cambridge History of Science*. Vol. 7: *The Modern Social Sciences*, edited by Theodore M. Porter and Dorothy Ross, 100–112. Cambridge: Cambridge University Press, 2003.

Lincoln, Margarette. "Shipwreck Narratives of the Eighteenth and Early Nineteenth Century: Indicators of Culture and Identity." *British Journal for Eighteenth-Century Studies* 20, no. 2 (1997): 155–72.

Lister, William B. C. *A Bibliography of Murray's Handbooks for Travellers and Biographies of Authors, Editors, Revisers and Principal Contributors*. Dereham: Dereham Books, 1993.

Livingstone, Justin. "The Meaning and Making of *Missionary Travels*: The Sedentary and Itinerant Discourses of a Victorian Bestseller." *Studies in Travel Writing* 15, no. 3 (2011): 267–92.

———. "*Missionary Travels*, Missionary Travails: David Livingstone and the Victorian Publishing Industry." In *David Livingstone: Man, Myth and Legacy*, edited by Sarah Worden, 33–51. Edinburgh: National Museums Scotland, 2012.

Lockhart, Jamie Bruce. "Documents Relating to the Lander Brothers' Niger Expedition of 1830." Accessed December 16, 2011, http://www.tubmaninstitute.ca/documents_relating _to_the_lander_brothers_niger_expedition_of_1830.

———. "In the Raw: Some Reflections on Transcribing and Editing Lieutenant Hugh Clapperton's Writings on the Borno Mission of 1822–25." *History in Africa* 26 (1999): 157–95.

——. *A Sailor in the Sahara: The Life and Travels in Africa of Hugh Clapperton, Commander RN*. London: I. B. Tauris, 2008.

Lockhart, Jamie Bruce, and Paul E. Lovejoy, eds. *Hugh Clapperton into the Interior of Africa: Records of the Second Expedition, 1825-1827*. Leiden: Brill, 2005.

Lloyd, Christopher. *Mr Barrow of the Admiralty: A Life of Sir John Barrow, 1764-1848*. London: Collins, 1970.

Lockhart, John G. *Memoirs of the Life of Sir Walter Scott, Bart*. 7 vols. Edinburgh: Robert Cadell; London: John Murray / Whittaker, 1838.

Lushington, Charles. *The History, Design, and Present State of the Religious, Benevolent and Charitable Institutions, Founded by the British in Calcutta and Its Vicinity*. Calcutta: Hindostanee Press, 1824.

Lutyens, Mary. "Murrays of Albemarle Street." *Times Literary Supplement*, October 24, 1968, 198-99.

Lydon, Ghislaine. "Writing Trans-Saharan History: Methods, Sources and Interpretations across the African Divide." *Journal of North African Studies* 10, nos. 3-4 (2005): 293-324.

Lyons, Martyn. *A History of Reading and Writing in the Western World*. Basingstoke: Palgrave Macmillan, 2010.

——. *Reading Culture and Writing Practices in Nineteenth-Century France*. Toronto: University of Toronto Press, 2008.

——. *The Writing Culture of Ordinary People in Europe, c. 1860-1920*. Cambridge: Cambridge University Press, 2013.

MacFarlane, Charles. *Constantinople in 1828*. London: Saunders & Otley, 1829.

——. *Reminiscences of a Literary Life*. London: John Murray, 1917.

Mackay, Mercedes. *The Indomitable Servant*. London: Rex Collins, 1978.

Mackenzie, Alexander. *Voyages from Montreal, on the River St. Laurence, through the Continent of North America, to the Frozen and Pacific Oceans; in the Years 1789 and 1793*. London: T. Cadell, Jun. & W. Davies / Corbert & Morgan; Edinburgh: W. Creech, 1801.

MacLaren, Ian S. "Exploration/Travel Literature and the Evolution of the Author." *International Journal of Canadian Studies* 5 (1992): 39-68.

——. "Explorers' and Travelers' Narratives: A Peregrination through Different Editions." *History in Africa* 30 (2003): 213-22.

——. "From Exploration to Publication: The Evolution of a 19th-Century Arctic Narrative." *Arctic* 47, no. 1 (1992): 43-53.

——. "In Consideration of the Evolution of Explorers and Travellers into Authors: A Model." *Studies in Travel Writing* 15, no. 3 (2011): 221-41.

——. "Samuel Hearne's Accounts of the Massacre at Bloody Fall, 17 July 1771." *Ariel* 22, no. 1 (1991): 25-51.

Maidment, Brian. "Periodicals and Serial Publications." In *The Cambridge History of the Book in Britain*. Vol. 5: *1695-1830*, edited by Michael F. Suarez and Michael L. Turner, 498-512. Cambridge: Cambridge University Press, 2009.

Malley, Shawn. *From Archaeology to Spectacle in Victorian Britain: The Case of Assyria, 1845-1854*. Farnham: Ashgate, 2012.

Mancall, Peter C. "Introduction: Observing More Things and More Curiously." *Huntington Library Quarterly* 70, no. 1 (2007): 1-10.

Martels, Zweder R. W. M. von, ed. *Travel Fact and Travel Fiction: Literary Tradition, Scholarly Discovery, and Observation in Travel Writing*. Leiden: Brill, 1994.

Martin, Alison E. "'These Changes and Accessions of Knowledge': Translation, Scientific Travel Writing and Modernity—Alexander von Humboldt's *Personal Narrative*." *Studies in Travel Writing* 15, no. 1 (2011): 39-51.

Martineau, Harriet. *How to Observe. Morals and Manners*. London: Charles Knight, 1838.

[———]. "Travel during the Last Half Century," Review of *The Principal Navigations, Voyages, Traffiques and Discoveries of the English Nation,* by Richard Hakluyt, and *The English Cyclopædia,* by Charles Knight. *Westminster Review* 70, no. 138 (October 1858): 236–58.

Marx, Karl. *Grundrisse: Foundations of the Critique of Political Economy.* Translated by Martin Nicolaus. Harmondsworth: Penguin, 1973.

Matthews, Gelien. *Caribbean Slave Revolts and the British Abolitionist Movement.* Baton Rouge: Louisiana State University Press, 2006.

Mayhew, Robert J. "Mapping Science's Imagined Community: Geography as a Republic of Letters, 1600–1800." *British Journal for the History of Science* 38, no. 1 (2005): 73–92.

———. 2007. "Materialist Hermeneutics, Textuality and the History of Geography: Print Spaces in British Geography, c.1500–1900." *Journal of Historical Geography* 33, no. 3 (2007): 466–88.

———. "Printing Posterity: Editing Varenius and the Construction of Geography's History." In *Geographies of the Book,* edited by Miles Ogborn and Charles W. J. Withers, 157–87. Farnham: Ashgate, 2010.

McClay, David. "John Murray and the Publishing of Scottish Polar Explorers." *Geographer: The Newsletter of the Royal Scottish Geographical Society* (Winter 2011-12), 7.

[McCoy, Rebecca.] *The Englishwoman in Russia; Impressions of the Society and Manners of the Russians at Home.* London: John Murray, 1855.

McEwan, Cheryl. *Gender, Geography and Empire: Victorian Women Travellers in West Africa.* Aldershot: Ashgate, 2000.

McGann, Jerome. *The Textual Condition.* Princeton, NJ: Princeton University Press, 1991.

McKitterick, David. "Introduction." In *The Cambridge History of the Book in Britain.* Vol. 6: *1830–1914,* edited by David McKitterick, 1–74. Cambridge: Cambridge University Press, 2009.

———. *Print, Manuscript and the Search for Order, 1450–1830.* Cambridge: Cambridge University Press, 2003.

McMurtry, R. Gerald. "The Influence of Riley's *Narrative* upon Abraham Lincoln." *Indiana Magazine of History* 30, no. 2 (1934): 133–38.

Meadows, Jack. *The Victorian Scientist: The Growth of a Profession.* London: British Library, 2004.

Medeiros, Michelle. "Crossing Boundaries into a World of Scientific Discoveries: Maria Graham in Nineteenth-Century Brazil." *Studies in Travel Writing* 16, no. 3 (2012): 263–85.

Miller, David Philip, and Peter Hanns Reill, eds. *Visions of Empire: Voyages, Botany, and Representations of Nature.* Cambridge: Cambridge University Press, 1996.

Millikin, Richard. *The River-Side: A Poem, in Three Books.* Cork: Printed by J. Conner, 1807.

Mills, Anthony J. "Sir Archibald Edmonstone: A Scottish Traveller Asserts a Claim." In *Egyptian Encounters,* edited by Jason Thompson, 43–49. Cairo: American University in Cairo Press, 2002.

"Mr. Layard's Second Expedition to Assyria." *Athenaeum,* no. 1325 (March 19, 1853), 339.

Moore, Thomas. *The Works of Lord Byron: With His Letters and Journals, and His Life.* 17 vols. London: John Murray, 1834.

Morrison-Low, A. D. *Making Scientific Instruments in the Industrial Revolution.* Aldershot: Ashgate, 2007.

Mr. Murray's List of New Works. London: John Murray, November 1861.

Mr. Murray's List of New Works Now Ready. London: John Murray, January 1852.

Murray, Hugh. 1817. *Historical Account of Discoveries and Travels in Africa.* 2 vols. Edinburgh: Archibald Constable; London: Longman, Hurst, Rees, Orme, & Brown, 1817.

"Murray's Colonial and Home Library." *Examiner,* October 28, 1843, 677.

Napier, William. *The Life and Opinions of General Sir Charles James Napier, G.C.B.* 4 vols. London: John Murray, 1857.

Neilsen, Kristian H., Michael Harbsmeier, and Christopher J. Ries, eds. *Scientists and Scholars in the Field: Studies in the History of Fieldwork and Expeditions.* Aarhus: Aarhus University Press, 2012.

Nelson, Thomas. *A Biographical Memoir of the Late Dr W. Oudney, and Captain Hugh Clapperton*. Edinburgh: Waugh & Innes; London: Whittaker, Treacher, 1830.

Neumann, Birgit, and Frederik Tygstrup. "Travelling Concepts in English Studies." *European Journal of English Studies* 13, no. 1 (2009): 1–12.

"Noctes Ambrosianae." *Blackwood's Edinburgh Magazine* 25, no. 153 (June 1829): 787–803.

North, Marcy L. *The Anonymous Renaissance: Cultures of Discretion in Tudor-Stuart England*. Chicago: University of Chicago Press, 2006.

Ogborn, Miles, and Charles W. J. Withers. "Introduction: Book Geography, Book History." In *Geographies of the Book*, edited by Miles Ogborn and Charles W. J. Withers, 1–25. Farnham: Ashgate, 2010.

Ogborn, Miles. "*Geographia*'s Pen: Writing, Geography and the Arts of Commerce, 1660–1760." *Journal of Historical Geography* 30, no. 2 (2004): 294–315.

———. *Indian Ink: Script and Print in the Making of the English East India Company*. Chicago: University of Chicago Press, 2007.

O'Hara, Kieron. *Trust: From Socrates to Spin*. London: Icon Books, 2004.

Ong, Walter J. *Ramus Method and the Decay of Dialogue: From the Art of Discourse to the Art of Reason*. Cambridge, MA: Harvard University Press, 1958.

Otness, Harold M. "Passenger Ship Libraries." *Journal of Library History* 14, no. 4 (1979): 486–95.

Outram, Dorinda. "On Being Perseus: New Knowledge, Dislocation, and Enlightenment Exploration." In *Geography and Enlightenment*, edited by David N. Livingstone and Charles W. J. Withers, 281–94. Chicago: University of Chicago Press, 1999.

Paston, George. *At John Murray's: Records of a Literary Circle, 1843–1892*. London: John Murray, 1932.

Pérez-Majía, Ángela. *A Geography of Hard Times: Narratives about Travel to South America, 1780–1849*. Translated by Dick Cluster. Albany: State University of New York Press, 2004.

Pococke, Richard. *A Description of the East, and Some Other Countries*. 2 vols. London: Printed for the author and sold by J. & P. Knapton; W. Innys; W. Meadows; G. Hawkins; S. Brit; T. Longman; C. Hitch; R. Dodsley; J. Nourse; J. Rivington, 1743.

Porter, Robert Ker. *Travels in Georgia, Persia, Armenia, Ancient Babylonia, &c. &c.* 2 vols. London: Longman, Hurst, Rees, Orme, & Brown, 1821–22.

Pratt, Mary Louise. *Imperial Eyes: Travel Writing and Transculturation*. 2nd ed. New York: Routledge, 2008.

"Presents Made to the Royal Society from November 1790 to June 1791." *Philosophical Transactions of the Royal Society of London* 81, no. 1 (1791): 423–27.

Price, Leah. *How to Do Things with Books in Victorian Britain*. Princeton, NJ: Princeton University Press, 2012.

Prinsep, Henry T. *Origin of the Sikh Power in the Punjab, and Political Life of Muha-Raja Runjeet Singh*. Calcutta: Military Orphan Press, 1834.

"Prospectus and Specimen of Mr. Murray's Colonial and Home Library." *Quarterly Literary Advertiser* (August 1843): unpaginated.

Raban, Jonathan. *For Love and Money: Writing, Reading, Travelling, 1969–1987*. London: Collins Harvill, 1987.

Raj, Kapil. *Relocating Modern Science: Circulation and the Construction of Knowledge in South Asia and Europe, 1650–1900*. Basingstoke: Palgrave Macmillan, 2006.

———. "When Human Travellers Become Instruments: The Indo-British Exploration of Central Asia in the Nineteenth Century." In *Instruments, Travel and Science: Itineraries of Precision from the Seventeenth to the Twentieth Century*, edited by Marie-Noëlle Bourguet, Christian Licoppe, and H. Otto Sibum, 156–88. London: Routledge, 2002.

Raven, James. "The Anonymous Novel in Britain and Ireland, 1750–1830." In *The Faces of Anonymity: Anonymous and Pseudonymous Publication from the Sixteenth to the Twentieth Century*, edited by Robert J. Griffin, 141–66. Basingstoke: Palgrave Macmillan, 2003.

——. *The Business of Books: Booksellers and the English Book Trade, 1450–1850*. New Haven, CT: Yale University Press, 2007.

——. "The Promotion and Constraints of Knowledge: The Changing Structure of Publishing and Bookselling in Victorian Britain." In *The Organisation of Knowledge in Victorian Britain*, edited by Martin Daunton, 263–86. Oxford: Oxford University Press for the British Academy, 2005.

Regard, Frederic. "Introduction: Articulating Empire's Unstable Zones." In *British Narratives of Exploration*, edited by Frederic Regard, 1–17. London: Pickering & Chatto, 2009.

Reidy, Michael S. *Tides of History: Ocean Science and Her Majesty's Navy*. Chicago: University of Chicago Press, 2008.

Rennell, James. *A Bengal Atlas: Containing Maps of the Theatre of War and Commerce on That Side of Hindoostan*. London, 1780.

——. *The Geographical System of Herodotus, Examined; and Explained, by a Comparison with Those of Other Ancient Authors, and with Modern Geography*. London: Printed for the author and sold by G. & W. Nichol, 1800.

——. *Illustrations, (Chiefly Geographical,) of the History of the Expedition of Cyrus, from Sardis to Babylonia; and the Retreat of the Ten Thousand Greeks, from Thence to Trebisonde, and Lydia*. London: G. & W. Nicol, 1816.

——. "On the Rate of Travelling, as Performed by Camels; and Its Applications, as a Scale, to the Purposes of Geography." *Philosophical Transactions of the Royal Society of London* 81 (1791): 129–45.

Reynolds, K. D. "FitzClarence, George Augustus Frederick, First Earl of Munster (1794–1842)." In *Oxford Dictionary of National Biography*. Oxford University Press, 2004–13. doi:10.1093/ref:odnb/9542.

Richards, Thomas. *The Imperial Archive: Knowledge and the Fantasy of Empire*. London: Verso, 1993.

Richardson, Alan. *Literature, Education, and Romanticism: Reading as Social Practice, 1780–1832*. Cambridge: Cambridge University Press, 1994.

Richardson, John. *Arctic Searching Expedition: A Journal of a Boat-Voyage through Rupert's Land and the Arctic Sea, in Search of the Discovery Ships under Command of Sir John Franklin*. 2 vols. London: Longman, Brown, Green, & Longmans, 1851.

[Rigby, Elizabeth.] *A Residence on the Shores of the Baltic: Described in a Series of Letters*. 2 vols. London: John Murray, 1841.

Riley, James. *An Authentic Narrative of the Loss of the American Brig* Commerce. New York: T. & W. Mercein, 1817.

Robertson, George. *Charts of the China Navigation, Principally Laid Down upon the Spot, and Corrected from the Latest Observations, with Particular Views of the Land*. London: J. Murray, 1788.

——. *Memoir of a Chart of the China Sea*. London: Printed for the author and sold by J. Murray, 1791.

Robinson, Edward. *Physical Geography of the Holy Land*. London: John Murray, 1865.

Robinson, Howard. *The British Post Office: A History*. Princeton, NJ: Princeton University Press, 1948.

Rogers, Shef. "Enlarging the Prospects of Happiness: Travel Reading and Travel Writing." In *The Cambridge History of the Book in Britain*. Vol. 5: *1695–1830*, edited by Michael F. Suarez and Michael L. Turner, 781–90. Cambridge: Cambridge University Press, 2009.

Rojek, Chris, and John Urry, eds. *Touring Cultures: Transformations of Travel and Theory*. London: Routledge, 1997.

Ross, Maurice J. *Polar Pioneers: A Biography of John and James Clark Ross*. Montreal: McGill-Queen's University Press, 1994.

"Royal Premium." *Journal of the Royal Geographical Society of London* 2 (1832): vii.

Rubiés, Joan-Pau. "Instructions for Travellers: Teaching the Eye to See." *History and Anthropology* 9, nos. 2–3 (1996): 139–90.

———. *Travellers and Cosmographers: Studies in the History of Early Modern Travel and Ethnology.* Aldershot: Ashgate, 2007.

Rudwick, Martin J. S. *Bursting the Limits of Time: The Reconstruction of Geohistory in the Age of Revolution.* Chicago: University of Chicago Press, 2005.

———. "The Emergence of a Visual Language for Geological Science, 1760–1840." *History of Science* 14 (1976): 149–95.

Ruland, Richard. "Melville and the Fortunate Fall: Typee as Eden." *Nineteenth-Century Fiction* 23, no. 3 (1968): 312–23.

Rupke, Nicolaas A. *Alexander von Humboldt: A Metabiography.* Chicago: University of Chicago Press, 2008.

———. "A Geography of Enlightenment: The Critical Reception of Alexander von Humboldt's Mexico Work." In *Geography and Enlightenment*, edited by David N. Livingstone and Charles W. J. Withers, 319–39. Chicago: University of Chicago Press, 1999.

Sabine, Edward. *Remarks on the Account of the Late Voyage of Discovery to Baffin's Bay, Published by Captain J. Ross, R. N.* London: John Booth, 1819.

Safier, Neil. "'Every Day That I Travel . . . is a Page that I Turn': Reading and Observing in Eighteenth-Century Amazonia." *Huntington Library Quarterly* 70, no. 1 (2007): 103–28.

Said, Edward. *Orientalism.* Harmondsworth: Penguin, 1978.

———. *The World, the Text, and the Critic.* Cambridge MA: Harvard University Press, 1982.

Sandwith, Humphry. *The Hekim Bashi; or, The Adventures of Giuseppe Antonelli, a Doctor in the Turkish Service.* London: Smith, Elder, 1864.

Secord, Anne. "Botany on a Plate: Pleasure and the Power of Pictures in Promoting Early Nineteenth-Century Scientific Knowledge." *Isis* 93, no. 1 (2002): 28–57.

———. "Pressed into Service: Specimens, Space, and Seeing in Botanical Practice." In *Geographies of Nineteenth-Century Science*, edited by David N. Livingstone and Charles W. J. Withers, 283–310. Chicago: University of Chicago Press, 2011.

Secord, James A. "Knowledge in Transit." *Isis* 95, no. 4 (2004): 654–72.

———. *Victorian Sensation: The Extraordinary Publication, Reception, and Secret Authorship of "Vestiges of the Natural History of Creation."* Chicago: University of Chicago Press, 2000.

Schaffer, Simon. "Easily Cracked: Scientific Instruments in States of Disrepair." *Isis* 102, no. 4 (2011): 706–17.

———. "'On Seeing Me Write': Inscription Devices in the South Seas." *Representations* 97, no. 1 (2007): 90–122.

Schaffer, Simon, Lissa Roberts, Kapil Raj, and James Delbourgo, eds. *The Brokered World: Go-Betweens and Global Intelligence, 1770–1820.* Sagamore Beach, MA: Science History Publications, 2009.

Schiffer, Reinhold. *Oriental Panorama: British Travellers in 19th Century Turkey.* Amsterdam: Editions Rodopi, 1999.

Schwartz, Joan M., and Terry Cook. "Archives, Records, and Power: The Making of Modern Memory." *Archival Science* 2, no. 1–2 (2002): 1–19.

Scott, John. *A Visit to Paris.* 2nd ed. London: Longman, Hurst, Rees, Orme, & Brown, 1815.

Scott, Walter. *The Life of Napoleon Buonaparte, Emperor of the French.* 9 vols. Edinburgh: Cadell; London: Longman, Rees, Orme, Brown, & Green, 1827.

Sell, Jonathan P. A. *Rhetoric and Wonder in English Travel Writing, 1560–1613.* Aldershot: Ashgate, 2006.

Shapin, Steven. "The House of Experiment in Seventeenth-Century England." *Isis* 79, no. 3 (1988): 373–404.

———. "The Image of the Man of Science." In *The Cambridge History of Science.* Vol. 4: *Eighteenth-Century Science*, edited by Roy Porter, 159–83. Cambridge: Cambridge University Press, 2003.

———. "The Man of Science." In *The Cambridge History of Science.* Vol. 3: *Early Modern Science*, edited by Katharine Park and Lorraine Daston, 179-91. Cambridge: Cambridge University Press, 2006.

——. *Never Pure: Historical Studies of Science as If It Was Produced by People with Bodies, Situated in Time, Space, Culture, and Society, and Struggling for Credibility and Authority*. Baltimore: Johns Hopkins University Press, 2010.

——. "Pump and Circumstance: Robert Boyle's Literary Technology." *Social Studies of Science* 14, no. 4 (1984): 481-520.

——. "Rarely Pure and Never Simple: Talking about Truth." *Configurations* 7, no. 1 (1999): 1-14.

——. "'A Scholar and a Gentleman': The Problematic Identity of the Scientific Practitioner in Early Modern England." *History of Science* 29 (1991): 279-327.

——. *The Scientific Revolution*. Chicago: University of Chicago Press, 1996.

——. *A Social History of Truth: Civility and Science in Seventeenth-Century England*. Chicago: University of Chicago Press, 1994.

Shapiro, Barbara J. "Testimony in Seventeenth-Century English Natural Philosophy: Legal Origins and Early Development." *Studies in History and Philosophy of Science* 33, no. 2 (2002): 243-63.

Sher, Richard B. *The Enlightenment and the Book: Scottish Authors and Their Publishers in Eighteenth-Century Britain, Ireland, and America*. Chicago: University of Chicago Press, 2006.

Sherman, Stuart. *Telling Time: Clocks, Diaries, and English Diurnal Form, 1660-1785*. Chicago: University of Chicago Press, 1996.

Sherman, William H. "Stirrings and Searchings (1500-1720)." In Hulme and Youngs, *The Cambridge Companion to Travel Writing*, 17-36.

Shteir, Ann B. *Cultivating Women, Cultivating Science: Flora's Daughters and Botany in England, 1760-1860*. Baltimore: Johns Hopkins University Press, 1996.

Skallerup, Harry R. *Books Afloat and Ashore: A History of Books, Libraries, and Reading among Seamen during the Age of Sail*. Hamden, CT: Archon Books, 1974.

Skinner, Kathleen L. "Ships, Logs, and Voyages: Maria Graham Navigates the Journey of H.M.S. *Blonde*." Honor's thesis, University of Texas at Austin, 2010.

Smiles, Samuel. *A Publisher and His Friends: Memoir and Correspondence of the Late John Murray, with an Account of the Origin and Progress of the House, 1768-1843*. 2 vols. London: John Murray, 1891.

Smith, Amy Elizabeth. "Travel Narratives and the Familiar Letter Form in the Mid-Eighteenth Century." *Studies in Philology* 95, no. 1 (1998): 77-96.

Smith, Bernard. *European Vision and the South Pacific*. 2nd ed. New Haven, CT: Yale University Press, 1985.

——. *Imagining the Pacific: In the Wake of the Cook Voyages*. New Haven, CT: Yale University Press, 1992.

Snader, Joe. *Caught between Worlds: British Captivity Narratives in Fact and Fiction*. Lexington: University Press of Kentucky, 2000.

Southey, Robert. *History of Brazil*. 3 vols. London: Longman, Hurst, Rees, & Orme, 1810-19.

Spix, Johann Baptist von, and Carl Friedrich Philipp von Martius. *Travels in Brazil, in the Years 1817-1820*. 2 vols. London: Longman, Hurst, Rees, Orme, Brown, & Green, 1824.

Spufford, Francis. *I May Be Some Time: Ice and the English Imagination*. London: Faber, 1996.

Stafford, Barbara Maria. *Voyage into Substance: Art, Science, Nature, and the Illustrated Travel Account, 1760-1840*. Cambridge, MA: MIT Press, 1984.

Stagl, Justin. *A History of Curiosity: Theory of Travel, 1550-1800*. Chur, Switzerland: Harwood Academic, 1995.

Staunton, George. *An Authentic Account of an Embassy from the King of Great Britain to the Emperor of China*. 3 vols. London: G. Nicol, 1797.

St. Clair, William. *The Reading Nation in the Romantic Period*. Cambridge: Cambridge University Press, 2004.

Steedman, Carolyn. *Dust*. Manchester: Manchester University Press, 2001.

[Sterne, Lawrence], *A Sentimental Journey through France and Italy*. 2 vols. London: T. Becket / P. A. DeHondt, 1768.

Stevens, Henry N. *Ptolemy's "Geography": A Brief Account of All Printed Editions Down to 1730*. 2nd ed. London: Henry Stevens, Son & Stiles, 1908.

Stewart, Larry. "Global Pillage: Science, Commerce, and Empire." In *The Cambridge History of Science*. Vol. 4: *Eighteenth-Century Science*, edited by Roy Porter, 825–44. Cambridge: Cambridge University Press, 2003.

Suarez, Michael F. "Introduction." In *The Cambridge History of the Book in Britain*. Vol. 5: *1695–1830*, edited by Michael F. Suarez and Michael L. Turner, 1–35. Cambridge: Cambridge University Press, 2009.

"Summary." *Literary Gazette*, no. 1878 (January 15, 1853), 63–64.

Sutherland, Kathryn. "Jane Austen's Dealings with John Murray and His Firm." *Review of English Studies* 64, no. 263 (2013): 105–25.

Tallmadge, John. "From Chronicle to Quest: The Shaping of Darwin's 'Voyage of the Beagle.'" *Victorian Studies* 23, no. 3 (1980): 325–45.

Taylor, Stephen. *The Caliban Shore: The Fate of the* Grosvenor *Castaways*. London: Faber, 2004.

Thompson, Carl. "Earthquakes and Petticoats: Maria Graham, Geology, and Early Nineteenth-Century 'Polite' Science." *Journal of Victorian Culture* 17, no. 3 (2012): 329–46.

———. *The Suffering Traveller and the Romantic Imagination*. Oxford: Oxford University Press, 2007.

Titlestad, Michael. "'The Unhappy Fate of Master Law': George Carter's *A Narrative of the Loss of the* Grosvenor, *East Indiaman* (1791)." *English Studies in Africa* 51, no. 2 (2008): 21–37.

Twyman, Michael. "The Illustration Revolution." In *The Cambridge History of the Book in Britain*. Vol. 6: *1830–1914*, edited by David McKitterick, 117–43. Cambridge: Cambridge University Press, 2009.

Vincent, David. *Literacy and Popular Culture: England 1750–1914*. Cambridge: Cambridge University Press, 1989.

Wahlgren-Smith, Lena. "On the Composition of Herbert Losinga's Letter Collection." *Classica et Mediaevalia* 55 (2004): 229–46.

Waterfield, Gordon. *Layard of Nineveh*. London: John Murray, 1963.

Waterston, Charles D., and Angus Macmillan Shearer. *Biographical Index of Former Fellows of the Royal Society of Edinburgh, 1783–2002*. Edinburgh: Royal Society of Edinburgh, 2006.

Wheatley, Kim. "The Arctic in the *Quarterly Review*." *European Romantic Review* 20, no. 4 (2009): 465–90.

White, Hayden. *Content of the Form: Narrative Discourse and Historical Representation*. Baltimore: Johns Hopkins University Press, 1987.

———. *Tropics of Discourse: Essays in Cultural Criticism*. Baltimore: Johns Hopkins University Press, 1985.

Williams, Bernard. *Truth and Truthfulness: An Essay in Genealogy*. Princeton, NJ: Princeton University Press, 2002.

Williams, Michael. 2001. *Problems of Knowledge: A Critical Introduction to Epistemology*. Oxford: Oxford University Press, 2001.

Wilson, Eric G. *The Spiritual History of Ice: Romanticism, Science, and the Imagination*. New York: Palgrave Macmillan, 2003.

Withers, Charles W. J. *Geography and Science in Britain, 1831–1939: A Study of the British Association for the Advancement of Science*. Manchester: Manchester University Press, 2010.

———. "Geography, Enlightenment and the Book: Authorship and Audience in Mungo Park's African Texts." In *Geographies of the Book*, edited by Miles Ogborn and Charles W. J. Withers, 191–220. Farnham: Ashgate, 2010.

——. *Geography, Science and National Identity: Scotland since 1520*. Cambridge: Cambridge University Press, 2001.

——. "Mapping the Niger, 1798–1832: Trust, Testimony and 'Ocular Demonstration' in the Late Enlightenment." *Imago Mundi* 56, no. 2 (2004): 170–93.

——. "Memory and the History of Geographical Knowledge: The Commemoration of Mungo Park, African Explorer." *Journal of Historical Geography* 30, no. 2 (2004): 316–39.

——. "On Enlightenment's Margins: Geography, Imperialism and Mapping in Central Asia, *c.*1798–*c.*1838." *Journal of Historical Geography* 39 (2013): 3–18.

——. *Placing the Enlightenment: Thinking Geographically about the Age of Reason*. Chicago: University of Chicago Press, 2007.

——. "Science, Scientific Instruments and Questions of Method in Nineteenth-Century British Geography." *Transactions of the Institute of British Geographers* 38, no. 1 (2013): 167–79.

——. "Travel, *En Route* Writing, and the Problem of Correspondence." In *Routes, Roads and Landscape*, edited by Mari Hvattum, Brita Brenna, Beate Elvebakk, and Janike Kampevold Larsen, 83–94. Farnham: Ashgate, 2011.

——. "Voyages et Credibilité: Vers une Géographie de la Confiance," *Géographie et Cultures* 33 (2000): 3–17.

——. "Writing in Geography's History: *Caledonia*, Networks of Correspondence and Geographical Knowledge in the Late Enlightenment." *Scottish Geographical Journal* 120, nos. 1–2 (2005): 33–45.

Withers, Charles W. J., and Innes M. Keighren. "Travels into Print: Authoring, Editing and Narratives of Travel and Exploration, *c.*1815–*c.*1857." *Transactions of the Institute of British Geographers* 36, no. 4 (2011): 560–73.

Wood, John. *A Description of Bath*. London: Printed for J. [*sic*] Murray, 1769.

Woodmansee, Martha. "On the Author Effect: Recovering Collectivity." In *The Construction of Authorship: Textual Appropriation in Law and Literature*, edited by Martha Woodmansee and Peter Jaszi, 127–56. Durham, NC: Duke University Press, 1994.

Yeo, Richard. "Between Memory and Paperbooks: Baconianism and Natural History in Seventeenth-Century England." *History of Science* 45, no. 1 (2007): 1–46.

Yothers, Brian. *The Romance of the Holy Land in American Travel Writing, 1790–1876*. Aldershot: Ashgate, 2007.

Youngs, Tim. "Where Are We Going? Cross-Border Approaches to Travel Writing." In Hooper and Youngs, *Perspectives on Travel Writing*, 168–80.

Zachs, William, Peter Isaac, Angus Fraser, and William Lister. "Murray Family." In *Oxford Dictionary of National Biography*. Oxford: Oxford University Press, 2004–13. doi:10.1093/ref:odnb/64922.

Zachs, William. *The First John Murray and the Late Eighteenth-Century Book Trade: With a Checklist of His Publications*. Oxford: Oxford University Press for the British Academy, 1998.

Index

Page numbers in italics refer to figures.

in epistolary form, 12; genre conventions of, 4, 31, 74–75, 124, 190, 213, 215–16, 220; as mediated process, 212; plain style in, 2, 35, 49, 50, 65, 86, 90, 106–8, 112, 116, 123–24, 220; practicalities of in-the-field, 20, 37, 49, 51, 54, 59–60, 79–80, 118, 127, 130, 160, 162, 192, 222; as profession, 107–8, 112; as record of experience, 3, 7; role of memory in, 41–43, 59, 62, 118; as source for later travelers, 4, 47, 56, 69, 75–82; stylistic deficiencies in, 4, 31; typology of, 37

Treatise in Judicial Evidence, A (Bentham), 220

Trebeck, George, *Travels in the Himalayan Provinces*, 265

Tremenheere, Hugh, 52–53; *Notes on Public Subjects*, 279

Trent, Thomas, *Two Years in Ava*, 253

trust, 11–16, 27, 68–69, 72–77, 87–89, 107, 114–21, 218–19

truth: authorial claims to, 3, 12–13, 70, 75, 88, 218–22; and lies, 11, 73; as plain and unvarnished, 2, 104, 106–8

truthfulness: contradictory nature of, 12; in relation to genre, 12, 74, 99; in relation to social status, 12, 17, 86, 99; role of first-hand experience in demonstrating, 13, 59, 73, 82, 220; role of instrumentation in demonstrating, 13, 91–98

Tuckey, James, 48, 66, 149, 297n24; *Narrative of an Expedition to Explore*, 28, 41, 43–44, 121, 217, 240

Tudor (anatomist on Tuckey's expedition), 41

Tulišen, *Narrative of the Chinese Embassy*, 245

Turkey (MacFarlane), 102, 277

Turner, William, *Journal of a Tour in the Levant*, 244

Tuweileb (guide), 89

Two Years in Ava (Trent), 253

Twyman, Michael, 155, 157

Typee (Melville, Herman), 200, 204

typography, 135, *136*, 185, 198

United States, 35, 53, 119, 172, 185

University of Berlin, 188

Vaillant, François le, 113–14

Vancouver, George, 1, 28

Vassa, Gustavus. *See* Equiano, Olaudah

Venables, Richard, *Domestic Scenes in Russia*, 263

Victoria, Queen, 155, 171

Views in Egypt (Cooper), 135, 248, 317n81

Visits to Monasteries (Curzon), 274–75

Vitruvius, 86

Vowell, Richard, *Campaigns and Cruises*, 142

Voyage de l'Arabie Pétrée (Laborde, Léon), 80, 138

Voyage of Discovery, Made under the Orders of the Admiralty, A (Ross, John), 23, 45, 47, 94–95, 242

Voyage of Discovery and Research, A (Ross, James Clark), 38, 45, 189, 273

Voyage of H.M.S. Blonde (Graham), 150, 250–51

Voyage of the 'Fox', The (McClintock), 6, 22, 45, 207–8, 211, 285

Voyage Round the World (Kruzenshtern), 235

Voyages and Descriptions (Dampier), 93

Voyages from Montreal (Mackenzie), 92

Voyages of Discovery and Research (Barrow, John), 137, 271

Voyage to Africa, A (McLeod), 244

Voyage up the River Amazon, A (Edwards), 35, 272–73

Waddington, George, *Journal of a Visit to Some Parts*, 246

Wakefield, Edward, *Adventure in New Zealand*, 271

Walker, J. & C., 160–61, 163

Wanderings in North Africa (Hamilton, James), 142, 184, 283

Weld, Charles, *Arctic Expeditions*, 278

Wellsted, James, 63, 161; *Travels in Arabia*, 57, 154–55, 262

Western Barbary (Drummond-Hay), 61, 142, 204–5, 269

Westminster Review, 221

What to Observe (Jackson), 62, 214

Wheatley, Kim, 29, 45

Whewell, William, 98, 301n101

White, Hayden, 12

Wilbraham, Richard, 36, 60, 161, *163*, 264

Wild Sports, The (Harris), 263

Wilkinson, George: *South Australia*, 137–38, 274; *The Working Man's Handbook*, 138, 277

Wilkinson, John, 27–28; *Modern Egypt*, 169, 268; *Topography of Thebes*, 260

William IV (king), 178–79